The Deep Ocean

The Deep Ocean
Life in the Abyss

Michael Vecchione
Louise Allcock
Imants Priede
Hans van Haren

PRINCETON UNIVERSITY PRESS
PRINCETON AND OXFORD

Published in 2023 by Princeton University Press
41 William Street, Princeton, New Jersey 08540
99 Banbury Road, Oxford OX2 6JX
press.princeton.edu

Library of Congress Control Number 2022944027

ISBN 978-0-691-22681-1
Ebook ISBN 978-0-691-23040-5

Typeset in Aktiv Grotesk and Dolly Pro

Printed and bound in China
10 9 8 7 6 5 4 3 2 1

British Library Cataloging-in-Publication
Data is available

This book was conceived, designed, and produced by
UniPress Books Limited
Publisher: Nigel Browning
Commissioning editor: Kate Shanahan
Project manager: Caroline Earle
Designer: Sandra Zellmer assisted by Katrin Zellmer
Illustrators: Sandra Zellmer, Carolina Thorbert,
and Ellis Muddle
Editor: Amy K. Hughes
Picture researcher: Alison Stevens

Cover designer: Wanda España
Cover image: Unidentified deep water tripodfish larva
© BluePlanetArchive / Steve Kovacs

Page 2: Murray's Abyssal Anglerfish
Melanocetus murrayi

Introduction   6

INTRODUCTION TO THE DEEP OCEAN   10

OCEANOGRAPHY   42

DEEP-SEA ORGANISMS   76

HABITATS   122

GLOBAL PATTERNS   178

HUMANITY AND THE DEEP OCEAN   230

Classification of deep-sea species   278

Glossary   280

Resources   282

Notes on contributors   283

Index   284

Acknowledgments   288

INTRODUCTION

Various dictionaries define "abyss" as some version of an immeasurably deep gulf or great space. We consider the abyss from an oceanic perspective. If an extraterrestrial civilization sent a probe to Earth to collect a random sample of multicellular life, the sample would very likely be a gelatinous organism found swimming in the deep sea. The "deep sea" means different things to different people, depending on their experience and intentions. For example, "deep-sea fishing" means getting in a boat and going far enough offshore to catch fishes that are different from those caught near the shore; "deep-sea diving" sometimes just means scuba diving in the ocean but more accurately refers to technically complex scuba to depths greater than about 200 ft. (60 m), where special gases may be required; whereas "deep-sea salvage" may describe recovery operations in difficult offshore conditions.

Those of us who are scientists studying the deep sea usually define it based on a specific set of conditions. One such condition is how sunlight changes with depth: where the ocean waters are so deep that penetration of sunlight is not sufficient to support photosynthesis, the basic physical/chemical process that converts the energy of the sun, together with carbon dioxide and water, into carbohydrates that can be used to fuel the functions of living organisms. This diminished power of the sun occurs at about the same depth at which water is too deep to be stirred by the passage of seasonal storms. This depth is also coincident with the depth at the edges of most continental shelves, the undersea extensions of the continents. In most places around the world, these conditions co-occur at about 660 ft. (200 m) deep.

We therefore define the deep ocean as extending from depths of about 660 ft. (200 m) down to the sediment and rocks of the seafloor and into the sediment where animals can live; this importantly includes the huge volume of seawater below 660 ft. (200 m) deep. Depending on assumptions about microbial distribution and high-flying animals, the deep ocean makes up about three-quarters of the space on our planet where life exists. Although we like to think of the earth as the familiar two-dimensional dry land, the biosphere is mostly abyss.

The average depth of the ocean is a little less than 2.5 mi. (4,000 m), but the deepest trench is almost 6.8 mi. (11,000 m) deep. Life throughout the deep sea has evolved in a dark, cold world of pressures much greater than we experience on land. Food availability is usually quite limited but sometimes varies between scarcity and plentitude, which may be limited in time. People often wonder how deep-sea organisms can tolerate such conditions. We should remember that pressurized dark and cold are their normal conditions. If you carefully and gently catch a deep-sea animal but bring it up through the warm surface waters and try to keep it alive at atmospheric pressure, the process will kill it.

When we explore this world and encounter the animals (there are no plants, but many microbes), some appear bizarre, even "alien," whereas others may seem familiar but either gigantic or miniature. Compared with land or shallow waters, we haven't had the ability until quite recently to explore and begin to understand the animals and ecosystems of the deep. Even now, surprising and at times startling discoveries are not unusual.

Chrysogorgid octocoral
Iridogorgia sp.
The spiral form of a deep-sea coral colony. Although quite different from familiar shallow-water reef corals, deep-sea corals come in many growth forms. Some create forests of treelike structures whereas others make massive hard mounds, providing habitat for many other animals. Some colonies are thousands of years old.

For most people, the deep ocean is out of sight and out of mind. Not long ago, common (even academic) thinking was that the ocean was so vast that humans were not capable of changing it substantially. This was especially true for the deep sea away from coastal human concentrations. The mistakenness of this idea is becoming increasingly clear. As we deplete coastal resources such as fisheries and fossil hydrocarbons, we have moved into increasingly deeper water to continue exploitation. Also, new methods are being developed for extracting from the deep ocean new resources, such as important minerals rare on the continents, or for exploiting unusual biochemical characteristics of deep-sea organisms. Meanwhile, as we explore, we are increasingly finding the results of human activities, including plastics and other pollution as well as temperature changes, oxygen depletion, and acidification from increasing carbon dioxide concentrations. We are also learning how very long it would take ecosystems of the deep ocean to recover from damages we may cause.

What we know about the deep ocean is increasing rapidly. This book summarizes our current state of knowledge, in hopes that readers will share our fascination with this largely unknown part of our planet and the life in its waters.

Crystal jelly
Aequorea sp.
This is called the crystal jelly because it is so transparent. In this photo you can barely see the upper edge of the bell. Although found in coastal and near-surface water, its distribution extends down into the mesopelagic. This hydrozoan jellyfish is both bioluminescent and fluorescent and is the source of the luminescent protein aequorin and the fluorescent molecule GFP (green fluorescent protein) used in biomedical research. The discovery and development of these proteins earned a Nobel Prize in 2008.

"We haven't had the ability until quite recently to explore and begin to understand the animals and ecosystems of the deep. Even now, surprising and at times startling discoveries are not unusual."

Ocean vs. oceans

A conceptual controversy among marine scientists asks whether we should continue to use the traditional names for the five major oceans (Atlantic, Pacific, Indian, Southern—the latter sometimes thought of as Antarctic, and Arctic Oceans) along with many marginal seas, as opposed to emphasizing that the connections among all of these form a single global ocean. When discussing the named oceans within the "global ocean" concept, each of the oceans is treated as a distinct basin connected to all other basins within the entirety. While this is physically correct, it leads to some confusion for issues such as biogeography and evolution and causes problems with wording. For example, the Indian Ocean becomes the Indian Ocean basin, over and over throughout the text. This increases wordiness without increasing information content. It even confuses the information somewhat. The basin concept works well for the Pacific, is adequate for the Indian, and somewhat less so for the Arctic with its distinct subbasins separated by transverse ridges, but does not work well for the Atlantic (which is really four basins: northeast, northwest, southeast, and southwest), and is terrible for the Southern Ocean because there is no Antarctic basin. The Southern Ocean comprises the southern extremes of the Pacific, Indian, and Atlantic basins, cut off from them by the dynamic Antarctic Circumpolar Current. While recognizing the global importance of the interconnected ocean system, we have chosen to use the traditional names of the major oceans and seas.

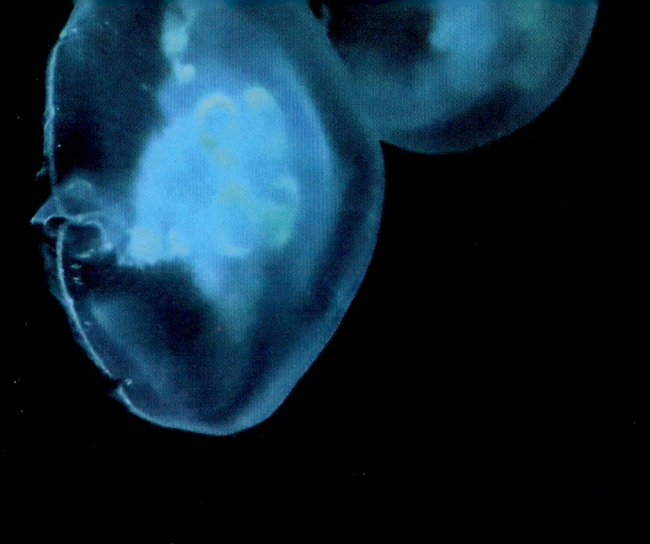

INTRODUCTION
TO THE
DEEP OCEAN

NOT YOUR FAMILIAR PLANET

Some differences between land and water-based environments

Previous pages: **Moon jellies** *Aurelia* sp.
Although common in epipelagic and even coastal waters, moon jellies are also found in the mesopelagic zone.

Go outside and look around. There is air to breathe, wind moving that air, and you can see many plants of various sizes including, perhaps, some very large old trees. Although trees seem tall from a human perspective, land-based ecosystems are essentially two-dimensional because all components are bound to the surface. You can't swim up in the air, although some animals can fly, glide, or even float on the wind, and people have invented tools for similar activities. However, all those animals, including us, having gone up, must come down. Gravity ties pretty much everything to the surface in terrestrial environments. Primary production (photosynthesis)—combining water, carbon dioxide, and energy from the sun to make complex carbohydrates and the oxygen component of the air in a form you and other animals can use—is based for all terrestrial food webs on large plants at the air–land interface.

All aquatic environments, both freshwater and marine, are fundamentally different from the familiar terrestrial ecosystems. Although some shallow-water plants and many animals are confined to underwater bottom areas (called benthic environments), other organisms permanently inhabit the entire volume of water above the bottom. This adds a third dimension to the two horizontal dimensions of the planet's surface. Water currents and animal movements can result in rapid changes to the distribution of life and physical properties in the ocean, especially in the vast volume of water. These rapid changes make life in the three-dimensional open ocean, most of which is deep sea, essentially four-dimensional.

Because water is so much denser than air, flotation (positive buoyancy, resulting from being lighter than the amount of fluid displaced by the volume of the body) is easy to achieve. The viscosity of water allows animals that are heavier than water to overcome gravity and swim far above the bottom (many never encountering the bottom) and slows the sinking of very small organisms. In the ocean, these small, usually single-celled organisms include most of the primary producers, called phytoplankton. The phytoplankton, which produce almost all the food in the ocean and about half of the molecular oxygen on the earth, have to stay close enough to the water surface that sunlight is adequate to support photosynthesis.

If we exclude land and the waters where sunlight is sufficient to drive photosynthesis, what remains is the deep sea, which by far encompasses most of the space on the earth that is habitable by multicellular life. Below the maximum penetration of sunlight, all is not totally dark. Biological light (bioluminescence) is produced by many deep-sea animals and is used for many purposes. Light in the deep sea ranges from dim, with day-night changes below the photosynthetic surface layer, to permanently almost nonexistent except for flashes and glows of bioluminescence. In addition to the differences in light dynamics, general characteristics of this world include pressures that are frighteningly intense from a terrestrial perspective as well as almost universally low temperatures, ranging from a few degrees above the freezing point of water to (in some places) a few degrees below it. Thermal anomalies exist in some basins isolated by bottom topography and in many small places where water heated by geothermal processes leaks from the rocks.

Heteropod
Carinaria lamarckii

This snail spends its entire life swimming with the shell downward and the belly fin up. Although primarily epipelagic it is found down into the upper bathypelagic zone (surface to at least 4,900 ft./1,500 m) and grows to at least 8.7 in. (220 mm) long. The small, transparent, keeled shell covering its viscera is sought after by shell collectors.

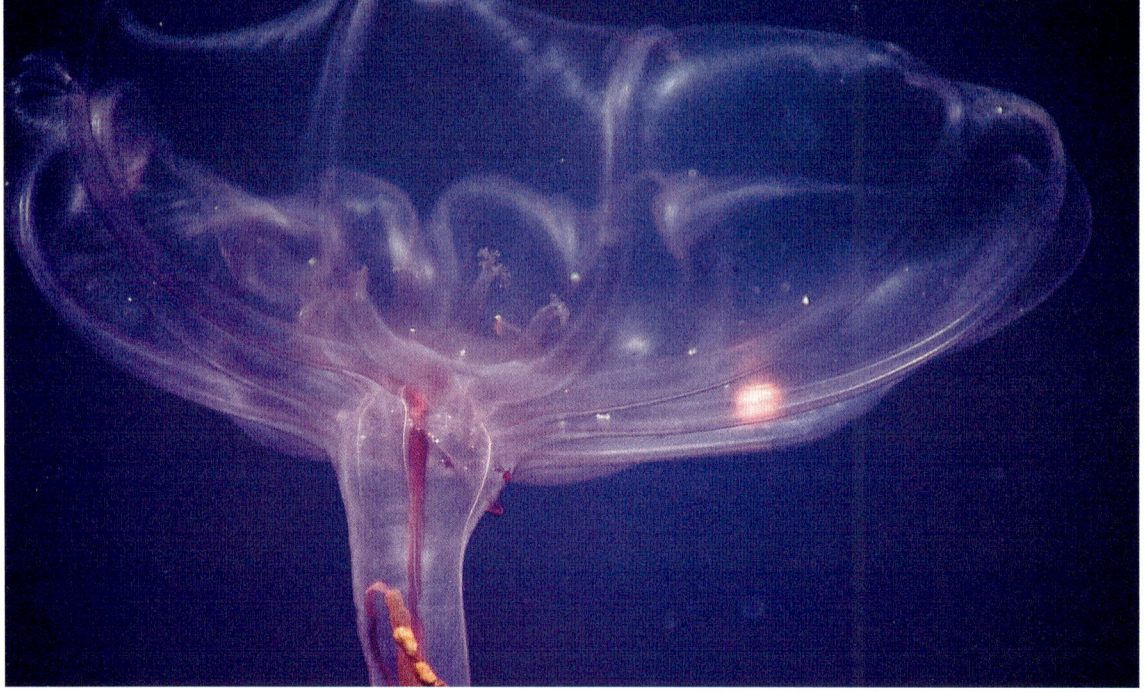

Swimming in the deep

Shortfin squids (*Illex* spp.) spotted during a dive on Pourtalès Terrace, Florida. Many larger mobile species from the continental shelf can move down the slope into deep water —descending to 3,280 ft. (1,000 m) depth.

Swimming cucumber

Pelagothuria natatrix

Deep-sea swimming cucumber found at 4,600 ft. (1,400 m) depth, Central Pacific Basin.

Important deep-sea characteristics

Many of the physical characteristics that make the deep sea, and therefore most of the livable space on the earth, such an alien environment for us will be discussed in detail in the sections that follow. Here we present a brief summary to set the stage. One characteristic, how salty the water is (salinity), can be so important to organisms in the coastal ecosystems with which we are most familiar, that it is a primary control on where they can live. However, away from the rivers of the coasts and from the surface where rain falls, salinity doesn't usually vary enough to be physiologically important to the organisms, although small salinity changes are very important in the density of seawater. Because these differences in density affect the vertical stability of the layers of water, they are among the important drivers of ocean currents.

On land, wind can move physical stuff such as dust and topsoil and can assist or divert the flight of various organisms. In aquatic environments, transport by currents is extremely important for the global distribution of dissolved chemicals, suspended sediments, heat transfer, and for the dispersal stages of most organisms. Current transport may be horizontal (over short or long distances) or vertical. The latter is very important for the deep sea because many of the chemical and biological materials in the deep originate from the surface or from the continents. Other processes contribute to the movement of material into the deep sea as well. These processes include simple sinking, turbulent mixing, turbidity currents (similar to underwater avalanches, but with sediment instead of snow), and active transport associated with the movements of swimming animals.

Another important characteristic of seawater is how much oxygen is dissolved in it and available for respiration by organisms. If, when you walked outside at the beginning of this section, you came into an area lacking oxygen, your distress would be immediate. In some areas of the deep sea, the dissolved oxygen content of the water can be low enough to be distressing or even lethal to local animals. Such areas, known as "oxygen-minimum zones," are forecast to become more intense and more widespread with continuing global climate change.

Human effects

The ocean may seem so vast, and the deep sea so remote, that even humans overpopulating the land could not cause widespread and lasting damage to it. This once was the prevailing thought in marine science, but now we know better. The effects of modern civilization, known as the Anthropocene era, are easily found in even the most remote parts of the deep sea. In addition to expanding oxygen-minimum zones, these effects include elevated temperatures, increased dissolved carbon dioxide and acidity (decreased pH), dumping of chemical pollutants, the presence of plastics essentially everywhere, and the physical and chemical impacts from extraction of resources such as fisheries, petrochemicals, and minerals. To develop ways to deal with such human-caused changes to the deep sea, we must understand the patterns and processes of the ecosystems of the deep, including the resilience of its inhabitants.

Studying the deep sea is just plain difficult, and comparable to exploring outer space. Therefore, even though the deep sea is most of the living space on our planet, our understanding of it lags far behind our knowledge of Earth's other ecosystems. We invite you to dive deeper with this book into our current understanding of life in the deep sea.

Oceanic zones

How depth zones are defined differs slightly depending on whether the focus is on bottom (or benthic) ecosystems or on the life up in the water column (pelagic). The names of the zones are different, but the depths of transitions are often similar because of physics and the physiological adaptations of the inhabitants. Oceanic depths are generally measured in meters, and throughout the book we have converted those numbers to the nearest 10 feet, and as such, measurements should not be considered precise. The depth transitions are generally considered to be at about 660 ft., 3,280 ft.,13,125 ft., and 19,700 ft. (200 m, 1,000 m, 4,000 m, and 6,000 m).

Marine biomes
Terminology of the depth divisions of bottom (benthic) and water-column (pelagic) ecosystems. The solid orange line is a generalized bottom profile. Benthic zones are labeled to the left of the orange line and pelagic zones to the right. The transitions between zones are more of a gradient than an abrupt demarcation.

Deep-sea anglerfish
Himantolophus sp.
A female deep-sea anglerfish or footballfish. This species lives in the bathypelagic, attracting prey with its luminescent lure. Lines of sensors on the skin detect vibrations. Males are much smaller and more mobile.

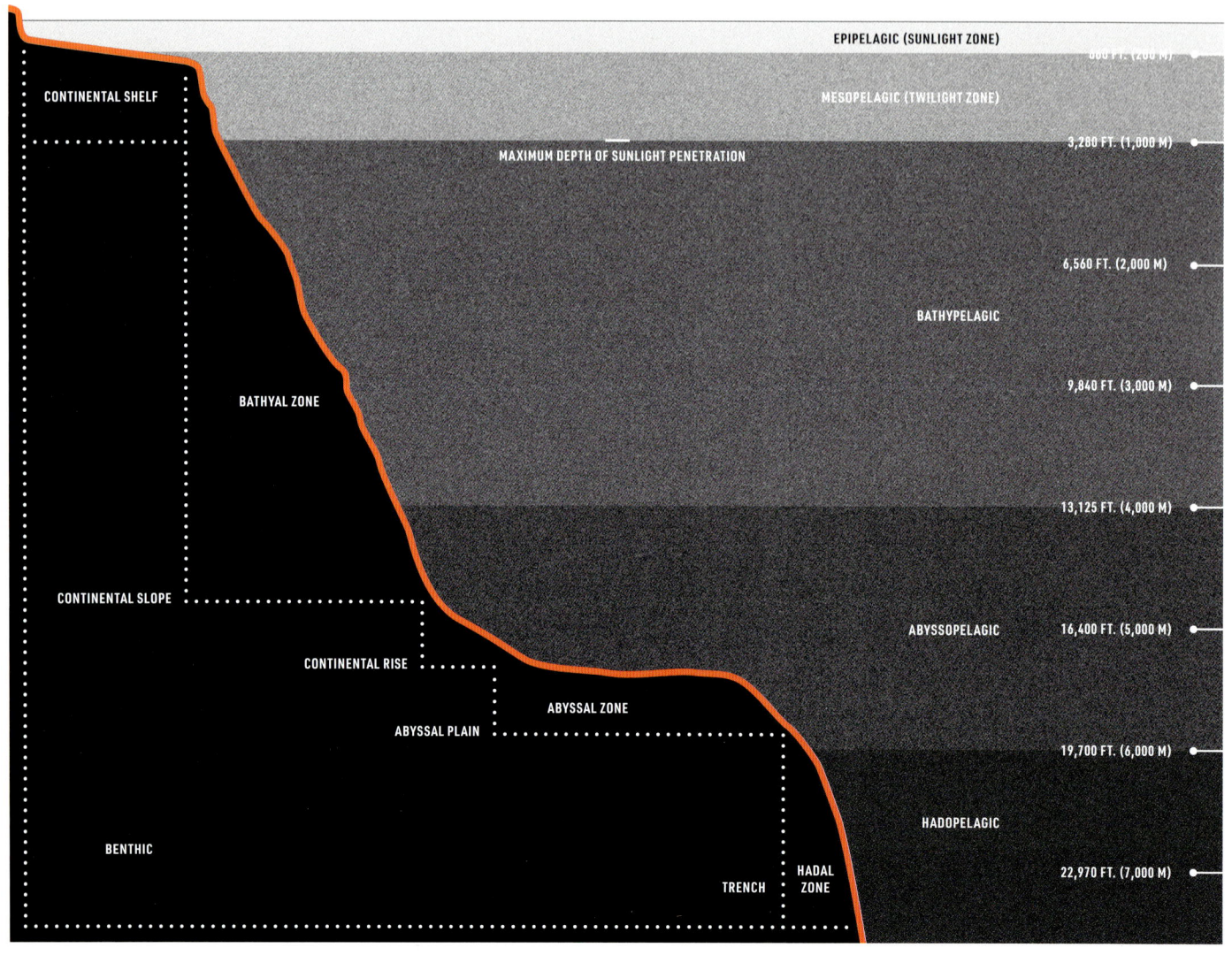

CONTINENTAL SHELF

BATHYAL ZONE

CONTINENTAL SLOPE

CONTINENTAL RISE

ABYSSAL ZONE

ABYSSAL PLAIN

BENTHIC

TRENCH | HADAL ZONE

EPIPELAGIC (SUNLIGHT ZONE)

660 FT. (200 M)

MESOPELAGIC (TWILIGHT ZONE)

MAXIMUM DEPTH OF SUNLIGHT PENETRATION

3,280 FT. (1,000 M)

6,560 FT. (2,000 M)

BATHYPELAGIC

9,840 FT. (3,000 M)

13,125 FT. (4,000 M)

ABYSSOPELAGIC

16,400 FT. (5,000 M)

19,700 FT. (6,000 M)

HADOPELAGIC

22,970 FT. (7,000 M)

THE CONDITIONS THAT DEFINE THE DEEP-SEA ENVIRONMENT

Important physical, chemical, and geological properties

What factors (preferably measurable) that define an ecosystem are relevant for general understanding of life in the deep sea? Here we consider the physical, chemical, and geological parameters of this environment. If we define the deep sea as the waters starting at 660 ft. (200 m) below the sea surface—beyond any continental shelf and below the well-lit zone—most parameters related to exchange with the atmosphere are excluded, except when their effects are transported into the deep.

Factors to consider

Although all deep sea is connected to shallower waters, even to those near the surface, the questions of how, when, where, and to what extent, are important and inform our parameters. Direct wind effects, such as turbulent mixing near the surface, do not reach the deep sea. However, indirectly storm-generated "inertial" internal waves (subsurface waves in the density stratification of the ocean that have a large enough wavelength to be affected by the earth's rotation) and large-scale wind-driven currents do penetrate into the deep sea (see pages 54–55). Similarly, although the bulk of the direct heat storage of solar energy is absorbed in the upper 330 ft. (100 m) of water below the surface, the transport of some of this heat downward into the deep sea cannot be excluded. Likewise, chemical parameters related to direct air–sea exchange, such as oxygen and carbon dioxide uptake by the ocean from the atmosphere, are excluded, but the transport of these chemical substances downward is not. Geological parameters that provide input at the surface but which cannot be excluded altogether include volcanic eruption and dust-storm sediments blown over the surface and sinking into the deep sea.

Sunlight penetration

Temperature generally decreases with increasing depth. Most of the solar heat is absorbed up to 330 ft. (100 m) down in the ocean, but some heat may travel to the deep sea.

Chondrichthyes, cartilaginous fishes

The Chondrichthyan orders Chimaeriformes (chimaeras or ghost sharks) and Squaliformes (sleeper and dogfish sharks) have specialized in colonizing the deep-sea slopes down to a depth of around 9,840 ft. (3,000 m). Top: *Chimaera* sp. in the Sulawesi Sea. Left: Portuguese Dogfish (*Centroscymnus coelolepis*), the world's deepest-living shark in the northeast Atlantic Ocean.

Physical properties

The waters in the deep sea are not a static, stagnant pool. Like the sea surface, the deep sea is always in dynamic motion. An important physical parameter is pressure: every 33 ft. (10 m) of additional depth adds 1 atmosphere—a unit of pressure equal to the pressure of the air at sea level, or approximately 14.7 lb. (6.7 kg) per square inch, reaching 1,000 atmospheres (14,700 lb. [6,668 kg]/sq. in.) at around 33,000 ft. (10,000 m). The enormous pressure at great depths calls for evolutionary adaptations in deep-sea marine life. Pressure variations directly relate to the dynamics that are of great importance for all transport and life in the deep sea. Some of the dynamic water movements are driven by variations in density, another vital physical parameter.

Variations in density are mainly determined by variations in temperature and salinity (salt content). Warmer (or fresher) waters are less dense than colder (saltier) waters. While the effects of gravity on variations in density set the deep waters in motion, the fact that the deep sea is located on a rotating sphere heated by the sun dominates its motions. The direction of the earth's rotation causes the Coriolis effect (deflection of motion related to changes in latitude) and imposes the direction of large transport circulations, like the jet stream in the atmosphere (see also pages 52–53). Within deep-sea basins, the direction of Earth's rotation controls circulations on the large scale—for example, the Antarctic Circumpolar Current—and it forces and intensifies the surface circulation on the western sides of sea basins. This wind-driven circulation, responding to mid-latitude westerlies and equatorial trade winds, dominates the near-surface ocean transport of heat and salt from equator to poles along the western sides of large basins and a weaker return flow on the eastern sides. This near-surface circulation of major currents extends down into the deep sea. It may vary in intensity but will not change direction.

Secondarily, the redistribution of heat and salt initiates deep-sea return flows. One mechanism to sink surface waters is cooling and evaporation by exchange with the atmosphere, generating vertical turbulent transport/motion (convection) of newly formed dense waters. Such convection occurs every year in the polar regions and in localized mid-latitude regions, such as some parts of the Mediterranean Sea, typically reaching hundreds of meters deep. Episodically, only once every 5–10 years, convection reaches thousands of meters deep and spreads dense waters into the ocean interior. However, if such limited vertical exchange were the only water movement, the deep ocean would be filled with cold stagnant water. The deep-sea flow is thus not primarily density driven. A mechanical turbulent mechanism is needed to mix and transport heat and salt in the deep sea. Indirectly, the deep-sea flow and transport are regulated by wind and tides, which in the deep sea are transferred to internal waves within vertical density differences. Such internal waves can attain amplitudes of 330 ft. (100 m) or more, and cause most of the deep-sea turbulent mixing when they break on island and continental slopes, much like surface waves breaking on a beach, but at a much larger scale. These breaking internal waves create deep-sea unrest of permanent turbulent motion.

Chemical properties

Important chemical parameters in the deep sea include salinity, constituting typically about 35 g of salt per 1 kg (about 7 teaspoons per liter) of seawater, and pH, a measurement of how acidic (values smaller than 7) or basic/alkaline (values larger than 7) water is; seawater is typically about pH 8. Oxygen, carbonate content, and major nutrients such as nitrogen and phosphate are found in quantities of the order of milligrams per kilogram of seawater. Micronutrients like iron and other metals are found in quantities of only several tens of nanograms (billionths of a gram) per kilogram of seawater but can still be limiting for deep-sea life. Such low quantities may be difficult to understand, given the large iron content when oceans were formed and the fact that the earth's core still mainly consists of metals. Iron has mostly precipitated out of the ocean.

Seafloor topography

Bottom: the deep seafloor is not flat, but contains more hills and mountains than on land. This example from the North Atlantic Ocean shows the prominent Mid-Atlantic Ridge.

Manganese nodules

Below: at some abyssal deep seafloors, manganese nodules are abundant, such as these photographed in deep-water areas of the southeastern US continental margin.

Depth vs. height

Right: there is more topography underwater than on land. The deepest point of the seafloor (6.78 mi./10.92 km, the Challenger Deep of the Mariana Trench) dwarfs the height of Mount Everest (5.5 mi./8.85 km above sea level).

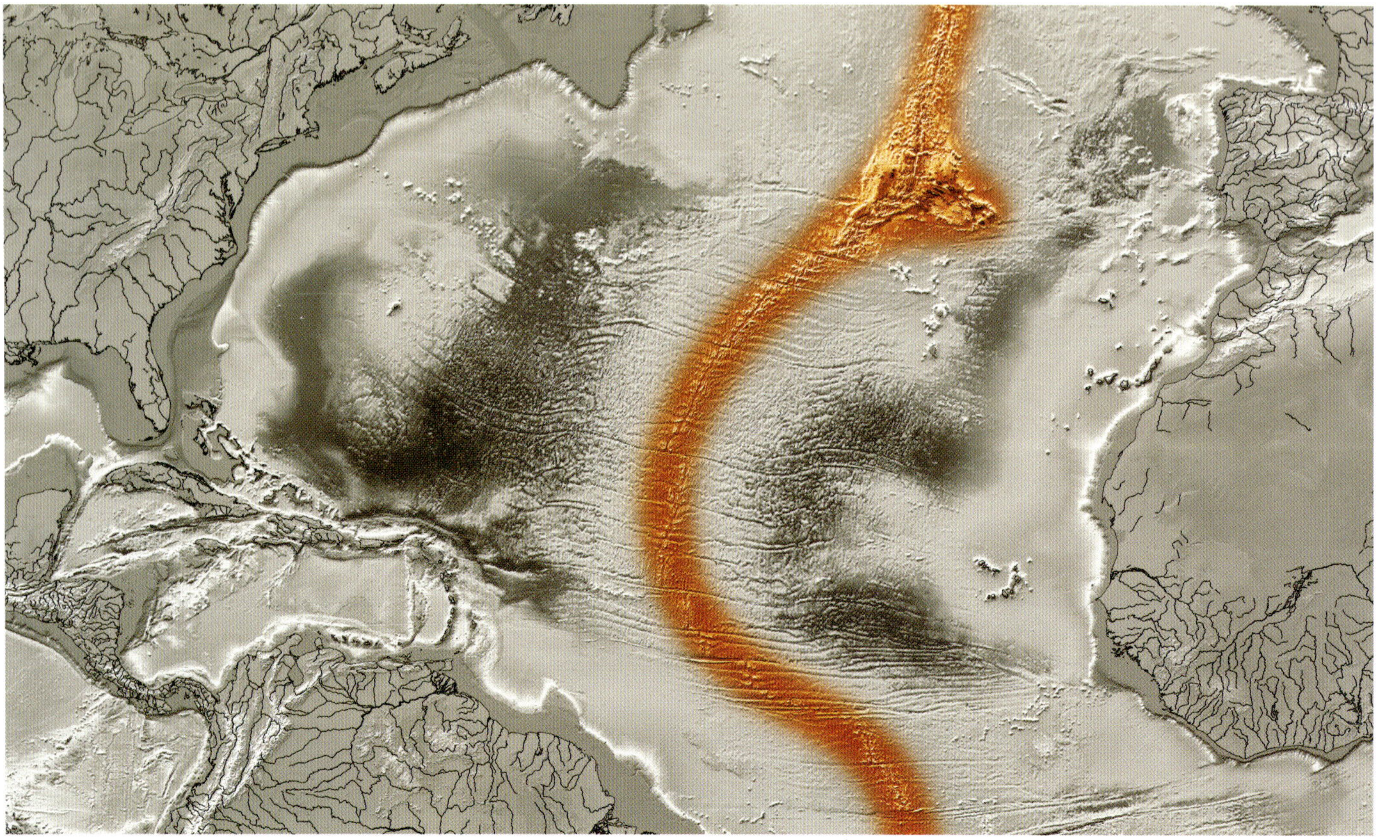

SEA LEVEL

3,280 FT. (1,000 M)

6,560 FT. (2,000 M)

9,840 FT. (3,000 M)

13,125 FT. (4,000 M)

MARIANA TRENCH

16,400 FT. (5,000 M)

19,700 FT. (6,000 M)

22,970 FT. (7,000 M)

MOUNT EVEREST
5.5 MI./8.85 KM ABOVE SEA LEVEL

26,250 FT. (8,000 M)

29,500 FT. (9,000 M)

32,800 FT. (10,000 M)

CHALLENGER DEEP
6.78 MI./10.92 KM BELOW SEA LEVEL

36,000 FT. (11,000 M)

Geological properties

As the earth's crust is not a smooth surface, the deep seafloor is also not flat, despite the millennia of precipitating sediments and biogenic matter. In fact, the deep seafloor can be very mountainous in places, making its topography a significant geological parameter. All deep-seafloor topography contributes to the steering of ocean currents and thus the transport of suspended matter, such as sediment particles and small organisms, dead or alive. As a boundary between the earth's crust and the deep-sea waters, topography's role in the resuspension of matter via turbulent mixing, bringing the matter back into the deep-sea waters from the seafloor, is vital for deep-sea life, but can also be problematic in terms of resuspending microplastics and pollutants.

The seafloor includes a large variety of substrates and textures that may interact with overlying water flows. Where internal waves break, they cause large-scale turbulence. This turbulence resuspends sediments with finest clays easily resuspended and coarser sands requiring a greater intensity of turbulence. In places, only hard substrates like bare rock exist. The earth's crust transfers relatively little heat to the deep-sea waters above, but in areas of underwater volcanic activity may transfer rare metals such as manganese and magnesium. Where tectonic plates meet and one plate subducts, or

slides beneath the other, deep trenches can be found, of which the deepest reach more than 6.2 mi. (10 km) below the sea surface. All such very deep spots are found in the Pacific Ocean; the deepest point of the Atlantic Ocean is about 5.2 mi. (8.4 km) from the sea surface.

Alongside deep trenches, volcanic arcs are formed, which are sources of many oceanic islands. Behind volcanic arcs, so-called back-arcs form part of the abyssal plain (a large, relatively flat expanse of seafloor). On some abyssal plains are found large loads of manganese nodules, which develop extremely slowly and are subject to impending deep-sea mining. Ferromanganese oxide crusts are also associated with submarine hydrothermal activity of volcanic origin, for example on the Mid-Atlantic Ridge. Outside subduction areas, volcanic activity can occur at hot spots and create submarine mountains (seamounts). Hot-spot volcanism is unique because it does not occur at the boundaries of the earth's tectonic plates where all other volcanism occurs. Instead, it occurs at abnormally hot centers known as mantle plumes. An example of hot-spot volcanism is the Hawaiian Islands.

INHABITANTS OF THE DEEP

An overview of the life-forms of the deep ocean

The organisms of the deep sea include representatives of almost all branches of the tree of life, with a major exception. Because the deep sea is, by definition, where the ocean is too deep for sufficient sunlight penetration to power photosynthesis, the organisms that can live there exclude plants and plantlike photosynthetic microbes (phytoplankton). Although not as well known as terrestrial and freshwater organisms, living things in the deep represent almost all phyla of animals (which actually form a very small branch on the tree), some fungi, and very diverse groups of microbes (many branches of the tree). Their place on the tree of life may be controversial, but viruses are definitely part of the deep-sea biota, including so-called giant viruses, the size of bacteria, with unusually complex genomes.

People have known about life in the deep sea since about the middle of the 19th century—and often ask how animals can live in such an inhospitable place. But conditions in the deep sea are not inhospitable for something that evolved under the conditions of darkness, high pressures, and low temperatures. Bring such creatures up to the surface, however, and they will die from the harsh (to them) conditions there. Many people, even now, tend to think of deep-sea animals as weird, alien, and monstrous. Although some are quite large, none are monsters. But many inhabitants of the deep certainly appear strange from a human perspective, with characteristics such as lures and long, sharp teeth that seem scary if we consider them through the lens of imagination rather than knowledge. Perspective is important, though—many animals portrayed as monsters in popular media are modeled after the appearance, and sometimes the behavior, of deep-sea animals. Folks are often surprised to learn that the model animals are actually quite small in comparison to a human. Even the very large animals, such as the Giant Squid (*Architeuthis dux*) or the sea-serpentine Oarfish (*Regalecus glesne*), normally live so far from places of usual human activities that chances of encounters wth healthy animals are very slim, and almost never threatening. (A threatening exception is a harpooned Sperm Whale, *Physeter macrocephalus*.)

> "*Many inhabitants of the deep certainly appear strange from a human perspective, with characteristics such as lures and long, sharp teeth that seem scary if we consider them through the lens of imagination rather than knowledge.*"

Sloane's Viperfish
Chauliodus sloani
Top left: a dragonfish that usually lives at 1,640–3,280 ft. (500–1,000 m) depth, preying on abundant fishes and crustaceans of the oceanic deep scattering layers.

Swordtail Squid
Chiroteuthis calyx
Top right: the family Chiroteuthidae includes some of the strangest-looking squids. With long complex tails in some developmental stages and unusual proportions of body and arm parts, they appear almost as a caricature of a more familiar squid. This adult is holding its long tentacles in a groove in the ventral arms.

Telescopefish
Gigantura sp.
Right: the name of the telescopefish refers to its eyes which, although they do not magnify, are tubular-like telecopes. The eyes on this species gaze forward but the animal probably orients vertically, head up, so that its silhouette is minimized while it searches for silhouettes of prey against the dim downward glow in the mesopelagic.

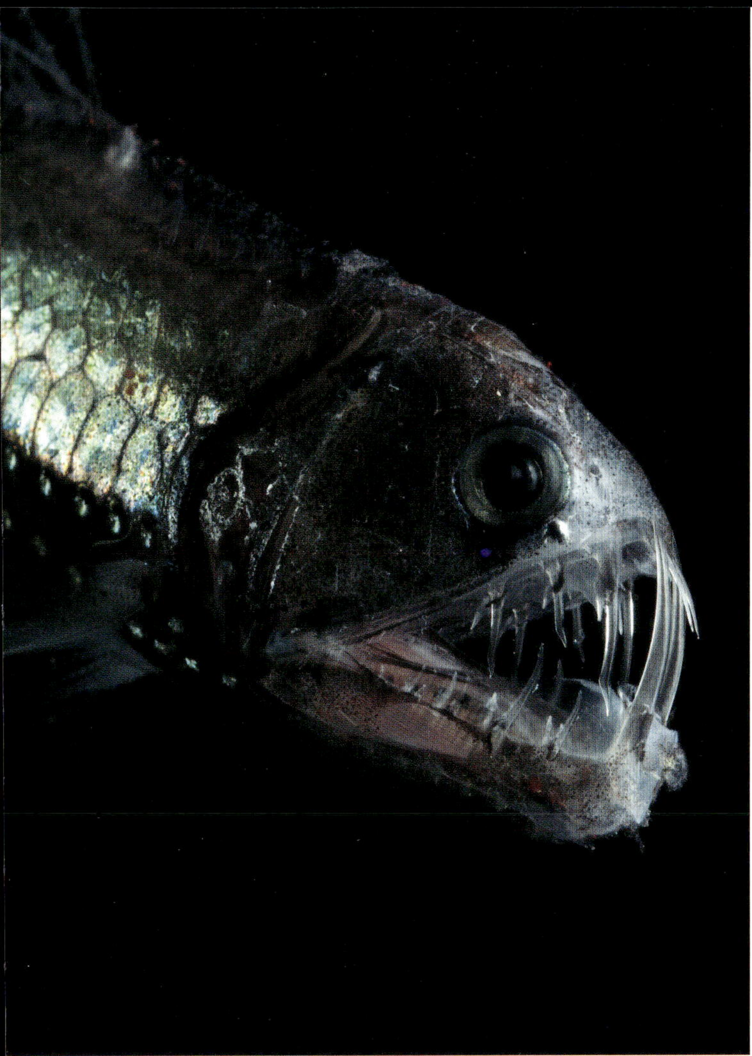
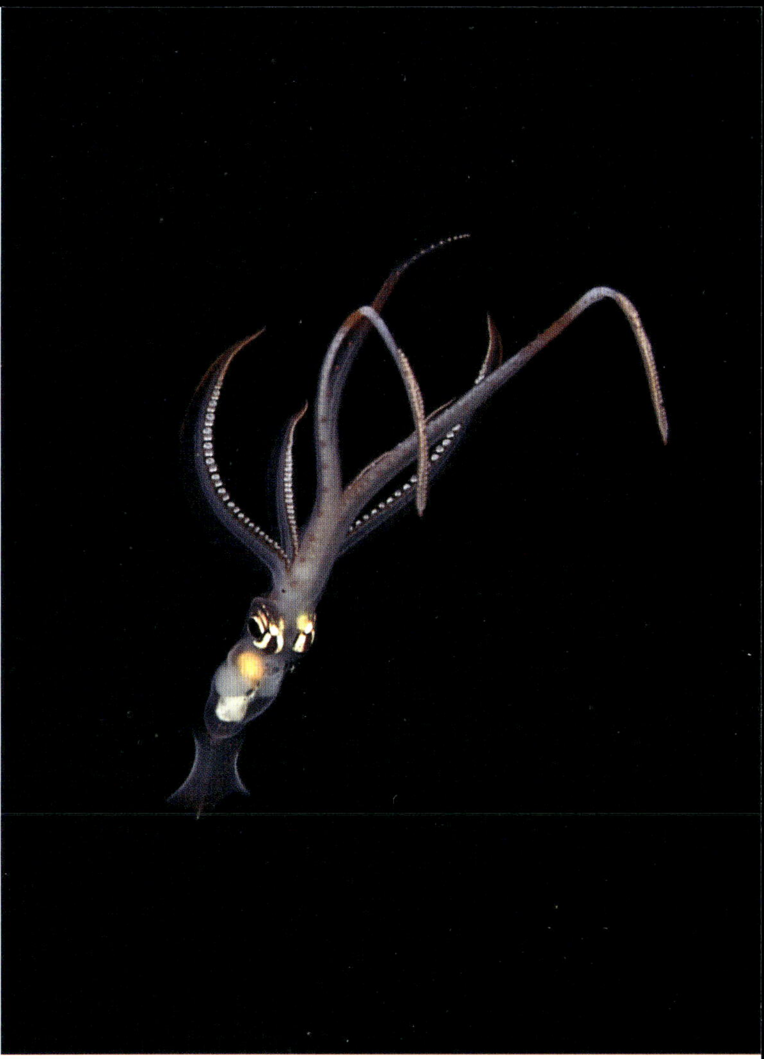
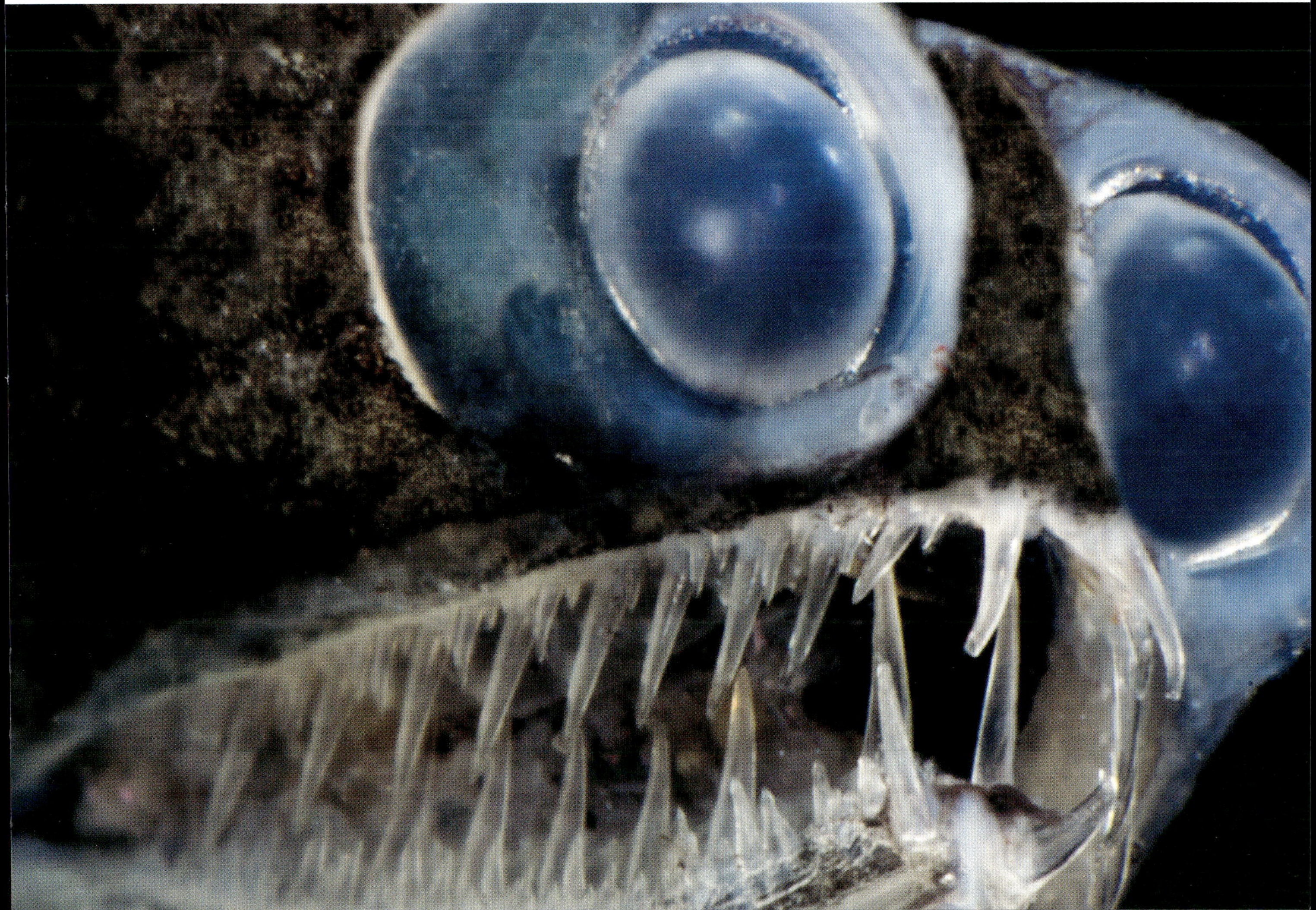

Habitat groups

Placing them in fundamental groupings can help us to understand the inhabitants of the deep sea, exploring what we do or don't know about them, and how we study them. One way to group them is by where they live: on or in the bottom (the inhabitants of which are known as benthic organisms) vs. up in the water, away from the bottom (pelagic organisms). Of course, transitions between benthic and pelagic environments are quite common and include benthic animals swimming high up into the water, vertically migrating pelagic animals encountering the bottom on the continental slope or an undersea mountain, and species that spend part of their lives in one habitat and another part in another habitat (e.g., pelagic larvae that settle onto the bottom to become benthic adults). Among the benthic animals, those that live on hard rocks are very different from those on or in soft sediments. Similarly, pelagic organisms differ greatly depending on their preferred depth and behavior. The vertical physical structure of the ocean (i.e., gradients of temperature, pressure, and light) results in clear differences in the communities of organisms, both benthic and pelagic, that live at different depths.

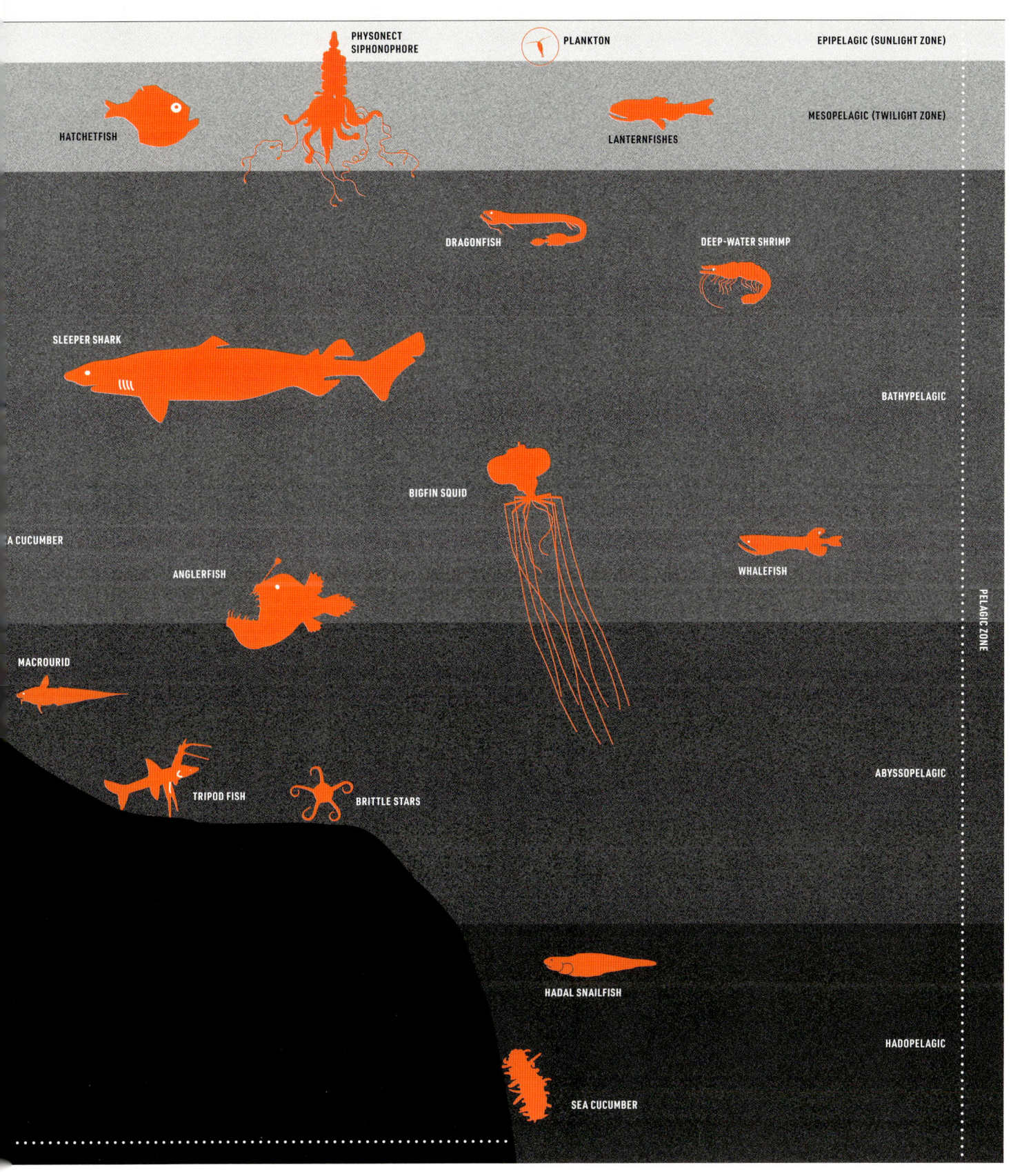

PHYSONECT SIPHONOPHORE

PLANKTON

EPIPELAGIC (SUNLIGHT ZONE)

HATCHETFISH

LANTERNFISHES

MESOPELAGIC (TWILIGHT ZONE)

DRAGONFISH

DEEP-WATER SHRIMP

SLEEPER SHARK

BATHYPELAGIC

BIGFIN SQUID

A CUCUMBER

ANGLERFISH

WHALEFISH

MACROURID

PELAGIC ZONE

TRIPOD FISH

BRITTLE STARS

ABYSSOPELAGIC

HADAL SNAILFISH

HADOPELAGIC

SEA CUCUMBER

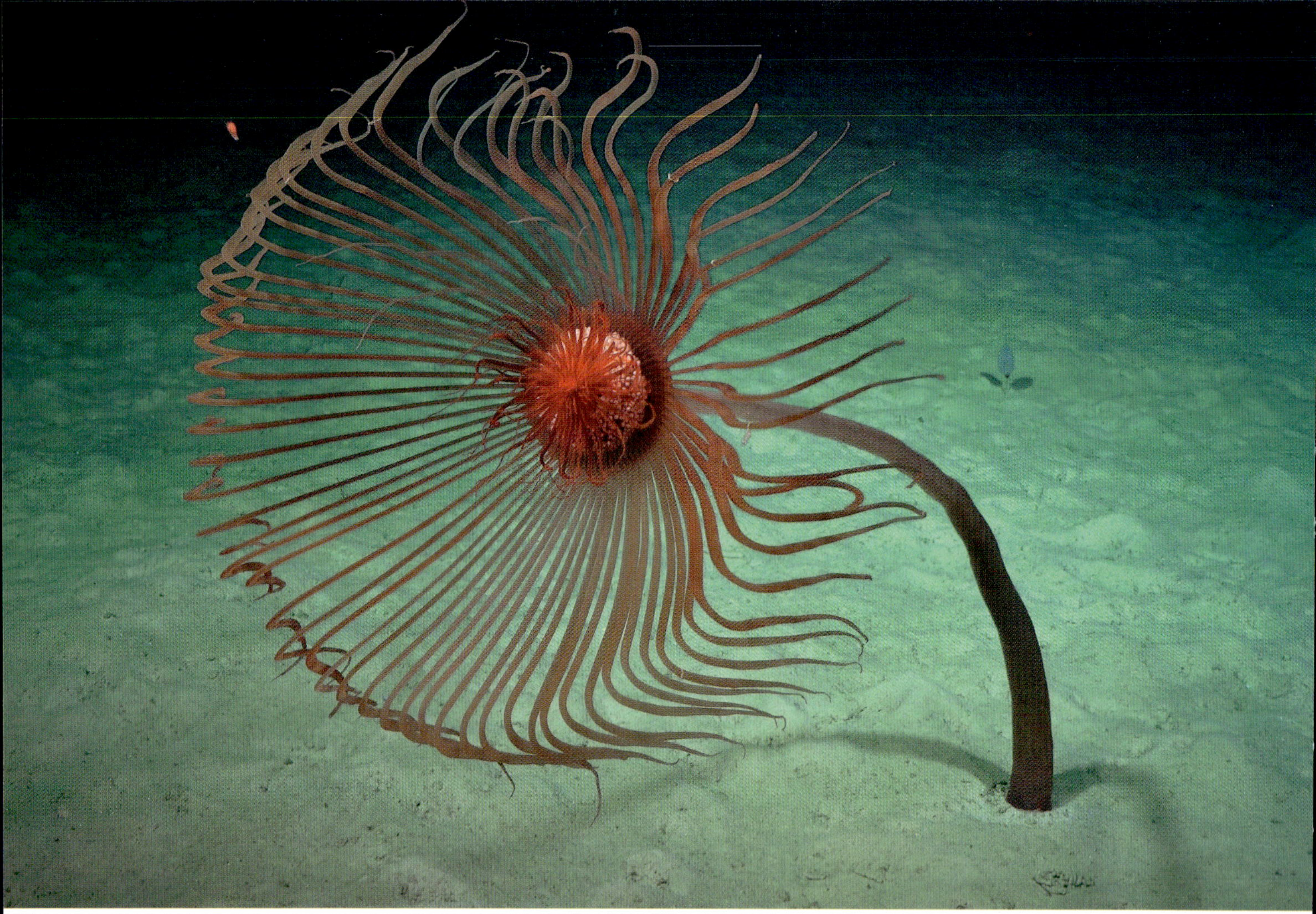

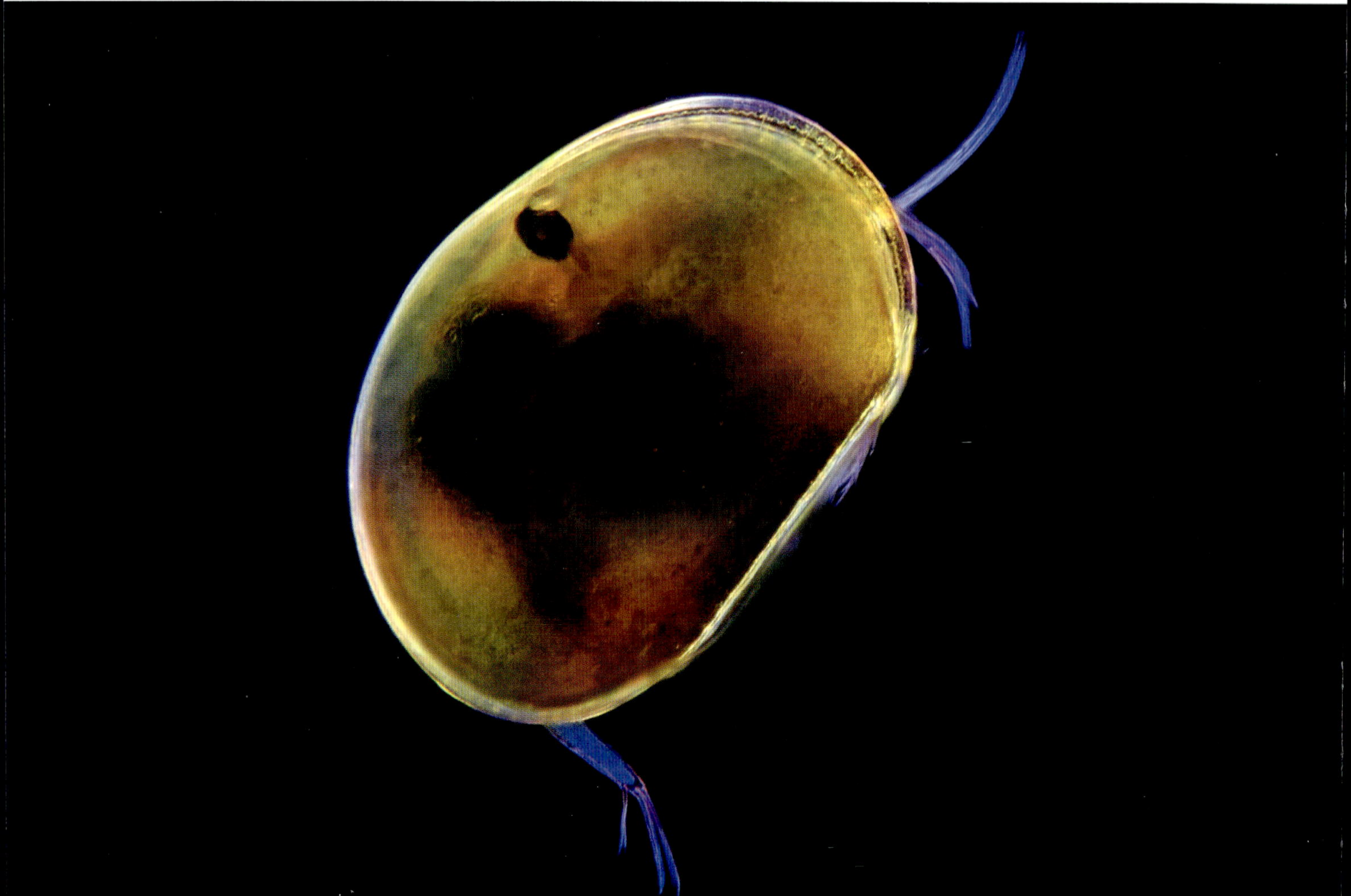

Size groups

Regardless of habitat, another way to think about deep-sea organisms is size groups. Very small organisms include various types of microbes (e.g., bacteria, archaea, viruses, and eukaryotes—single-cells with nuclei) as well as microscopic multicellular animals. The methods used to understand the biology and ecology of these very small things are quite different from those used to study larger animals. Even within the larger animals, size influences lifestyle, as does habitat. Among mobile species, whether swimming or crawling, large animals can generally move farther and faster than smaller ones, affecting their ability to search for food, mates, and so forth. Even sessile species or colonies, though fixed in place, can reach higher up above the bottom if they are particularly large, to filter food drifting or swimming in near-bottom currents. Of the familiar groups of animals found in the deep sea, species are often larger (exhibiting gigantism, with its mobility advantage) or smaller (dwarfism, requiring less food to complete the life cycle) than related shallow-living species we may know.

Feeding

In addition to the physical parameters (e.g., temperature, pressure, and light) characteristic of various deep ecosystems, availability of food—which may be continually sparse or only episodically available—is a critical and sometimes limiting factor throughout most of the deep sea. The exception is chemosynthetic ecosystems, in which microbes produce organic materials from chemicals and geothermal energy, sometimes in the absence of oxygen; several types of chemosynthetic ecosystems are known in many locations, but they are all very limited in size. Otherwise, the ultimate source of food in the deep is at the ocean's surface. Variability in the flow of surface production into the deep aligns with patterns of surface productivity driven by factors such as season, latitude, and availability of nutrients. Nutrient availability at the surface, in turn, is related to things like distance from continents and riverine input, upwelling of nutrient-rich deep water, and other physical transport by major currents or winds. Effective collection and efficient use of limited food resources are recurrent themes in the biology of deep-sea organisms.

Evolutionary relationships

Of course, studies of deep-sea organisms can be organized in many ways. An important one involves evolutionary relationships—who is related to whom, and how did their characteristics and variation evolve? Some evolutionary lines are quite diverse in the deep sea, whereas others have just a few known representatives, either because their global diversity is low or because most of the diversity is found in other ecosystems around the world. Some biologists focus on particular taxa (evolutionary groups), such as fishes, echinoderms (sea stars, urchins, and kin), or cephalopods (octopods, squids, and kin), regardless of their size or habitat. Some taxa (e.g., fishes) are widely studied and comparatively well known, while others (such as many groups of small worms, microbes, or jellyfishes) receive much less attention and are therefore very poorly known. Mysteries abound in the deep and await the attention of future investigators.

A giant solitary hydrozoan
Branchiocerianthus sp.

Top: this sessile (unmoving) species may appear somewhat like a skinny sea anemone, but it is more closely related to pelagic colonial siphonophores.

Ostracod (seed shrimp)
Gigantocypris sp.

Left: these microscopic crustaceans have a bivalved carapace, looking somewhat like a clam shell with a shrimp squeezed inside of it.

HOW DO WE KNOW ABOUT LIFE IN THE ABYSS?

Selecting the right tool for the job

Over nearly two centuries of oceanic exploration, dedicated tools have been developed to support the study of life in the deep sea. Equipment used to understand the physics, chemistry, and geology of the deep ocean provides information important to infer the distribution of deep-sea organisms. Briefly describing the many tools currently in use to study different aspects of the deep sea raises the questions of where to begin and which gear to include. Let's consider here a few categories most useful for deep-sea biology. Our section on history (see pages 36–41) includes a description of the chronology of interplay between new tool development and breakthroughs in our understanding of the deep ocean. It would be a mistake, though, to think that new types of tools replaced previous tools. After inception, different categories of tools have advanced technologically and kept a place in the oceanographic toolbox, each providing a different perspective on the deep ocean and its inhabitants. Additionally, new tools are under development as this text is being written and, depending on the success of those developments, will add to our toolbox and allow different perspectives on life in the deep.

Sounding the bottom and the water column

While light is rapidly absorbed by water, depending on color (wavelength), sound does a much better job of penetrating great distances of water, depending on frequency, and returning information in the form of echoes. Thus, sonars with different frequencies or with multiple frequencies began to be employed, replacing the weighted lines previously used for exploration. Sonars are used very commonly to determine the depth and topography of the seafloor as well as geological structure beneath the deep seafloor, even in the deepest trenches. How much detail we can infer depends on the sound frequency used and the distance to the target. Therefore, in addition to ship-mounted sonars, we lower transducers into the depths, either by towing them beneath and behind a ship or by mounting them on various types of submersibles.

Anti-submarine warfare during World War II demonstrated that the seafloor is not the only feature detected by sonar. Because many pelagic animals and some physical features like bubble streams also create strong echoes, sonar is also commonly used to determine depths of many things floating or swimming in the water between the surface and the bottom. To obtain a long temporal sequence of changes in what is in the water column, a sonar transducer can be mounted pointing upward from a benthic (bottom) lander or floating on a deep buoy.

Satellite-based measurements

Although radio, and therefore radar, signals do not penetrate through water, satellites are useful for acquiring large-scale overviews of the sea surface. Unfortunately, sea-surface parameters are of limited use for deep-sea biology. Temperature, and lately salinity, patterns are useful for inferring configurations of large currents, including the locations of eddies and oceanographic fronts, which can be very important for determining distributional aggregations and limits of biological communities, including those in the deep pelagic environment. Color at the surface can be very useful for describing patterns of phytoplankton productivity, which ultimately provides the food in the deep. A very recent development is the use of lidar (laser-based light detection and ranging) from a satellite to assess large-scale patterns of diel (on a 24-hour cycle) vertical migration of pelagic animals.

Lidar

Right: the space-based CALIPSO lidar (light detection and ranging) satellite measures the planet's largest animal migration, which takes place when small sea creatures swim up from the depths at night to feed on phytoplankton, then back down again just before sunrise.

Sonar measurements

Left: technicians prepare a customized REMUS autonomous underwater vehicle used to study sound-scattering layers in southern California.

Nets and water catchers

Among the oldest of tools for determining what lives where in the ocean are the nets developed originally to catch food. The antiquity of the concept, however, does not mean that towing nets is an outdated mode of studying life in the deep. Small nets with small mesh catch the small animals of the zooplankton. Larger nets not only catch larger individuals but also collect faster swimming things (e.g., fishes, squids, shrimps) that are represented in the zooplankton only as early life stages. Some specialized ships can tow extremely large nets that collect samples, giving very different perspective on the life cycles of many species that grow large and swim fast. The utility of pelagic net sampling has been enhanced by the development of complex gear with multiple nets, each of which samples within a different layer in the water, providing information about vertical distribution and movements.

It is also important to sample water, both for the microbes in it and for chemical properties, including environmental DNA. Every oceanographic research ship has the capability of lowering a sampler with multiple bottle-like devices that can be closed at target depths to bring water back for analysis. Such gear is usually also outfitted with additional sensors to measure, for example, temperature, conductivity (to determine salinity), pressure (to determine depth), and other things like light from stimulated bioluminescence.

Dredges, grabs, corers

To find out what lives on, or in, the bottom, we can either tow a dredge, scraping things from the bottom into a net, or we can lower a device to collect sediment and the organisms living in it. Varieties of the latter range from a box corer that takes a one-square-meter bite out of the sediment to a multi-corer that punches many small samples out of a limited area. For larger, faster animals (e.g., crabs, benthic fishes) a rugged, heavy net can be towed over fairly flat bottom areas, but such tows have been shown to do substantial damage to benthic communities, especially those attached to hard substrate.

Landers, baited cameras, traps, and longlines

It is also possible to attract mobile scavengers and predators to baits or lures, either to catch them with traps or hook and line or to photograph them. This method works both on the bottom and up in the water but is most often used in the former. Complex benthic landers with sensor packages and timing mechanisms have become a standard method for assessing mobile fauna in hard-to-reach areas (e.g., steep terrain and trenches).

Plankton nets

Top: researchers launch a moderate-sized and a small "bongo" sampler. The two sizes are to catch different size fractions of the plankton assemblage. Each sampler has two nets rigged such that there is no wire harness in front of the net mouth. This decreases avoidance of the nets by the planktonic organisms.

Benthic lander

Left: a benthic lander, in this case an early version of the "Eye in the Sea" camera system that can be rigged with either bait or an artificial lure to attract and photograph or video-record scavenging animals.

Box corer

A box corer grabs a sample of the muddy seafloor in the Chukchi Borderlands, north of the Arctic Chukchi Sea. Researchers can use these samples to study the marine organisms that live in the sediments on the ocean floor.

PRESSURE DROP SHIP

GUAM

660 FT. (200 M)
HUMPBACK WHALE

ACRYLIC-BUBBLE HUMAN-OCCUPIED VEHICLE (HOV)

3,280 FT. (1,000 M)
GLIDER

ELEPHANT SEAL

6,560 FT. (2,000 M)

CUVIER'S BEAKED WHALE

9,840 FT. (3,000 M)
SHARK

DEEP GLIDER

13,125 FT. (4,000 M)

16,400 FT. (5,000 M)
AUV (AUTONOMOUS UNDERWATER VEHICLE)

ROV (REMOTELY OPERATED VEHICLE)

19,700 FT. (6,000 M)
HOV (HUMAN-OCCUPIED VEHICLE)

22,970 FT. (7,000 M)

26,250 FT. (8,000 M)
MARIANA SNAILFISH

29,500 FT. (9,000 M)
AMPHIPOD

32,800 FT. (10,000 M)

FORAMINIFERA   EXTREME ROV   EXTREME HOV

CHALLENGER DEEP
36,000 FT. (11,000 M)

Relative maximum operating depths for some of the types of submersibles used to study and work in the deep ocean. Both crewed and crewless submersibles exist that can reach the deepest parts of the ocean, but most have shallower depth constraints. Note that if this were drawn to scale, the ship (224 ft./68 m, in length) would be barely visible, and the submersibles and animals would not be visible.

- **Guam**

  A Pacific Ocean island, Guam is the closest land to the Mariana Trench (about 200 mi./322 km away). With such islands, the land visible above the ocean surface is just the top of a huge mountain rising from the deep sea.

- **Mariana Trench (pictured)**

  The deepest of a few dozen geological gashes in the bottom of the ocean. Caused by the movement of tectonic plates covering the earth's surface, these small, ultra-deep areas are unique but very poorly known.

- ***Pressure Drop* ship**

  Studying the deep ocean requires investment in major equipment. For example, the *Pressure Drop* is one of many moderately large, specialized ships designed to operate in sometimes difficult conditions at sea while pursuing specific exploration and research goals.

- **Humpback Whale**

  Megaptera novaeangliae

  Humpback Whales dive to a depth of 660 ft. (200 m).

- **Acrylic-bubble human-occupied vehicle**

  A type of submersible in which people can go into the deep ocean. This type has been limited to a depth of about 3,280 ft. (1,000 m) because of the limited strength of the Plexiglas used to construct the high-visibility chamber in which the humans sit, but technological advancements are increasing this depth limit.

- **Glider**

  A special type of AUV (see below). Because gliders use hydrodynamics and changes in buoyancy rather than motors, they don't require large batteries and can independently pursue long-term missions.

- **Elephant seal**

  The largest and deepest-diving pinniped (seal). There are two species, Northern Elephant Seal (*Mirounga angustirostris*) and Southern Elephant Seal (*M. leonina*).

- **Cuvier's Beaked Whale**

  Ziphius cavirostris

  The deepest-diving whale—it has been recorded at a depth of 9,816 ft. (2,992 m) and routinely forages down to 3,280 ft. (1,000 m).

- **Shark**

  Many people are surprised to learn that sharks don't live in the deepest parts of the oceans. Although there are many species of deep-sea sharks, some quite large and long-lived, they are not found below about 9,840 ft. (3,000 m).

- **Deep glider**

  Although most gliders are limited to depths of 3,280 ft. (1,000 m), new developments are increasing the depth range of these and other technologies.

- **AUV**

  An autonomous underwater vehicle follows a preprogrammed set of commands for navigation and data collection. Because there is no tether, communications with the vehicle are very limited until it is recovered and the data downloaded.

- **ROV**

  A remotely operated vehicle is connected by a tether to a human operator, generally aboard a ship at the surface. The ROV can send video and other data up the tether in real time.

- **HOV**

  Human-occupied vehicles able to go deeper than 3,280 ft. (1,000 m) have a titanium sphere with small, very thick portholes for seeing out. Because the outside of the porthole is much larger than the inside, the great pressure at depth holds it in place.

- **Snailfish**

  The deepest-known fish, and therefore the deepest-living vertebrate, is the Mariana Snailfish (*Pseudoliparis swirei*), which lives down to about 26,250 ft. (8,000 m).

- **Extreme HOV and extreme ROV**

  Both HOVs and ROVs specially designed to withstand extreme pressure have explored some of the deepest parts of the deepest trenches.

- **Challenger Deep**

  At about 35,840 ft. (10,925 m), the Challenger Deep is the deepest part of the deepest trench, the Mariana, and therefore the maximum depth of the ocean.

- **Amphipod**

  Hirondellea gigas

  This large (1.2 in./3 cm) scavenging crustacean lives at the deepest depths of the Mariana Trench, protecting its exoskeleton with aluminum hydroxide gel.

Submersibles

Submersible vehicles come in a variety of configurations useful for various purposes. A fundamental dichotomy of submersible types is based on whether or not humans actually ride them into the deep. Human-occupied vehicles (HOVs) can be further separated based on the construction of the chamber into which the humans squeeze themselves. In the most comfortable (and most fun) versions, humans sit in a transparent sphere with visibility in almost every direction. The disadvantage in this design has been that the clear bubble is limited to a maximum depth of about 3,280 ft. (1,000 m). To go deeper requires a titanium sphere engineered with small portholes through which the pilot and observers can look. The depth limit on these varies among vehicles, depending on design and construction; it may be 3.7 mi. (6,000 m) or full-ocean depth of 6.8 mi. (11,000 m). Both kinds can be outfitted with a variety of exterior tools for photography, manipulation of objects, or collection of samples, in addition to environmental sensors.

Innovation is increasing the variety of robotic vehicles. These can also be divided into two fundamental types: those that are connected to, and controlled by, a human operator via a cable vs. those that operate independently following a program set before launching. Currently, the workhorse of deep-ocean submersible operations for both science and industry is the cabled remotely operated vehicle (ROV), usually controlled from a ship at the surface high above the ROV. This requires careful coordination between the ship operations and navigation of the ROV. One advantage of ROVs over human-occupied submersibles is that with an ROV, multiple scientists can watch and direct in real time from the ship, or even ashore via the Internet, whereas in occupied submersibles only those aboard the submersible can have that privilege during the dive. Again, these submersibles can be outfitted with tools, and their depth limit depends on design and construction, from near-surface to full-ocean depth.

The other type of robot is the autonomous underwater vehicle (AUV). These are preprogrammed and don't require human intervention during a dive. There are many types of AUV that are designed for a variety of missions. Depth capability also varies by design and construction. Some have been designed to go places inaccessible to other methods, such as under Antarctic ice shelves. AUVs have several advantages, including: (1) they can be launched from a ship without specialized positioning or winches with thousands of meters of cable; (2) the research vessel can pursue other objectives between launching and retrieval; and (3) multiple AUVs can be programmed to work together.

A BRIEF HISTORY OF DEEP-SEA EXPLORATION

A chronicle of study and discovery in the oceans

Seafaring people have probably had some awareness since prehistoric times that the oceans are deep and populated by strange animals. Art from classical civilizations includes recognizable representations of deep-sea animals. However, scientific exploration of the deep sea did not begin until the late 18th century when accurate maps of the world's oceans first became available. In 1775, the French mathematician Pierre-Simon Laplace measured tides on the coasts of Brazil and Africa, and his dynamic theory of tides predicted the average depth of the Atlantic Ocean to be about 13,000 ft. (3,962 m), within 10% of the modern accepted value, marking the beginning of science of the deep sea.

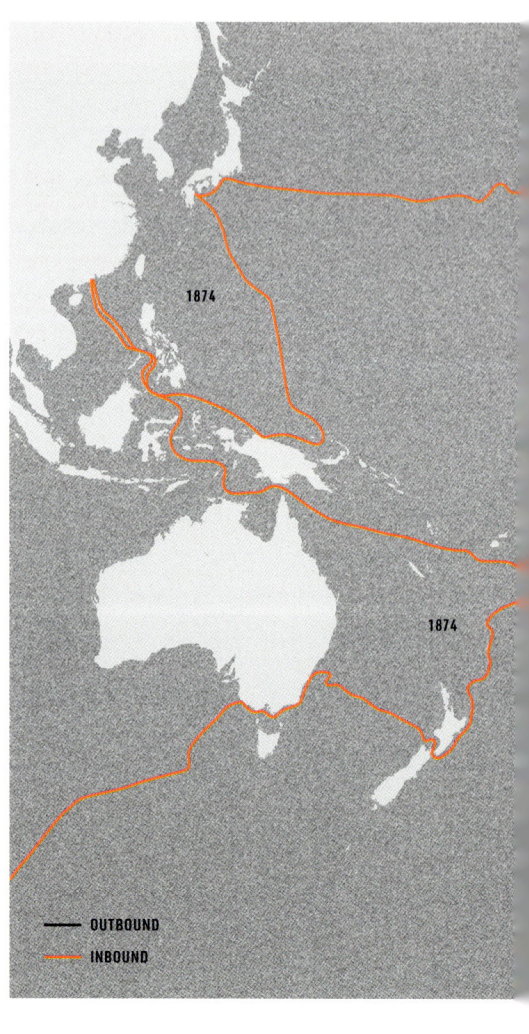

HMS *Challenger*

This 225 ft. (69 m) British Navy three-masted corvette had an auxiliary steam engine specially adapted for a round-the-world voyage of ocean exploration, with removal of most of the guns. The ship sailed from Portsmouth, England, on December 21, 1872, with 237 men on board, covered 67,000 nautical miles and returned on May 24, 1876.

Voyage of discovery

The track of *Challenger*'s voyage across all the major oceans. The crew sailed close to Antarctica and photographed icebergs, but did not sight land.

1874

1874

—— OUTBOUND
—— INBOUND

Early scientific exploration

In 1810, the naturalist Antoine Risso began describing fishes living down to 6,560 ft. (2,000 m) deep in the Mediterranean Sea off his native city of Nice, France. He named 21 species that had been caught by fishermen's nets, traps, and hooks, including species of sleeper sharks, lanternfishes, dragonfishes, slickheads, rattails, and spiny eels (see Benthic Fishes, pages 88–91 and Pelagic Fishes, 104–109). In 1818, British Navy officer John Ross retrieved animals on sounding lines from depths below 3,280 ft. (1,000 m), and in 1839–43, his nephew, Captain James Clark Ross, found evidence of life down to 5,900 ft. (1,800 m) off Antarctica. However, in 1843, the naturalist Edward Forbes of Edinburgh, finding that the abundance of life in bottom dredge samples decreased with depth, postulated that depths greater than 300 fathoms (1,800 ft./550 m) must be devoid of life. This "azoic hypothesis" of the deep sea briefly gained widespread attention.

But further dredge sampling by the zoologist Michael Sars during 1860–70 cataloged over 400 species living 2,690 ft. (820 m) deep off Norway, and the Scottish naturalist Charles Wyville Thompson found life down to 14,070 ft. (4,289 m) west of the British Isles. Thompson obtained funds and a British Royal Navy ship, the HMS *Challenger*, to lead a voyage around the world (1872–76) to investigate the physical conditions, chemistry, and biology of the deep-ocean basins. The voyage was an outstanding success, laying the foundations of modern deep-sea oceanography using specially developed new methods. Achievements included the definition of the general shape of the great ocean basins, the mapping of bottom deposits and currents, and the discovery of 4,717 new species. Thompson died shortly after the voyage, so the 50 volumes of the expedition report were edited by fellow researcher John Murray.

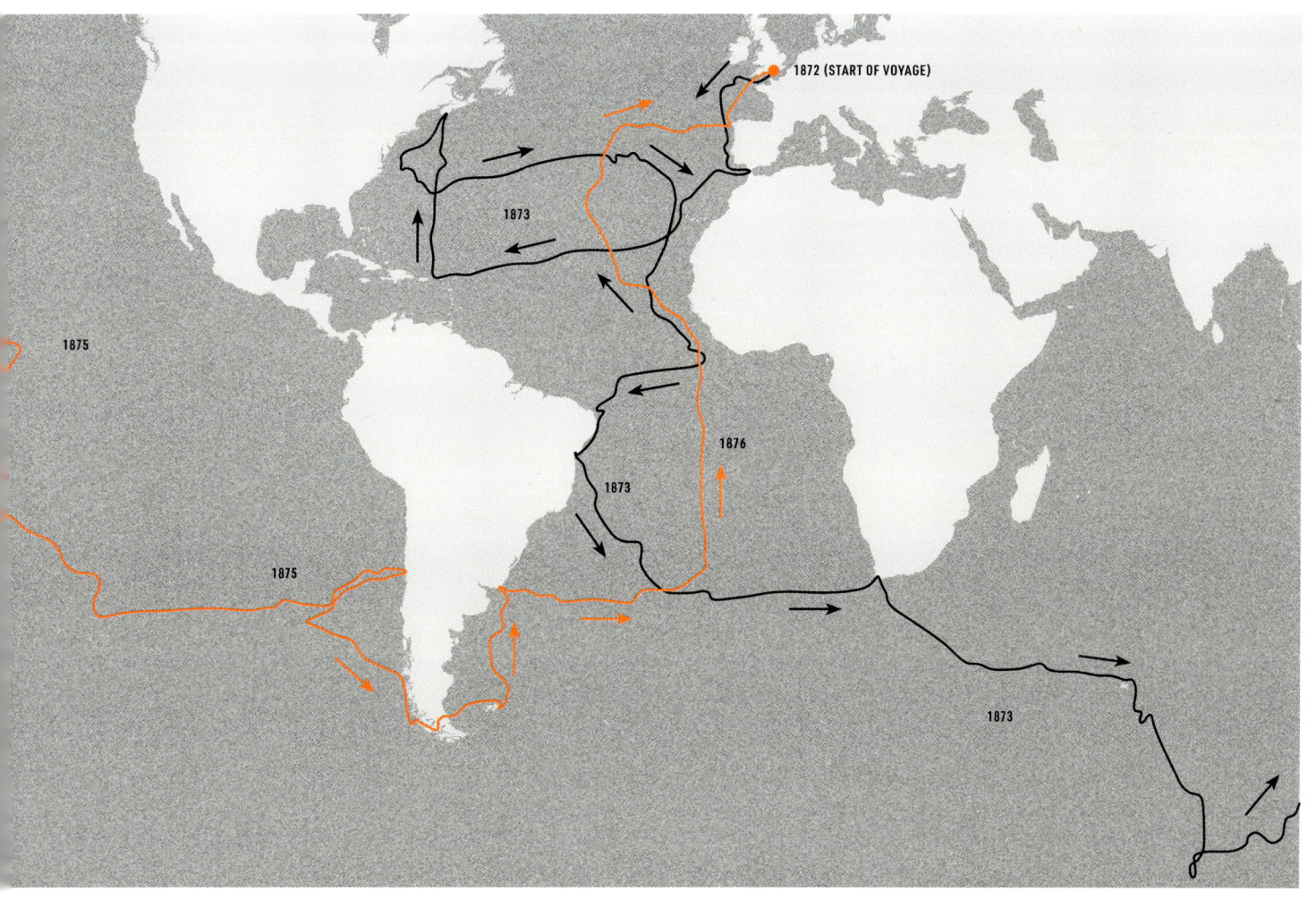

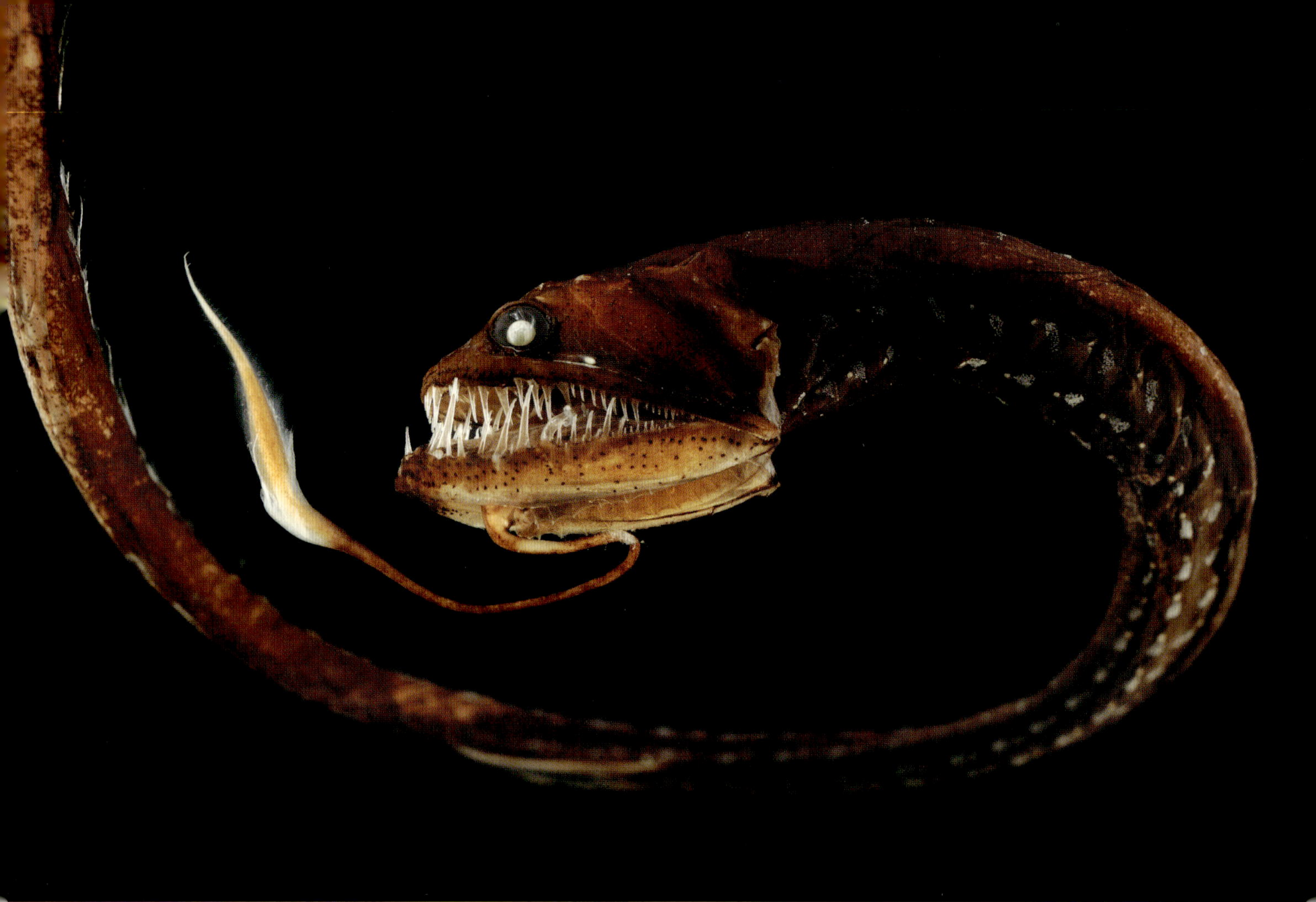

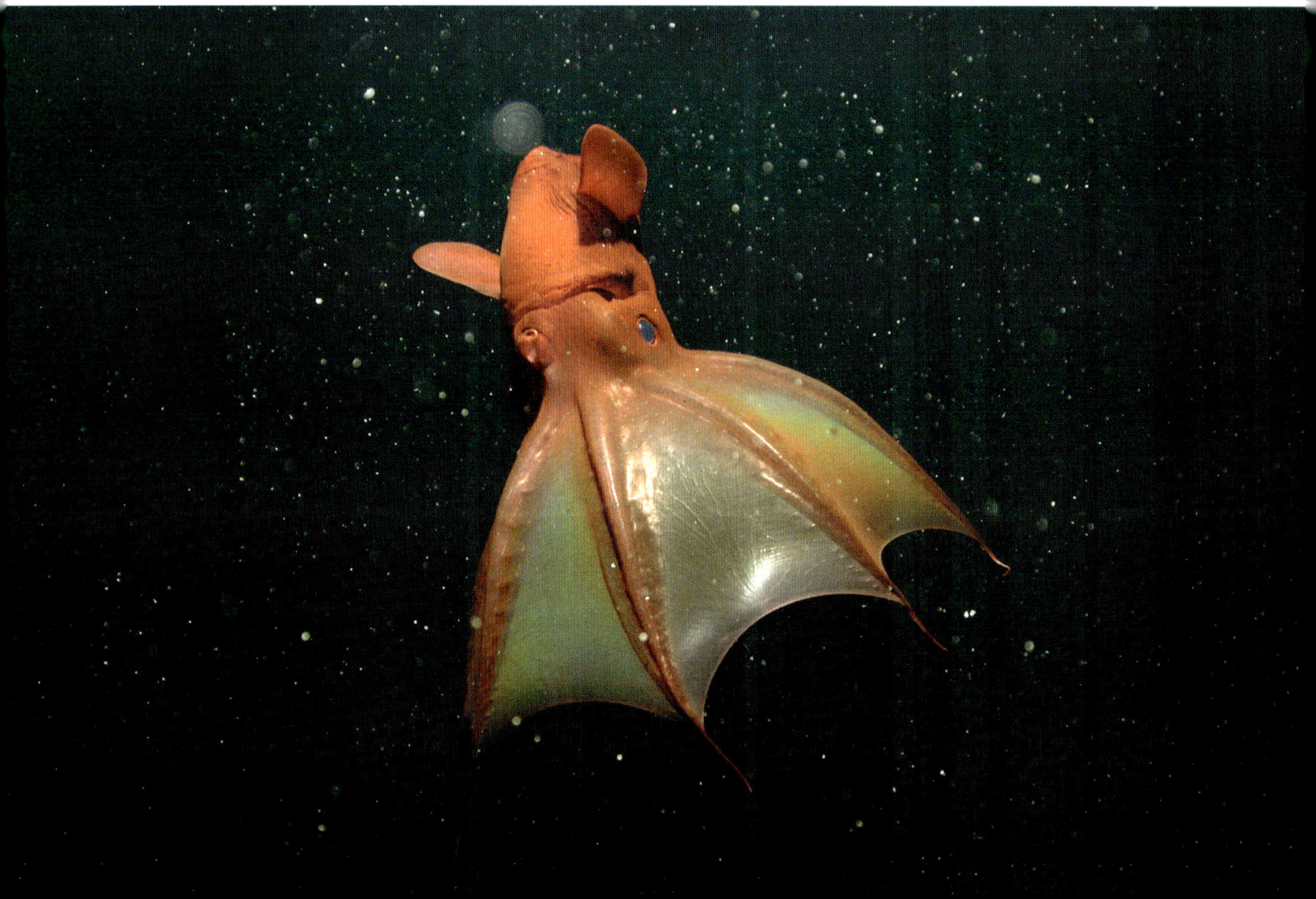

A golden age of deep-sea exploration

Following the *Challenger*, other nations launched expeditions, notably the *Gazelle* (Germany, 1874–76), investigating deep-water circulation; the *Travailleur* and *Talisman* (France, 1880–83), deep-sea fishes; voyages in the west Atlantic and the Pacific sponsored by Alexander Agassiz (United States, 1877–91); the *Vityaz* (Russia, 1886–89), which circumnavigated the globe taking physical measurements; the deep-sea expedition under zoologist Carl Chun (Germany, 1898–99), which made major advances in biology using large fine-mesh nets; and ocean exploration by Prince Albert I of Monaco (1885–1915), using a succession of research yachts. The first transatlantic underwater telegraph cable had been completed in 1858, and the surveys of the routes for this and other cables around the world further advanced knowledge of the deep sea.

In 1904, ocean physicist Vagn Walfrid Ekman of Sweden showed how kinetic energy is dissipated by internal waves at interfaces between water masses of different densities. Such waves are now thought to govern most turbulent mixing in the ocean interior (see pages 60–65). In 1912, the German geophysicist Alfred Wegener first proposed the theory of continental drift, important for understanding the long-term dynamics of deep ocean basins. But the onset of World War I, in 1914, ended a golden age of deep-sea exploration.

The postwar return to exploration lasted from 1918 to 1939, until the next global war. Danish expeditions under the biologist Johannes Schmidt (1920–30) discovered the oceanic oxygen-minimum zones (see pages 172–173) and the deep-sea spawning of eels. A German expedition in the Atlantic (1925–27) completed the first full-ocean-depth hydrographic mapping of temperature, salinity, and oxygen across the ocean. Between 1930 and 1934, US explorer William Beebe and inventor Otis Barton made the first human descents into the deep sea, reaching a depth of 3,028 ft. (923 m) in their submersible vessel, known as a bathysphere.

The end of World War II triggered a new explosion of exploratory activity, starting with the Swedish deep-sea expedition of 1947–48. This was followed by the Danish *Galathea* expedition, under the leadership of oceanographer Anton Brunn, who in 1952 discovered life at 33,430 ft. (10,190 m) in the Philippine Trench. Continuous echo-sounding profiles, initially spurred by military development of sonar for anti-submarine warfare, began to reveal seabed features in unprecedented detail as well as curious "deep scattering layers" that moved toward the surface at night.

From 1957 to 1960, the oceanographer Henry Stommel developed the theory of abyssal circulation, showing how the great ocean basins are ventilated by cold oxygenated water originating in polar regions. This circulation system was dubbed the "great ocean conveyor" by Wallace Broecker in 1987 and became the foundation of later major studies such as the World Ocean Circulation Experiment (WOCE).

Pacific Blackdragon
Idiacanthus antrostomus
Discovered in the late 19th century, a time of flourishing deep-sea exploration, this dragonfish has a luminescent barbel under its chin, which it uses to lure prey into its vast mouth gape. The spots along the body are light organs. It lives in the eastern Pacific down to 3,600 ft. (1,100 m) coming toward the surface at night to prey on midwater fishes and crustaceans.

Vampire "Squid"
Vampyroteuthis infernalis
The Vampire "Squid" in the eastern North Pacific. With a confusing name, this is neither a vampire nor a squid. It is an unusual species but most closely related to octopods. Although biology of the species is best known from around Monterey Canyon, where red color predominates, its distribution is global in non-polar deep seas and elsewhere the skin is black. This black color, together with the deep web looking like a cloak between the arms, reminded scientists who discovered and described it at the beginning of the 20th century of descriptions of Dracula and other vampires, resulting in both the common and scientific names.

Bathysphere
Engineer Otis Barton (left) and naturalist William Beebe (right) with the 4.7-ft. (1.4-m) diameter steel bathysphere in which they descended to over half a mile (0.8 km) depth off Bermuda in 1934, observing mesopelagic bioluminescent creatures through 8-in. (20-cm) diameter quartz windows which are on the far side of the sphere not visible in this image. The large circular opening is the crew access hatch.

Surveying and exploring the seafloor

Mapping of the physical geography of the ocean floors made great strides in the mid-20th century. In 1957, geologist Bruce Heezen and cartographer Marie Tharp began producing physiographic maps of all the oceans, showing the extensive mid-ocean ridge system. In 1963, paleogeomagnetic studies by geophysicists Lawrence Morley, Frederick Vine, and Drummond Matthews provided the first direct evidence of seafloor spreading from the mid-ocean ridges. By the end of the decade, the theories of continental drift and plate tectonics were generally accepted. In 1997, geophysicists Walter Smith and David Sandwell produced new global high-resolution maps of the ocean floor using satellite data.

The mid-20th century also saw development of two types of submersibles, human-occupied vehicles (HOVs) and remotely operated vehicles (ROVs). These vessels increasingly provided scientists with direct access to the deep sea and the seafloor, opening up the possibilities for discovery. In 1960, explorers Jacques Piccard and Don Walsh reached the deepest point in the world ocean, the Challenger Deep, in the bathyscaphe *Trieste*.

The Woods Hole Oceanographic Institution (WHOI) improved on the bathyscaphe model with *Alvin*, launched in 1964, a more maneuverable submersible that contained imaging and sampling gear that enabled scientists to make new discoveries in the deep sea. In 1966, *Alvin* allowed WHOI biologist Richard Backus and others to directly observe that deep scattering layers in the Pacific Ocean were made up of vast numbers of mesopelagic fishes (see Pelagic Fishes, pages 104–109). With the assistance of *Alvin*, a team of researchers including oceanographers Jack Corliss and Bob Ballard discovered chemosynthetic life at hydrothermal vents in the Pacific Ocean in 1977. Similar chemosynthetic organisms were discovered via submersibles at hydrocarbon cold seeps by marine geologist Charles Paull and others in 1984. With both these discoveries, scientists realized that the deep sea contains a variety of specialized communities thriving in the absence of energy input from the sun (see Chemosynthetic Ecosystems, pages 158–167).

Discoveries made on the sea surface also informed studies of the deep. In 1980, the space engineer Warren Hovis and others published the first satellite images of sea-surface chlorophyll, and it became possible to produce global maps of sea-surface primary production and to quantify the flux of organic matter to the deep sea. Geoscientist Susumu Honjo of WHOI, in 1976, pioneered complementary direct measurement of flux by time-series sediment traps moored at different depths below the surface. These initiatives resulted in global programs of remote sensing and flux studies of the deep sea.

Today, exploration and discovery continue, including recognition of new deep-sea species by numerous individuals and institutions. Recent international coordination efforts have included the Census of Marine Life (CoML, 2000–2010), which studied marine biodiversity and abundance in the world's seas and oceans.

Bathyscaphe _Trieste_

Preparations prior to the first dive to the deepest point in the world's oceans on January 23, 1960. Jacques Piccard, the designer, and US Navy officer Don Walsh spent 20 minutes at the bottom of the Mariana Trench.

DSV _Alvin_

The first of a new class of human-occupied deep submergence vehicles delivered on June 5, 1964. It was capable of powered maneuvering in three dimensions down to 7,870 ft. (2,400 m). Recent versions of _Alvin_ and other HOVs can reach much greater depths.

Mapping the ocean floor

Marie Tharp, geologist and oceanographer, constructing a map of the Atlantic Ocean floor in the 1950s. Discovery of the global mid-ocean ridge system contributed to the theories of continental and plate tectonics.

"Human-occupied vehicles (HOVs) and remotely operated vehicles (ROVs) provided scientists with direct access to the deep sea and the seafloor, opening up the possibilities for discovery."

OCEANOGRAPHY

UNDERWATER TOPOGRAPHY

Continents, plates, ridges, and subduction zones

Previous pages: **Internal waves**
Satellite image of the Andaman Sea, off the coast of Thailand. Internal waves can be hundreds of meters tall and tens to hundreds of kilometers long—and yet moving all the way to the deep sea beneath the sea surface. The sunlight helps make the internal waves visible (patterns and ripples in the center of the image)—the waves in the deep displace the surface by about 4 in. (0.1 m).

About 3.8 billion years ago, the ocean most likely formed from the escape of water vapor and other gases from molten rocks to the atmosphere surrounding the cooling planet. Debate is ongoing whether comets, specifically those of a type known as "hyperactive," contributed some of the earth's water. Once the earth cooled below 212°F (100°C), water vapor condensed into rain, which filled the basins that we know now as our world ocean. Gravity prevented the water from leaving the planet.

Continental and oceanic plates

The earth's surface is divided into eight major and several minor tectonic plates—large, rigid pieces of the planet's crust and upper mantle. These plates are of two primary types: continental plates and oceanic plates. Of the major plates, the largest, the Pacific Plate, is the only one that is nearly entirely oceanic; it supports some tropical islands as landmasses above water. However, as landmasses take up only 30% of the earth's surface, most of the continental plates are about half above water and half under, forming substantial expanses of deep seafloor. For example, both the North American and South American tectonic plates reach well eastward under the ocean to the Mid-Atlantic Ridge, forming the Atlantic Ocean floor. The Australian Plate contains portions of both the Indian Ocean floor and the Pacific Ocean floor.

While the two types of plates share many characteristics, several differences distinguish them. Oceanic plates average about 4–5 mi. (6–8 km) thick, compared to an average of 25 mi. (40 km) for continental plates. Oceanic plates are composed of basalt rich in iron, magnesium, and calcium. Continental plates are dominated by granitic rock with abundant silica, aluminum, sodium, and potassium. Because of their heavy ferromagnesian elements, oceanic plates are about one-fifth denser than continental plates. This difference in density causes oceanic plates to descend (or subduct) beneath the lighter continental plates when they meet. It also allows the denser oceanic plates to sink farther into the earth's upper mantle, causing them to lie below sea level. The more buoyant continental plates float higher on top of the mantle, resulting in dry land.

Major tectonic plates
The division of tectonic plates on the earth's surface. The Pacific Plate is the largest.

Seafloor spreading ridges
The topography of the earth reveals the seafloor spreading ridge system at a depth of 8,200 ft./ 2,500 m (yellow). Deep ocean trenches (purple) are the sites where the cool and dense plates sink into the earth. The abyssal plains are shown in dark blue.

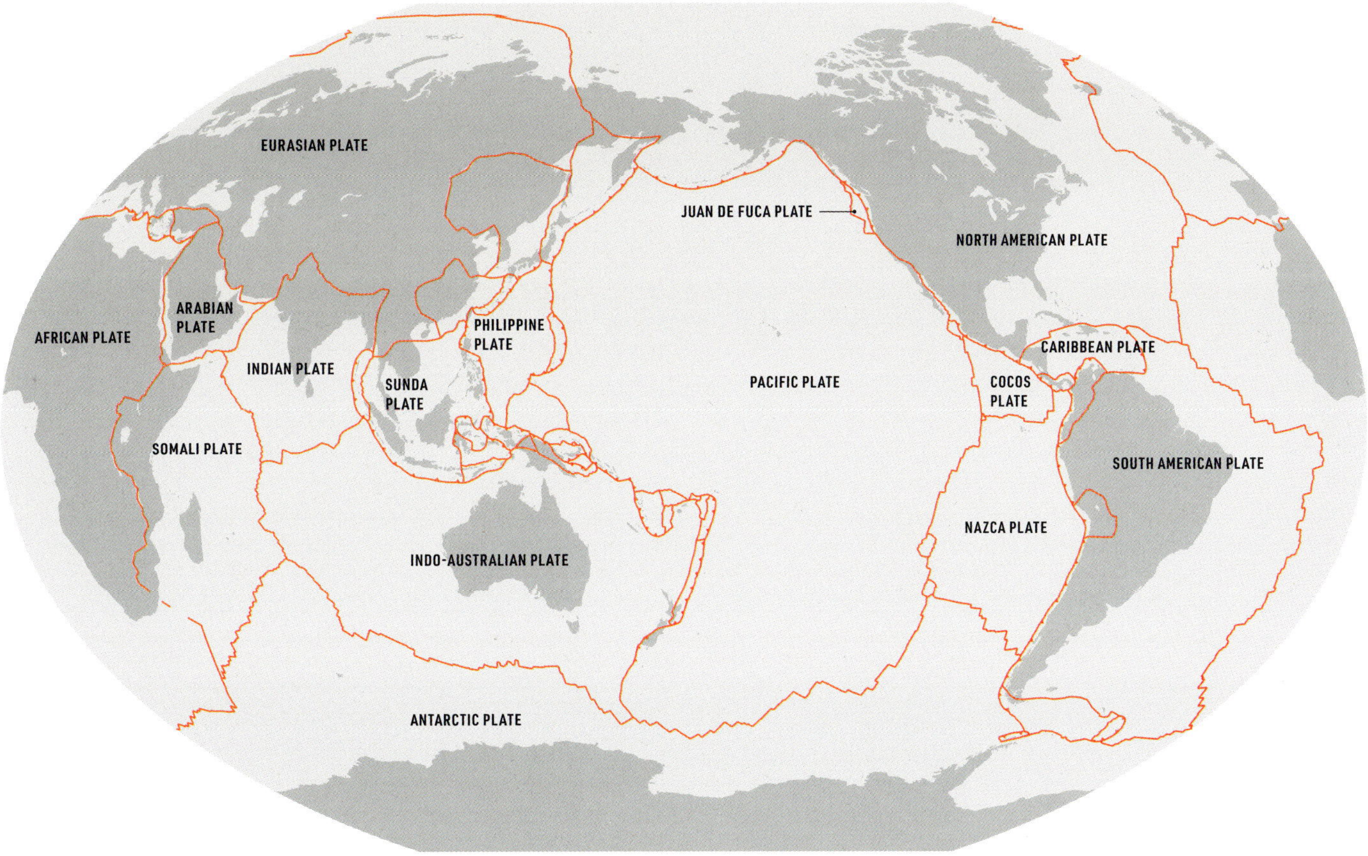
EURASIAN PLATE
JUAN DE FUCA PLATE
NORTH AMERICAN PLATE
ARABIAN PLATE
AFRICAN PLATE
PHILIPPINE PLATE
CARIBBEAN PLATE
INDIAN PLATE
SUNDA PLATE
PACIFIC PLATE
COCOS PLATE
SOMALI PLATE
SOUTH AMERICAN PLATE
NAZCA PLATE
INDO-AUSTRALIAN PLATE
ANTARCTIC PLATE

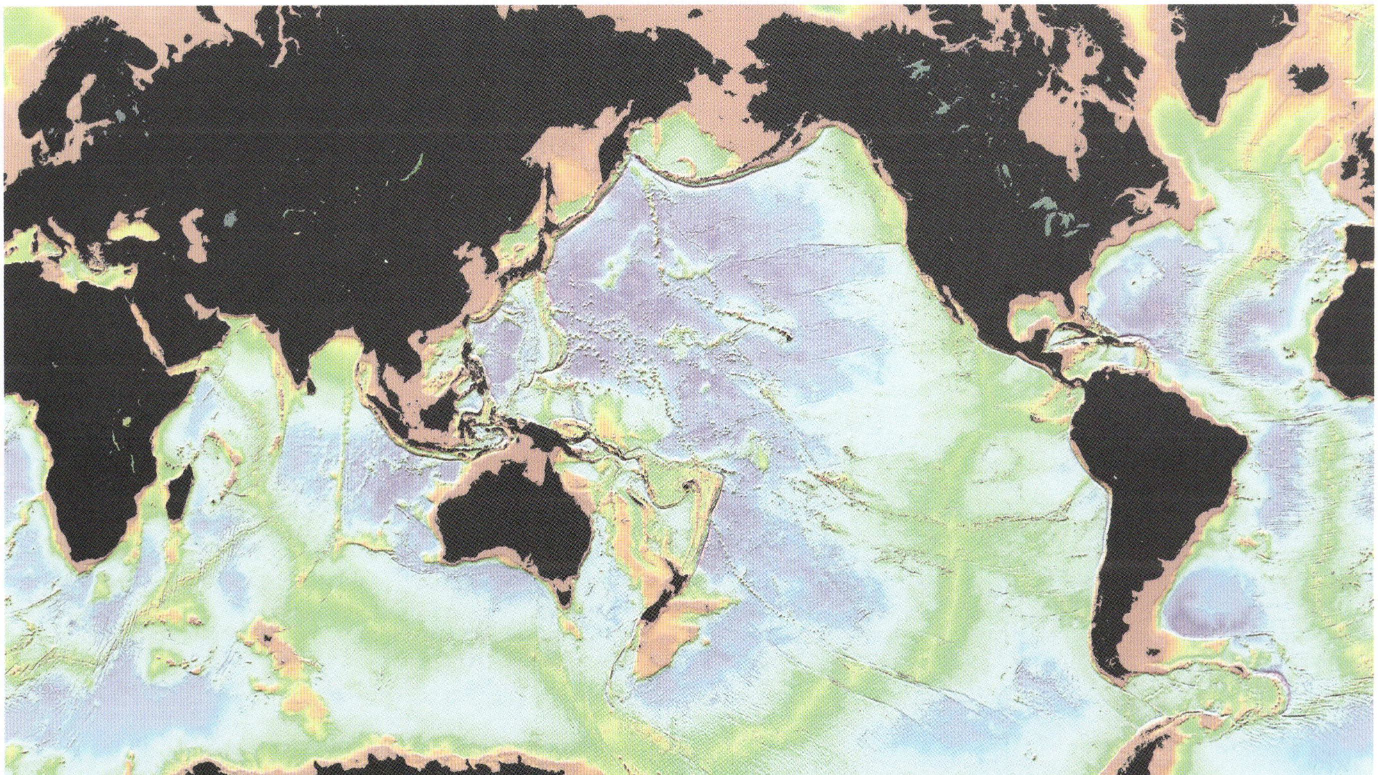

Ridges and subduction zones

Oceanic plates are formed at divergent plate boundaries along mid-ocean ridges, where upwelling magma creates new oceanic crust. As lava flows from these volcanic ridges, it quickly cools, forming igneous rock. The spread of the Atlantic Ocean at the Mid-Atlantic Ridge, a large chain of volcanic seamounts, all lying underwater except for a few islands, is about 1 in. (2.5 cm) per year. Continental plates are formed primarily by converging plate boundaries, where oceanic plates collide with, and plunge underneath, continental plates—a process called subduction. As oceanic plates subduct, they melt to form magma. This magma cools over millions of years, producing new continental crust.

Oceanic and continental plates differ considerably in age because of plate tectonics, processes that control the structure and properties of the earth's crust and its evolution through time. Divergent plate boundaries continually renew oceanic plates, while the subduction zones of convergent boundaries continually recycle them. As a result, the oldest oceanic rocks are less than 200 million years old. In contrast, continental plates take a long time to form but are rarely destroyed. Much of the continental crust exceeds 1 billion years in age, and its oldest rocks may be as old as 4 billion years.

The subduction zones lead to the deepest points on Earth: the hadal trenches, which exceed 19,700 ft. (6,000 m) in depth. Trenches deeper than 27,500 ft. (8,400 m) are found only in the Pacific Ocean. Along the perimeter of the Pacific oceanic plate are numerous deep spots; the deepest, down to around 32,800 ft. (10,000 m), are found only on its western side. Ocean trenches also show variation in topography and thus in depth. For example, the Mariana Trench is 1,550 mi. (2,500 km) long, but only a handful of spots, less than 30 mi. (50 km) long, are deeper than 32,800 ft. (10,000 m), including three in the Challenger Deep, one of which is the deepest place on Earth. The nearby continental plates are volcanic in nature and are known earthquake areas.

Abyssal plains

Shallower than the trenches, huge relatively flat seafloor areas are called abyssal plains. There, too, topography may show variability, in the form of abyssal hills and also isolated mountains. Such mountains can reach high up, even above the sea surface, forming islands. When such mountains remain under the surface, they are called seamounts, or guyots when their tops are flat. If they are part of a volcanic arc they may form hot spots of unusually warm and metal- and nutrient-rich waters.

All underwater topography is of importance for deep-sea life. In areas where minerals are spread into the ocean, waters are enriched, and particularly adapted life may thrive. Currents will redistribute the enriched waters. The topography also focuses deep-sea motions in various ways, and they may become so vigorous that turbulent mixing results (see pages 52–65).

Seafloor regions

The naming of various
seafloor regions. All occur
in the deep sea, except the
continental shelf.

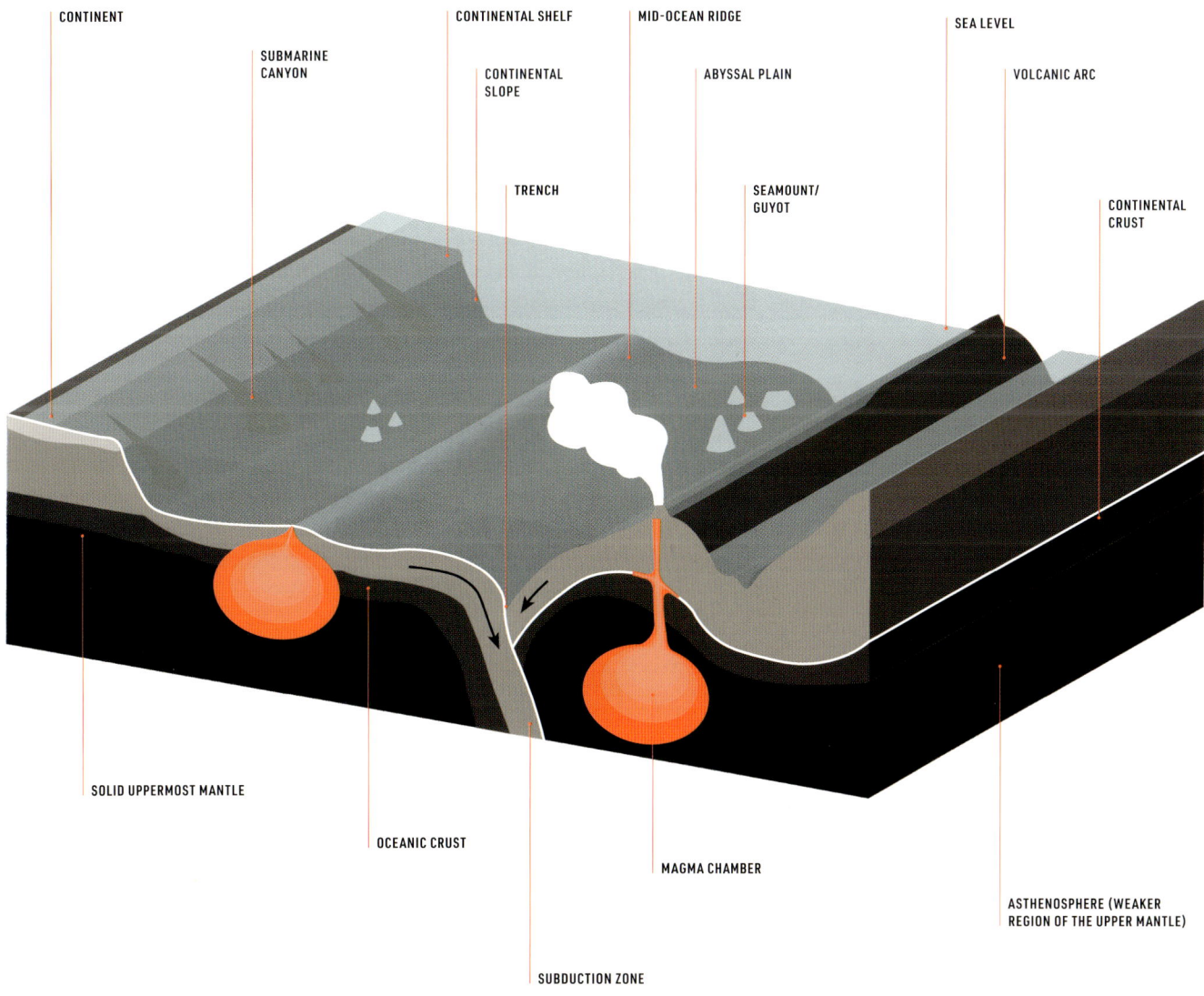

CONTINENT

SUBMARINE
CANYON

CONTINENTAL SHELF

CONTINENTAL
SLOPE

TRENCH

MID-OCEAN RIDGE

ABYSSAL PLAIN

SEAMOUNT/
GUYOT

SEA LEVEL

VOLCANIC ARC

CONTINENTAL
CRUST

SOLID UPPERMOST MANTLE

OCEANIC CRUST

MAGMA CHAMBER

ASTHENOSPHERE (WEAKER
REGION OF THE UPPER MANTLE)

SUBDUCTION ZONE

PROPERTIES OF SEAWATER

It's more than just water

Of course, ocean water is not just water. The average density of seawater at the surface is around 1,025 kg/m³; this is more than 800 times denser than air (which has a density of 1.2 kg/m³). Seawater is denser than fresh water (which has a density of 1,000 kg/m³ at 39°F/4°C) because the dissolved salts increase the mass by a larger proportion than their addition to volume. On average, seawater in the world's oceans has a salinity (the amount of dissolved inorganic matter) of about 3.5%. This used to be referred to as "35 per mil," or parts per thousand (ppt), but this is now considered a dimensionless number, as is "salinity of 35 g/kg." This means that every kilogram (roughly about a liter by volume) of seawater contains approximately 35 g (7 teaspoons) of dissolved salts. The dominant dissolved ions are sodium ($Na+$) and chloride ($Cl-$), with smaller amounts of magnesium, sulfate, and calcium. In comparison, human blood has a salinity of about 9 g/kg. The freezing point of seawater decreases as the salt concentration increases. At typical salinity, freezing occurs at about 28.4°F (−2°C), the lowest temperature of liquid seawater.

Speed of sound

In the deep sea, values of different parameters may vary from their near-surface values. Temperature generally decreases with increasing depth, while salinity shows a more variable vertical profile in the deep sea, depending on the location. In general, most vertical density variations are caused by differences in temperature. After correction for slight, but nonnegligible, compressibility effects, density increases seemingly monotonically with depth—a monotonic trend is either only increasing or only decreasing. Speed of sound, however, is not a monotonic function of depth. Minimum speeds occur at intermediate depths, between about 3,280 and 4,290 ft. (1,000 and 1,500 m), because temperature, pressure, and salinity affect the speed of sound. Near the surface, temperature effects dominate, with lower speeds at lower temperatures, while deeper down pressure dominates, with higher speeds at higher pressure. This results in the SOFAR (sound fixing and ranging) channel of minimum sound speed at intermediate depths, causing sound waves to be retained within that depth range while propagating long distances. Low-frequency sound waves within the channel may travel thousands of kilometers before dissipating.

Oxygen saturation

Similar to the profile of sound speed with depth, oxygen saturation in the deep sea shows variation, reaching a minimum between about 1,640 and 4,920 ft. (500 and 1,500 m), depending on local physical and biological circumstances. This oxygen-minimum zone (OMZ) results from an interplay of relatively low turbulent mixing that limits supply of fresh oxygen from waters above and below and large oxygen consumption by bacteria feeding on organic matter raining down from the surface. The source of ocean oxygen is the exchange with atmospheric air. Most of the organic matter is decomposed in the upper 4,920 ft. (1,500 m), and oxygen saturation is relatively high in deeper waters due to low consumption.

Water properties

Depth profiles showing
the temperature, salinity,
density, speed of sound,
and oxygen saturation in
the Puerto Rico Trench in
the western Atlantic Ocean.

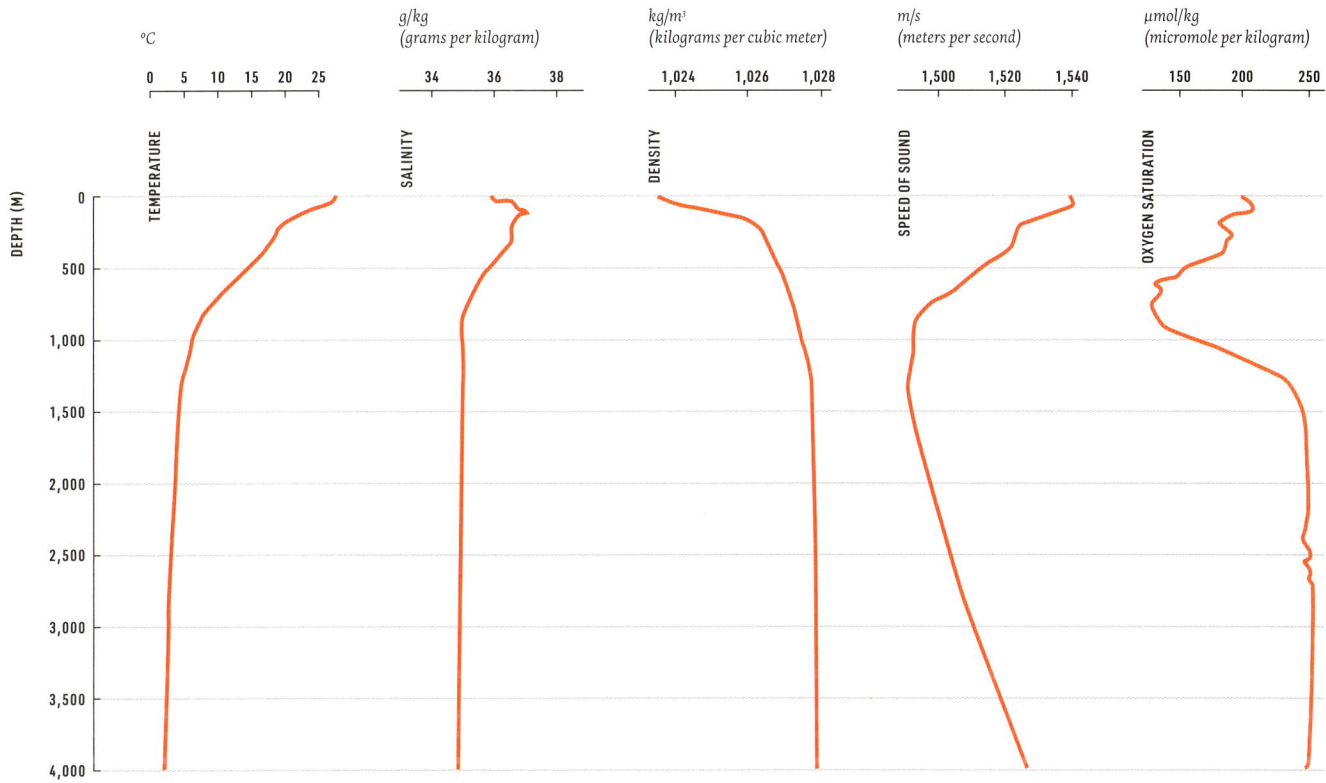

Sound waves

Fin Whales (*Balaenoptera
physalus*) are known to
dive down to the SOFAR
(sound fixing and
ranging) channel of
minimum sound speed at
intermediate depths
to communicate with
other Fin Whales large
distances away.

pH and salinity

Seawater is basic (alkaline), as indicated by typical pH values are larger than 7, limited to a range between 7.6 and 8.2. Generally the higher (relatively more basic) values are found near the surface, and the lower (relatively more acidic) values are found in the deep sea, reaching minimum values around 3,280 ft. (1,000 m), a depth profile that is very similar to the deep-sea oxygen profile.

Ocean salinity has been stable for billions of years, most likely a consequence of a chemical/tectonic system that removes as much salt as is added. Seawater contains more dissolved inorganic matter than fresh water, and the ratios of solutes differ dramatically. For example, seawater contains about 2.8 times more bicarbonate than river water, governing pH and alkalinity in seawater. However, the percentage of bicarbonate in seawater as a ratio of all dissolved ions is far lower than it is in river water. Differences like these are due to the varying times that seawater solutes remain in solution; sodium and chloride have very long residence times, while calcium (vital for carbonate binding—i.e., shell building by marine animals) tends to precipitate much more quickly. Small amounts of other substances are found, including amino acids at concentrations of up to 2 micrograms of nitrogen atoms per kilogram, which are thought to have played a key role in the origin of life.

Microbial life

Seawater also contains abundant amounts of microbial life in the form of bacteria, archaea (single-celled prokaryotic microorganisms), and viruses. DNA research has revealed a great diversity in microbial life forms, to the extent that a bucket of seawater may hold more than 20,000 species. The world's oceans are thought to contain over 10 million species. Bacteria are found at all depths in the deep sea as well as in seafloor sediments; some are aerobic (consuming oxygen), others anaerobic (not requiring oxygen for growth). Most are free-swimming, but many exist within other organisms—for example, bioluminescent light-emitting bacteria. Cyanobacteria (blue-green algae) played an important role in the evolution of ocean processes, enabling the development of stromatolites (microbial reefs) and oxygen in the atmosphere. Some marine bacteria survive in a pH range of 7.3–10.6, while other species will grow only at pH 10.0–10.6, exceptional and unusually alkaline local conditions compared with general ocean waters.

Archaea, which also live in the deep sea, may constitute as much as half the ocean's biomass. These bacteria-like organisms survive in extreme environments, such as the hot and sulfurous hydrothermal volcanic vents on the seafloor. Studies of seafloor sediments have revealed archaea that break down methane as well as bacteria that break down seafloor rocks, thereby influencing seawater chemistry.

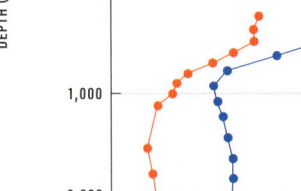

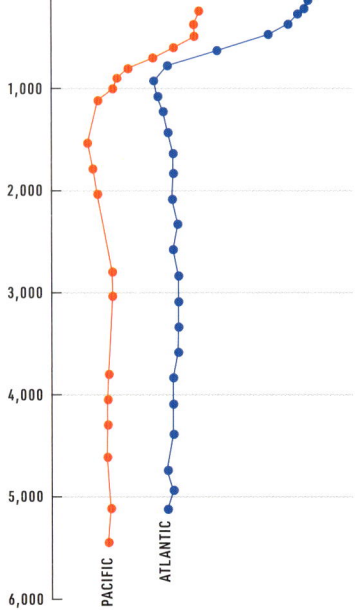

Deep ocean acidity

Ocean average profiles of pH as a function of depth. At all levels the Atlantic is less acid than the Pacific, because of shorter residence times of the water.

Bioluminescent bacteria

Top left: *Vibrio fischeri* is a bioluminescent bacterium that is found in many oceans around the world, working in a symbiotic relationship with certain deep-sea marine life.

Stromatolites

Bottom left: these microbial reef communities are formed from the precipitation of calcium carbonate by cyanobacteria.

Cyanobacteria

These single-celled organisms, which live in the deep sea, evolved to produce energy from sunlight and oxygenated the ocean. Top right: A micrograph of a diatom (oval) on cyanobacteria. Bottom right: A micrograph of *Prochlorococcus marinus*, a globally significant marine cyanobacterium

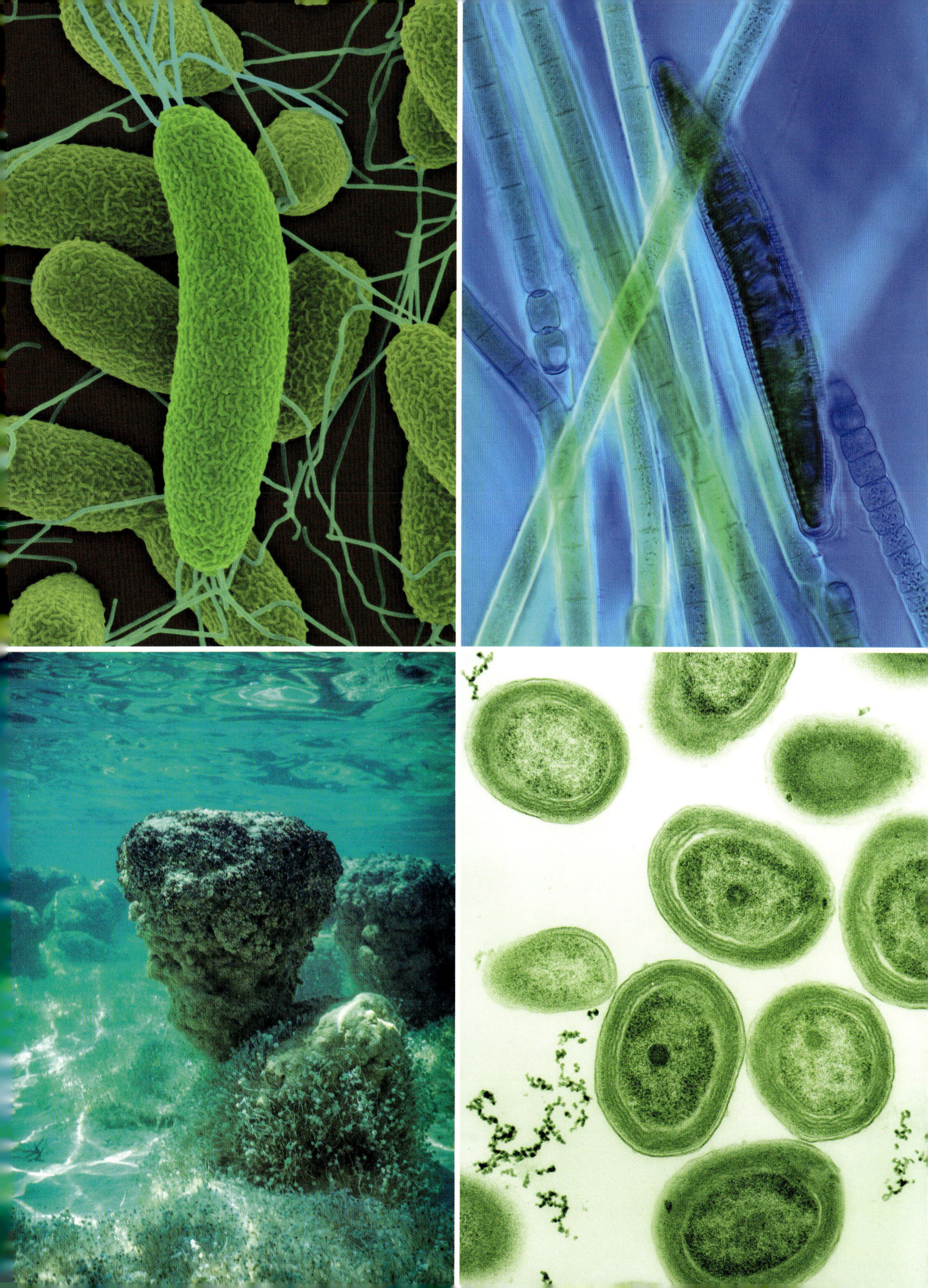

CURRENTS AND TURBULENCE

How ocean waters move

Water movements in the deep sea are not steady flows, but vary over multiple space and time scales. The spatial scales of variation range from ocean-basin scales for large flows of thousands of kilometers to millimeter scales for the smallest turbulence. The time scales vary between several hundreds of years of residence for ocean-basin flows, as determined from radiocarbon dating, to a hundredth of a second for turbulence-dissipation time scales. To date, measurement devices or numerical modeling capabilities have not been constructed to describe the dynamics of the deep sea at all scales. For example, large-scale ocean circulation models do not contain details of turbulent mixing processes. Whether quantifying these could ever be possible, given the complexity of intrinsic processes with various interactions between motions at different scales, is questionable. The dynamics of the ocean may prove to be inherently unpredictable and, until proven otherwise, they remain a mystery to solve, like the physics principles of turbulence. However, that does not preclude the study of limited-scale portions or specific dynamical processes in confined deep-sea basins.

On the large spatial scale, water motions in the deep sea follow those generated via the atmosphere near the sea surface, albeit commonly with smaller magnitudes. Westerlies, strong winds blowing from west to east around mid-latitudes in the Northern and Southern Hemispheres, drag water flows near the surface that eventually reach hundreds of meters into the deep sea. The winds are aided by equatorial trade winds blowing from east to west to set up gyres, large ocean-basin-scale circulations. Gyres consist of pressure-difference-driven flows that are intensified with the aid of the earth's rotation and steered by topography at the basin's western boundary. Examples of intensified wind-driven flows are the Agulhas Current in the Indian Ocean, the Kuroshio Current in the Pacific Ocean, and the Gulf Stream in the Atlantic Ocean. The world's largest flow (by volume of water transported) is the Antarctic Circumpolar Current, which is also mainly wind-driven. All these large-scale flows are diverted by the Coriolis effect, a pseudo-force resulting from the earth's rotation. In the Northern Hemisphere, a flow driven by a pressure difference is deflected to the right, up to the point that the Coriolis and pressure-difference forces balance and cancel each other out. Wind setup—winds blowing waters to pile up on one side—creates a surface pressure-difference force that is constant with depth. In the deep-sea interior, horizontal density differences (fronts) induce an independent pressure-difference force that varies with depth.

Coriolis effect

The Coriolis effect results in the deflection of major ocean currents to the right in the Northern Hemisphere and to the left in the Southern Hemisphere. The Coriolis force also results in major ocean spirals of ocean-circling currents called "gyres," directed clockwise in the Northern Hemisphere and counterclockwise in the Southern Hemisphere.

Ocean currents

The circulation on the ocean surface driven by westerly winds and the earth's rotation can reach deeper than 660 ft. (200 m).

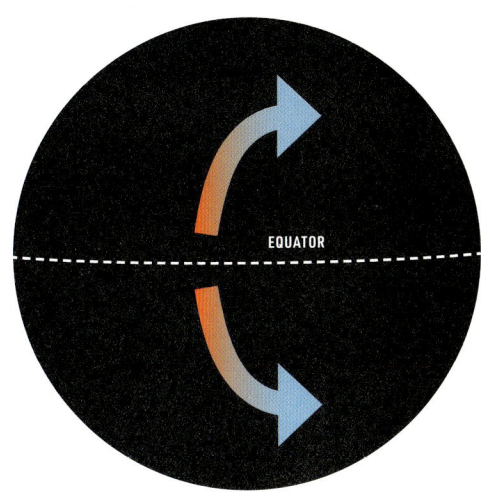

EQUATOR

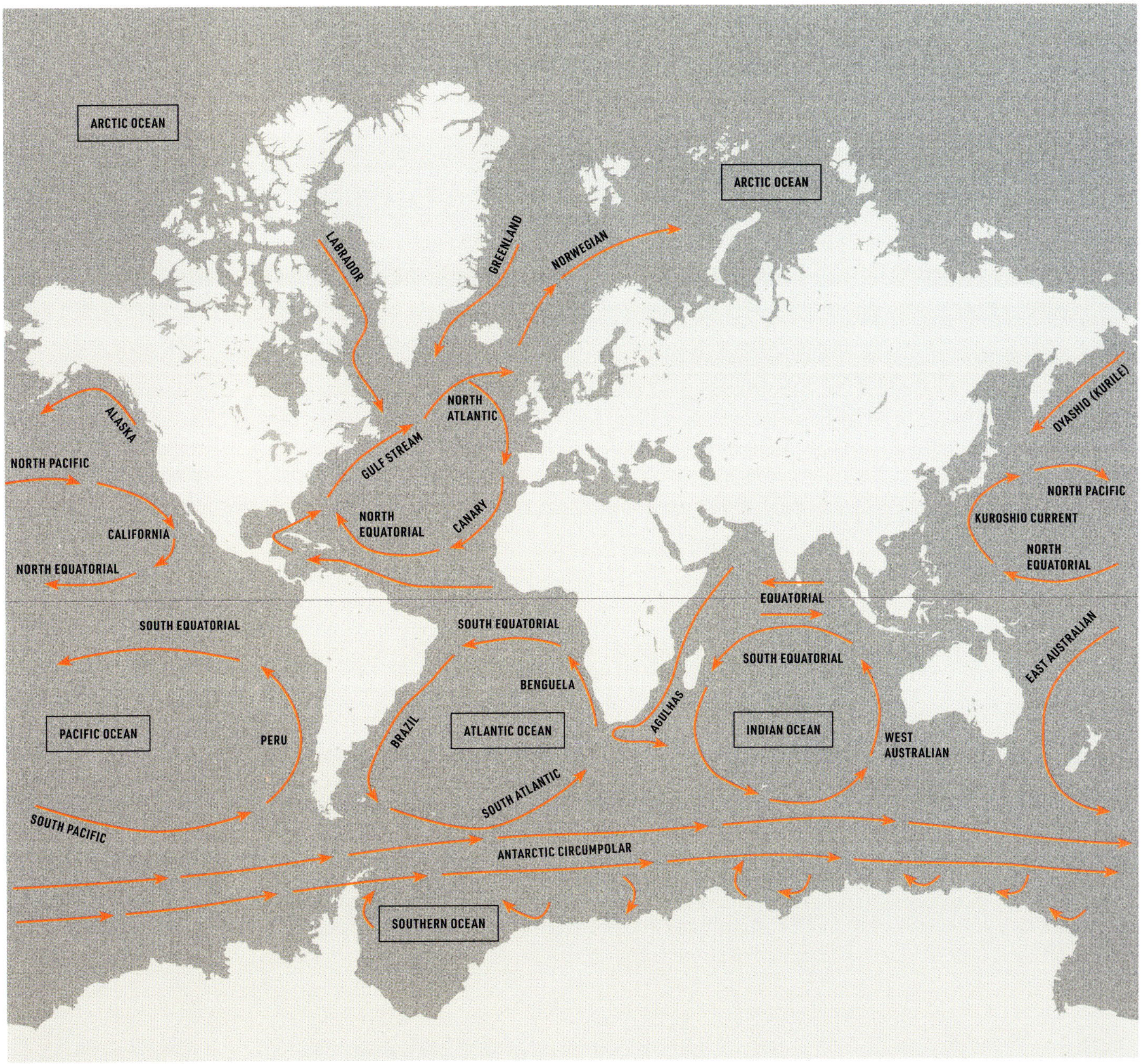

ARCTIC OCEAN

ARCTIC OCEAN

LABRADOR

GREENLAND

NORWEGIAN

OYASHIO (KURILE)

ALASKA

NORTH PACIFIC

NORTH
ATLANTIC

GULF STREAM

NORTH PACIFIC

KUROSHIO CURRENT

CALIFORNIA

NORTH
EQUATORIAL

CANARY

NORTH
EQUATORIAL

NORTH EQUATORIAL

EQUATORIAL

SOUTH EQUATORIAL

SOUTH EQUATORIAL

SOUTH EQUATORIAL

EAST AUSTRALIAN

BENGUELA

PACIFIC OCEAN

PERU

BRAZIL

ATLANTIC OCEAN

AGULHAS

INDIAN OCEAN

WEST
AUSTRALIAN

SOUTH ATLANTIC

SOUTH PACIFIC

ANTARCTIC CIRCUMPOLAR

SOUTHERN OCEAN

Deep-sea flows

These flows can often be found in opposite directions to those near the surface. Deep-sea flows over underwater topography like seamounts and mid-ocean ridges generate turbulent mixing.

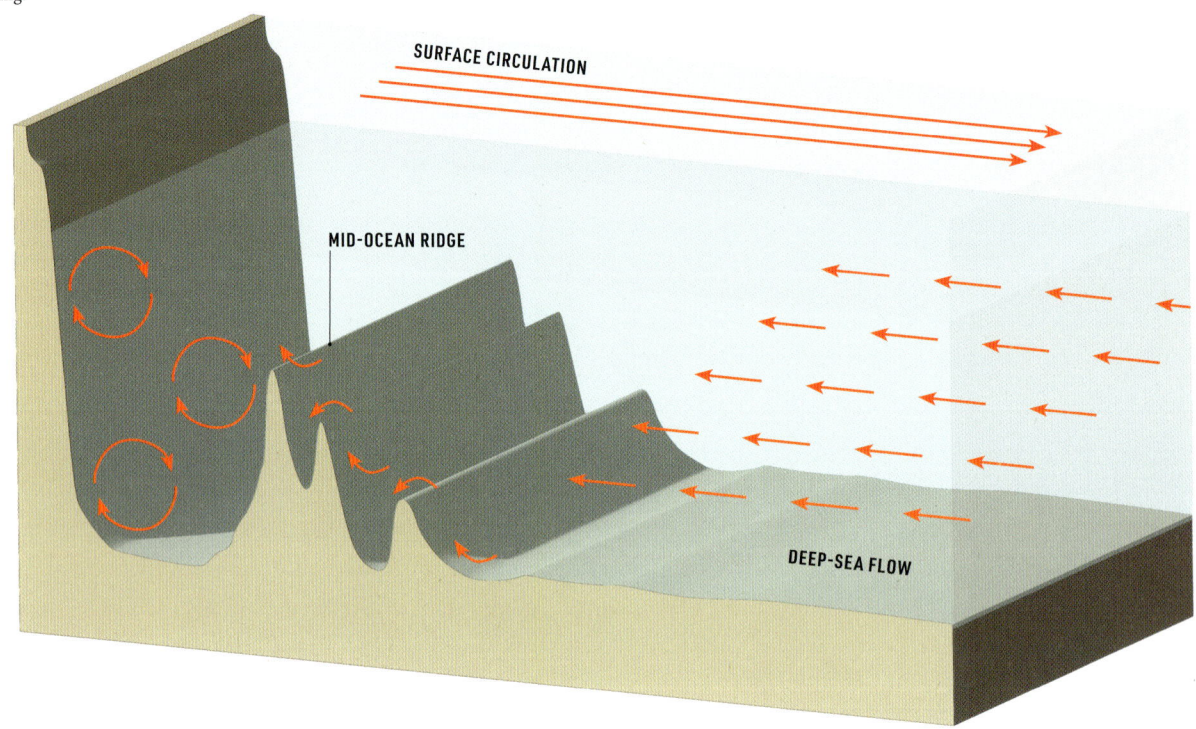

SURFACE CIRCULATION

MID-OCEAN RIDGE

DEEP-SEA FLOW

Deep-sea flows and density

The deep sea is generally stably stratified in density. While the less dense, warm, and relatively fresh waters are found mainly near the surface, some of the heat (and salt) is transported to greater depths. The deep sea is thus characterized by horizontal layers of constant density, of which the value steadily increases with depth, although at greater depths the pace of increase is less. When the density layers are not horizontal, they cause horizontal fronts. These horizontal density differences lead to pressure differences that vary with depth. The pressure differences result in water flow, which becomes directed along the front by the Coriolis effect. Examples of such density-driven flows on the large scale are along the boundaries of primarily wind-driven flows such as the Gulf Stream or Kuroshio Current. The Antarctic Circumpolar Current is accompanied by three large fronts, one of which is the polar front.

Although the large-scale, wind-driven flows are primarily in a balanced state, in which the pressure difference is balanced by the Coriolis effect, this does not mean that they are steady flows. Like the atmospheric winds, the flows show considerable variability in space and time. The flows are dynamically unstable; they maintain a steady direction, from which they repeatedly divert (or meander), like a river not flowing in a straight line. When the meandering becomes strong, the bend in the flow may pinch off to form an eddy, a circular movement of water causing a whirlpool, which can have a horizontal diameter of 60 mi. (100 km). The flow speed also varies, repeatedly, by more than two times. Large-scale flows thus give the impression of a strongly pulsed, variable phenomenon.

Large-scale flows driven by density differences are also found in the deep sea, where current flows are generally weaker than near the surface. These flows are intensified, and less attenuated, above topographic boundaries like continental slopes. Depending on the direction of the horizontal density difference, which may be different from the surface pressure difference, the deep-sea flows go in the opposite direction (as countercurrents) to the near-surface flows above. At depths greater than 3,280 ft. (1,000 m), such flows can still attain speeds of several tens of centimeters per second (0.1–0.2 m/s), up to ten times greater than flow speeds well away from topographic boundaries in the deep-sea interior.

Smaller topographic features than a continental slope may induce smaller-scale flows. Canyons incising a continental slope will deflect a boundary flow, which will be accelerated or decelerated accordingly. Submarine mountain chains like the Mid-Atlantic Ridge and the Central Indian Ridge are incised by numerous fracture zones, long and deep valleys between the ridge sections. Water flowing into the fracture zones will accelerate. As the fracture zones themselves are not flat, narrow basins, but also show variable topography with smaller-scale cross-channel ridges, the through-flow will be accelerated or decelerated during its passage. With the accelerations, more turbulent mixing with the overlying waters is generated. Although dense waters flowing over the seafloor are modified by the mixing and thus become less dense, the fracture zones are an important conduit for their passage between deep-sea basins (see pages 74–75).

Antarctic Circumpolar

A computer model of the flow around Antarctica. The flow shows numerous whirls and eddies. The most intense flow is colored red.

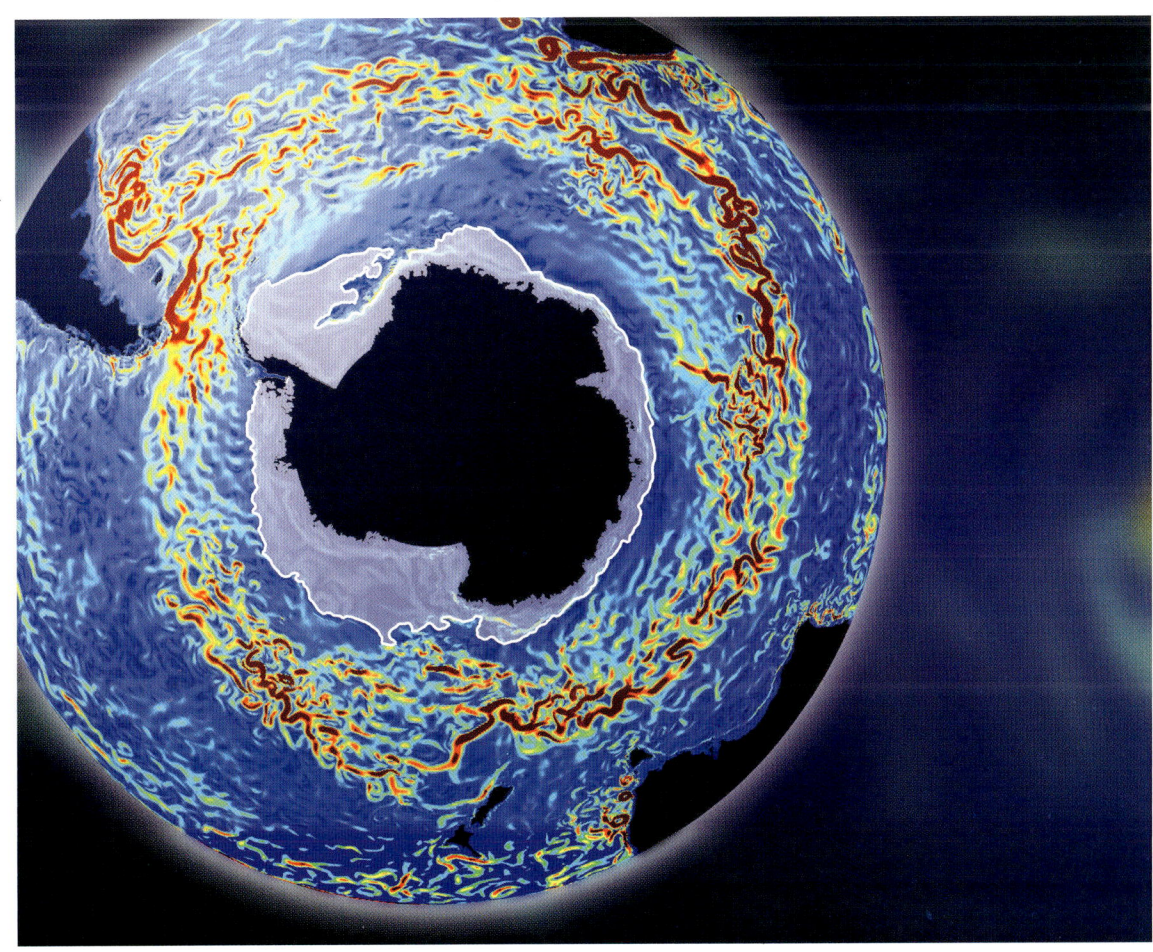

Density and the formation of water masses

Water masses are defined as large volumes of seawater that have distinctive properties, such as salinity, temperature, oxygen, or radio-isotope content, which each have values within particular bounds. Once formed, water masses will move away from the source, collide with other water masses, and eventually mix with them via turbulence processes. The deepest of these water masses, which also move and do not reside at a fixed spot, are denser than overlying water masses, consistent with stable vertical-density stratification in the deep sea. Deep, dense water masses newly form when the sea is unstably stratified in density, starting from the surface. This occurs only episodically, generally in localized areas near the poles as well as in a few mid-latitude sites in the Mediterranean Sea. Normally, in tropical, but also nearly all mid-latitude regions, near-surface density stratification by solar heating is so extensive that it strongly reduces turbulent mixing deeper than a few tens of meters. At greater depths, the stable stratification is thus not destroyed via instabilities by evaporation and nighttime cooling.

Under late-winter conditions in polar (and some Mediterranean) regions, however, overlying air may cool and evaporate sufficiently to result in relatively low temperature and high salinity content in near-surface waters compared to their deep-sea values. When the near-surface waters are cooler and/or saltier than waters below, the resulting unstable density stratification generates convective turbulent downward and upward motions, until reaching the depth level where the density values match those of the sinking waters. At their respective density level, the waters originating from the surface will spread into the deep interior by essentially horizontal transport.

Once every 5–10 years, such sinking reaches seafloor deeper than 3,280 ft. (1,000 m), and the newly formed deep, dense water replaces the resident deep-water masses. The replacement is one way to set the deep water in motion, and there will be a frontal horizontal density difference with the previous deep-water masses. From their sources, such as areas near the poles, the episodically formed deep, dense water masses will flow in pulses toward the equator. During the sinking and the equatorward flow, they will slowly but surely mix with overlying waters, due to interior friction via turbulent shear flow of waters moving at different speeds and/or in opposite directions.

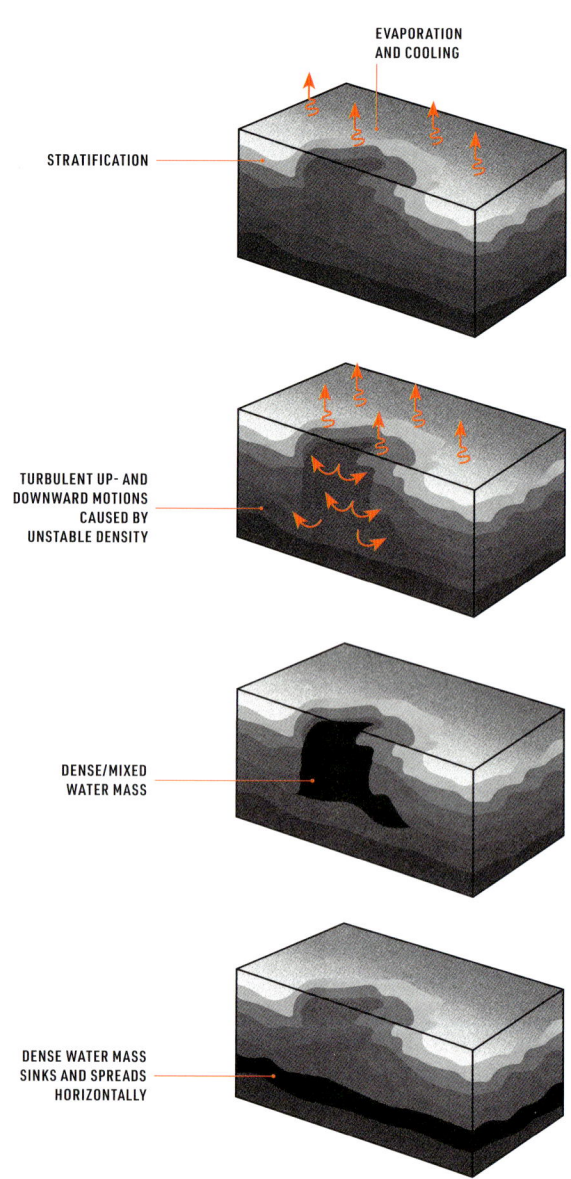

EVAPORATION
AND COOLING

STRATIFICATION

TURBULENT UP- AND
DOWNWARD MOTIONS
CAUSED BY
UNSTABLE DENSITY

DENSE/MIXED
WATER MASS

DENSE WATER MASS
SINKS AND SPREADS
HORIZONTALLY

Dense water-mass formation

Schematic diagrams of deep convection. From above to below: preconditioning by evaporation and cooling, deep convection in narrow tubes, mixing, and horizontal spreading after sinking (after Marshall and Schott, 1999).

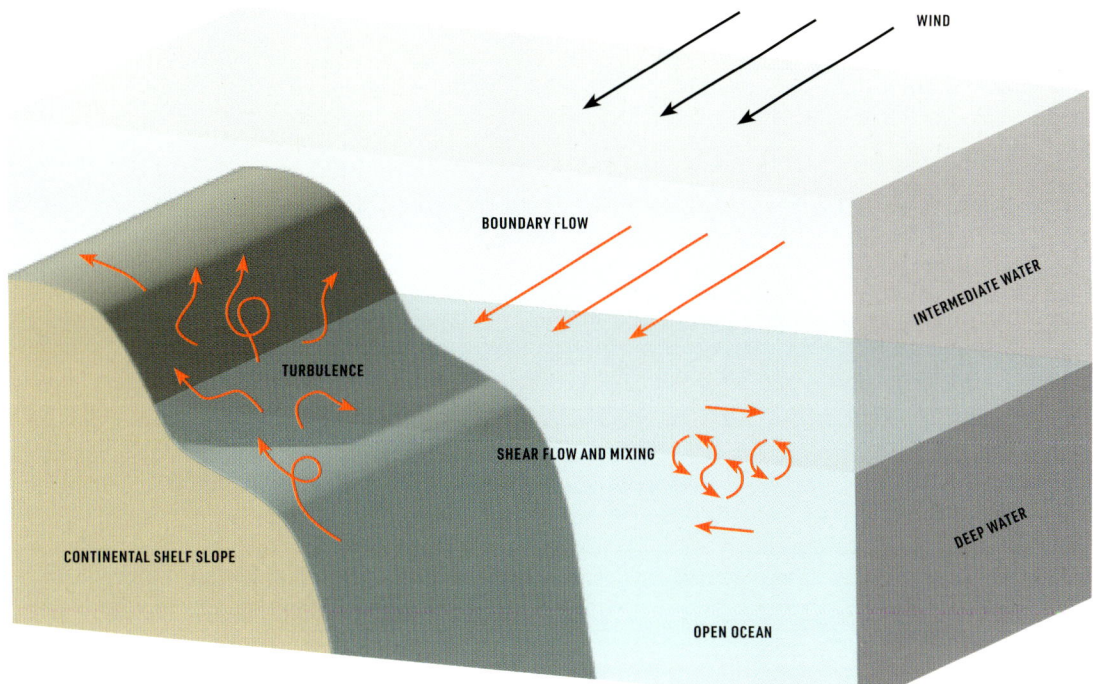

Mixing and turbulence
Open-ocean deep currents flow smoothly along constant density surfaces (isopycnals), where they mix via turbulence generated from flow-shear. Turbulent mixing by internal-wave breaking and boundary flows is 100–1,000 times more powerful above sloping topography than in the open ocean.

(labels in figure: WIND; BOUNDARY FLOW; INTERMEDIATE WATER; TURBULENCE; SHEAR FLOW AND MIXING; DEEP WATER; CONTINENTAL SHELF SLOPE; OPEN OCEAN)

Deep-sea flow and mixing

Although the dense deep-water formation may invoke the image of a continuous large-scale, density-driven flow from polar to equatorial regions and beyond, the reality is different. The deep, dense water formation occurs in pulses, and so does the flow of dense waters over the deep-sea topography. Dense water formation varies not only on seasonal scales, but also on decadal scales, with the real deep formation occurring once in 5–10 years. Within the seasons, formation pulses vary from daily scales to shorter-than-hourly scales. The episodic pulses, varying on so many scales, do not continuously cause a deep-sea river or conveyor belt. Envisioning the deep-sea flow using such metaphors may be tempting but does not correspond with the dynamical processes of the ocean.

Once a dense water mass is formed, its spread along the seafloor or, when not reaching to the seafloor, along layers of constant density (isopycnals), smooths the pulses somewhat. The smoothing is governed by interior slow, turbulent mixing processes, which may become stronger when the flow of interacting layers occurs at different speeds or directions (shear), resulting in friction and turbulent mixing. Thus, the denser waters near the deep seafloor will gradually become less dense during their transport equatorward (and overlying waters gradually denser). No matter how slow the modification, as water masses can be traced for decades, this interior exchange via turbulent mixing is still about 10–100 times faster than molecular exchange would be in a nonturbulent flow. More energetic, and thus faster, turbulent mixing occurs in some regions of the deep sea.

In the open ocean, the shear- and/or internal-wave-induced turbulent mixing is about 10 times more intense near a flat seafloor of a deep basin than turbulent mixing away from the bottom. Turbulent mixing is 100–1,000 times more powerful in a layer about 330–660 ft. (100–200 m) thick above sloping topography than in the open water column. We already noted the accelerating flows over small-scale topography in fracture zones of a mid-ocean ridge, but the largest turbulent mixing in the deep sea occurs via the breaking of internal waves. Such waves are ubiquitous in the density-stratified deep sea, as we elaborate next. All the turbulent mixing that touches the seafloor whirls up sediments, which are subsequently transported by the larger-scale flows. Outside the polar regions, all the turbulent mixing also ensures that the ocean is stratified by the heat transfer from surface to bottom, thereby compensating for the potential energy loss due to the sinking of newly formed dense—cold and salty—water. Without turbulent mixing, the deep ocean would become a stagnant pool of cold water, not driven at all by differences in density.

Due to their large scales, flows in the deep sea are turbulent everywhere and a laminar- (or smooth-) flow environment does not exist, except at the level (smaller than 0.5 in./1 cm) of water flowing past a very small swimming animal.

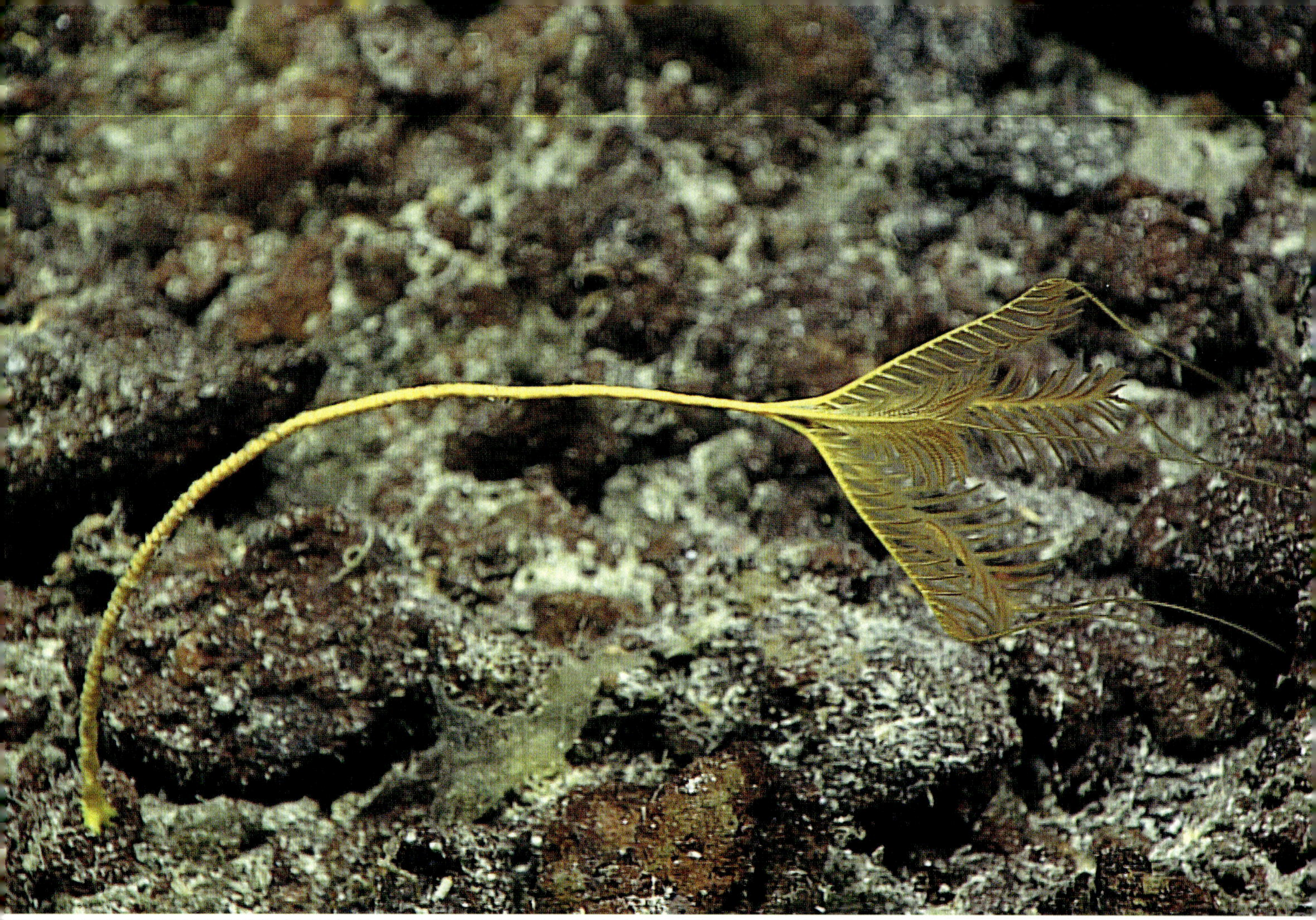

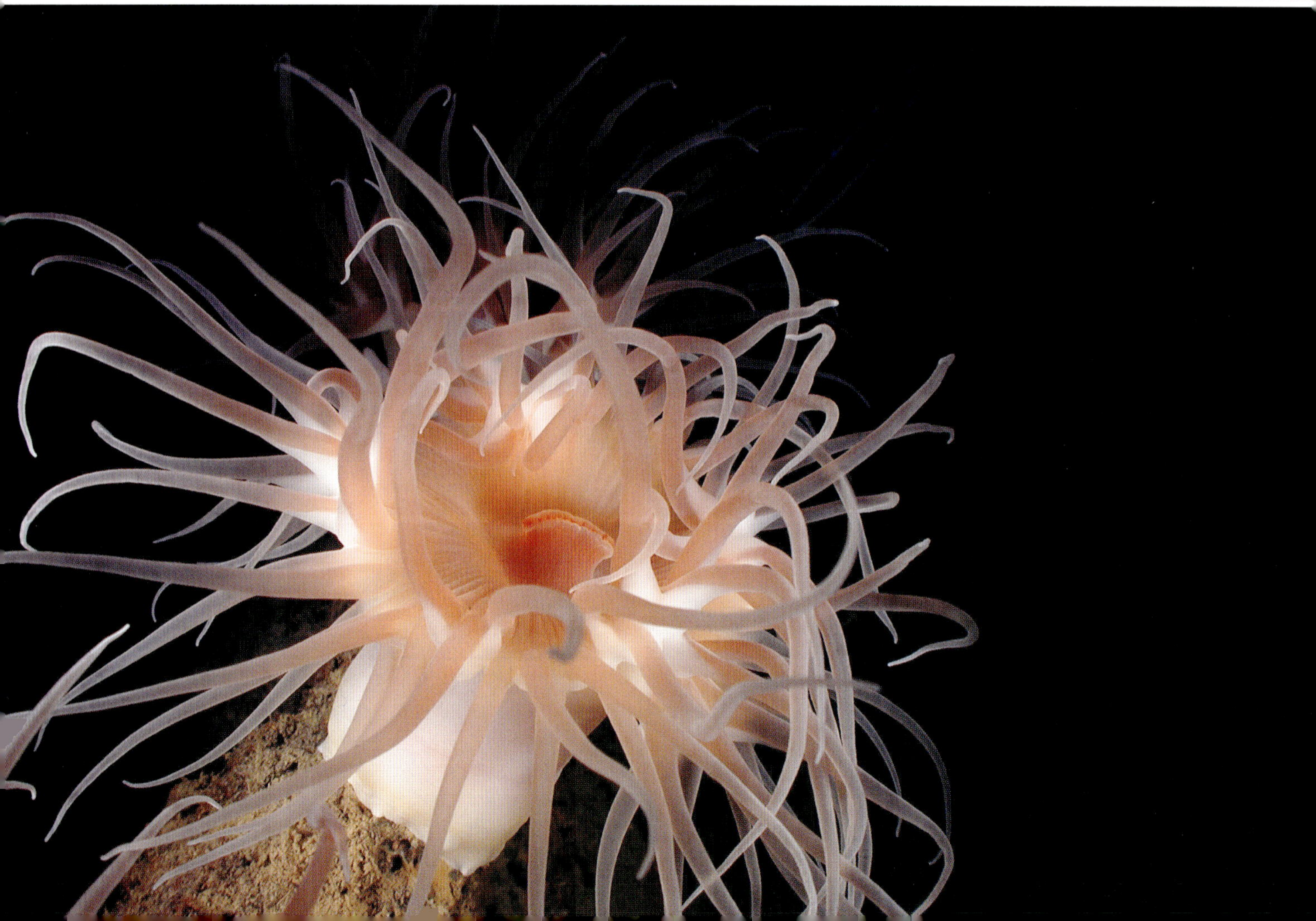

Going with the flow

Clockwise from top left: *Porphyrocrinus* sea lily, American Samoa; brown cerianthid tube anemone, Gulf of Mexico; *Relicanthus* sp., a new species from a new order of Cnidaria, Clarion-Clipperton Fracture Zone, eastern Pacific; large stalked anemone, Gulf of Mexico.

Tides, wind, and internal waves

A major source of turbulence in the deep sea is related to tides, the gravitational pull of the moon and sun on ocean water. If a current meter is held anywhere in the deep sea for some time, it will measure kinetic energy, the energy of motion. A spectrum—a graph displaying energy levels as a function of time—will generally show the largest peaks at (1) semidiurnal and (2) inertial frequencies. This implies that most energetic motions are related to (1) the tide—dominant semidiurnal variation shows tidal peaks that occur twice a day; and (2) the rotation of the earth, with the Coriolis effect resulting in an inertial motion with an oscillation frequency of about one day, depending on the latitude. These motions contain about 1–2 TW (terawatts) and 0.5–1.0 TW of energy, respectively, globally—more kinetic energy than the other large-scale flows in most of the deep sea. Such values may be compared with 16 TW in present consumption by humankind, about 2,000 TW of ocean heat transport, and roughly the same, or more, of atmospheric heat transport, and about 120,000 TW of solar radiation reaching the ocean surface.

Whittard Canyon internal tide

A 660 ft. (200 m) from-crest-to-trough tall internal tide detected by temperature sensors at Whittard Canyon in the northeast Atlantic. The graphic shows the peaks and troughs. Map shows location of Whittard Canyon, highlighted with the red circle.

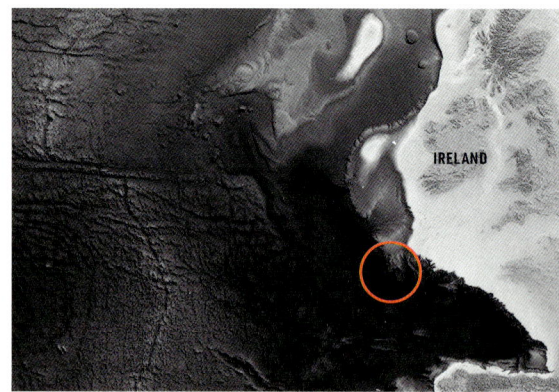

IRELAND

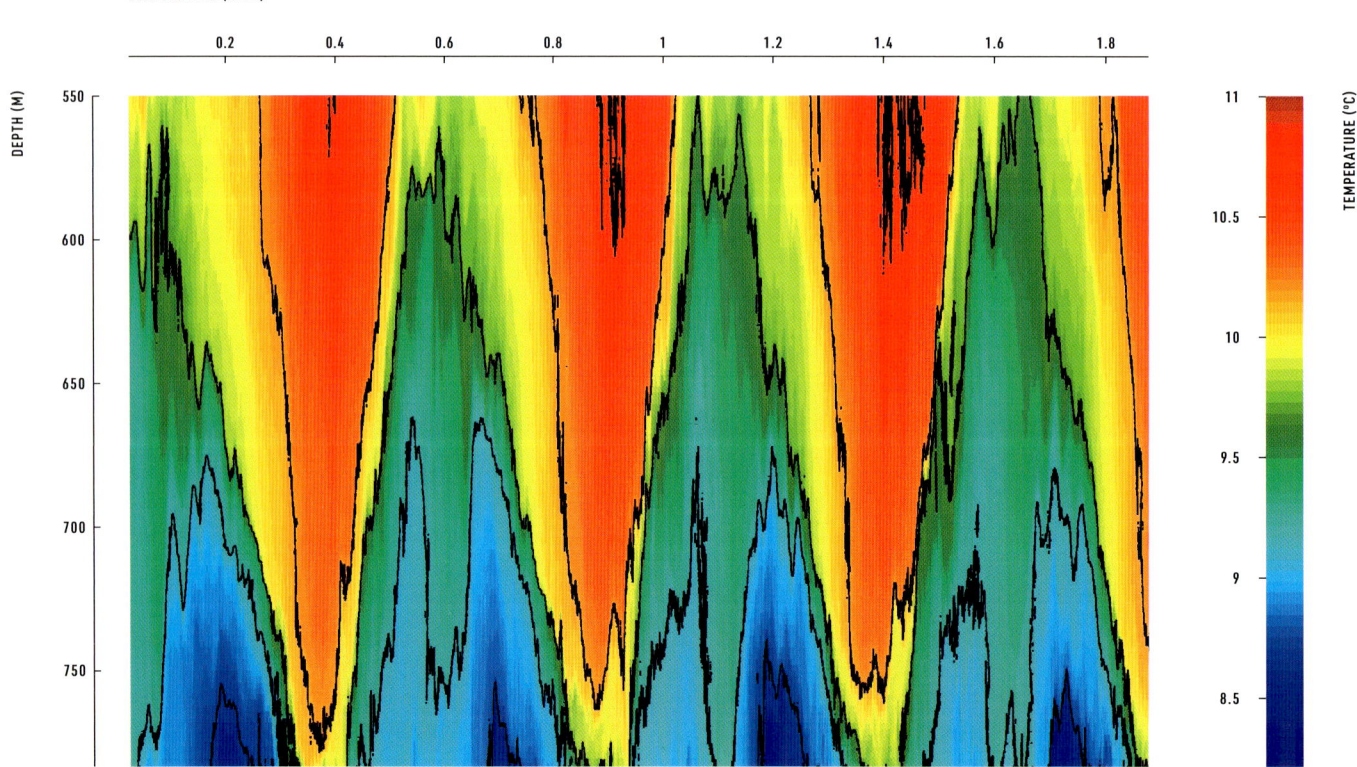

ELAPSED TIME (DAYS)

Why would the small amounts of ocean kinetic energy in tidal and inertial motions, compared with the huge amounts of heat transport, be important for deep-sea turbulence and, thereby, for deep-sea circulation? The two main kinetic energy peaks have different sources. The semidiurnal peak has its source in the surface pressure difference following the earth–moon (and sun) interaction. In the deep sea, the resulting relatively weak horizontal tidal motions may induce a vertical motion when they encounter underwater topography. The vertical tidal motion will displace horizontal density layers, which thereby create internal waves with tidal frequency.

As the vertical density differences are relatively weak in the deep sea, typically less than one one-thousandth of the density difference between air and water, internal waves are different from sea-surface waves. Internal-wave amplitudes can reach over 330 ft. (100 m) vertically, while they go nearly unnoticed at the sea surface, displacing only 4 in. (10 cm). Their typical wavelengths (the distance from peak to peak) are in the order of 0.6–6.2 mi. (1–10 km), while surface tides can have wavelengths of thousands of kilometers. Internal tidal-flow speeds can grow up to several tens of centimeters per second in the deep sea, whereas surface tidal flows reach only 0.4–0.8 in./s (1–2 cm/s) in the deep. Internal waves travel slowly, at speeds of typically 4–39 in./s (0.1–1.0 m/s), compared with speeds up to several hundred meters per second for ocean-surface tidal highs and lows. Even at their most rapid frequency, internal waves are still slow waves, having typical periods of an hour or more in the deep sea, compared to sea-surface wind waves typically having 10-second periods. The lowest-frequency internal waves, those with the longest period, have a period of typically a day, depending on the latitude of generation. Any moving disturbance over and in a fluid on a rotating sphere will leave behind inertial motions—transient motions that are deflected by the Coriolis effect. Think of an atmospheric storm passing over the sea or a density front collapsing in the deep sea. The inertial motions left behind by such passing disturbances will set up internal waves that move near horizontally. The larger their vertical component, the deeper they extend into the deep sea, where stratification will be weaker.

Internal-wave breaking

Internal waves keep the stably stratified deep-sea interior in permanent motion, everywhere. When waves are moving freely in the interior, very little wave breaking may occur. However, throughout the deep-sea interior, internal waves of different frequencies and originating from different sources may locally enhance their energy when the waves interact. The smooth waves may then deform and become unstable. When they pass through layers of different currents above and below, undergoing vertical shear, the deformation may grow to the point of breaking. Shear is found, for example, in localized large-scale current systems, like the Kuroshio or the Antarctic Circumpolar Current, which weaken and thus shear at greater depths. Shear occurs in other areas, such as density-driven flows down a canyon, as well as in internal waves.

Paradoxically, the internal-wave shear destabilizes the density stratification, so that internal waves apparently destroy their own support via generation of shear. Although turbulent, this mixing occurs sporadically, intermittently, in the deep-sea interior, with a puff here and a puff there. Internal-wave turbulence can become more than 1,000 times stronger than this interior turbulence in localized spots of the deep sea above sloping underwater topography.

Slope characteristics

Like surface waves breaking on a beach, internal waves can break above underwater topography depending on slope characteristics of both topography and internal waves. By propagating in three dimensions under water, internal waves will encounter the boundaries of a basin. For example, large inertial waves generated by a typhoon (hurricane) in the northwest Pacific caused turbulent breaking 660 ft. (200 m) tall at 9,840 ft. (3,000 m) depth on the continental slope, amid slightly smaller internal tides colliding with the topography. Upon such an encounter with sloping deep-sea topography, internal-wave energy accumulates, and the overgrowing internal waves must break into turbulent mixing. In practice, such spots are not fixed in space and time but vary, because breaking occurs where the slope of the internal wave matches the slope of the seafloor, and because the slope of internal waves depends on local water stratification, which varies in space and time.

**Turbulent breaking
on a continental slope**

Large inertial waves generated by a typhoon in the northwest Pacific caused turbulent breaking 660 ft. (200 m) tall at 9,840 ft. (3,000 m) depth on the continental slope, amid slightly smaller internal tides colliding with the topography. The white bar indicates the duration of the inertial period (about 1.25 days). The dotted line indicates the lower detail image of the internal wave breaking.

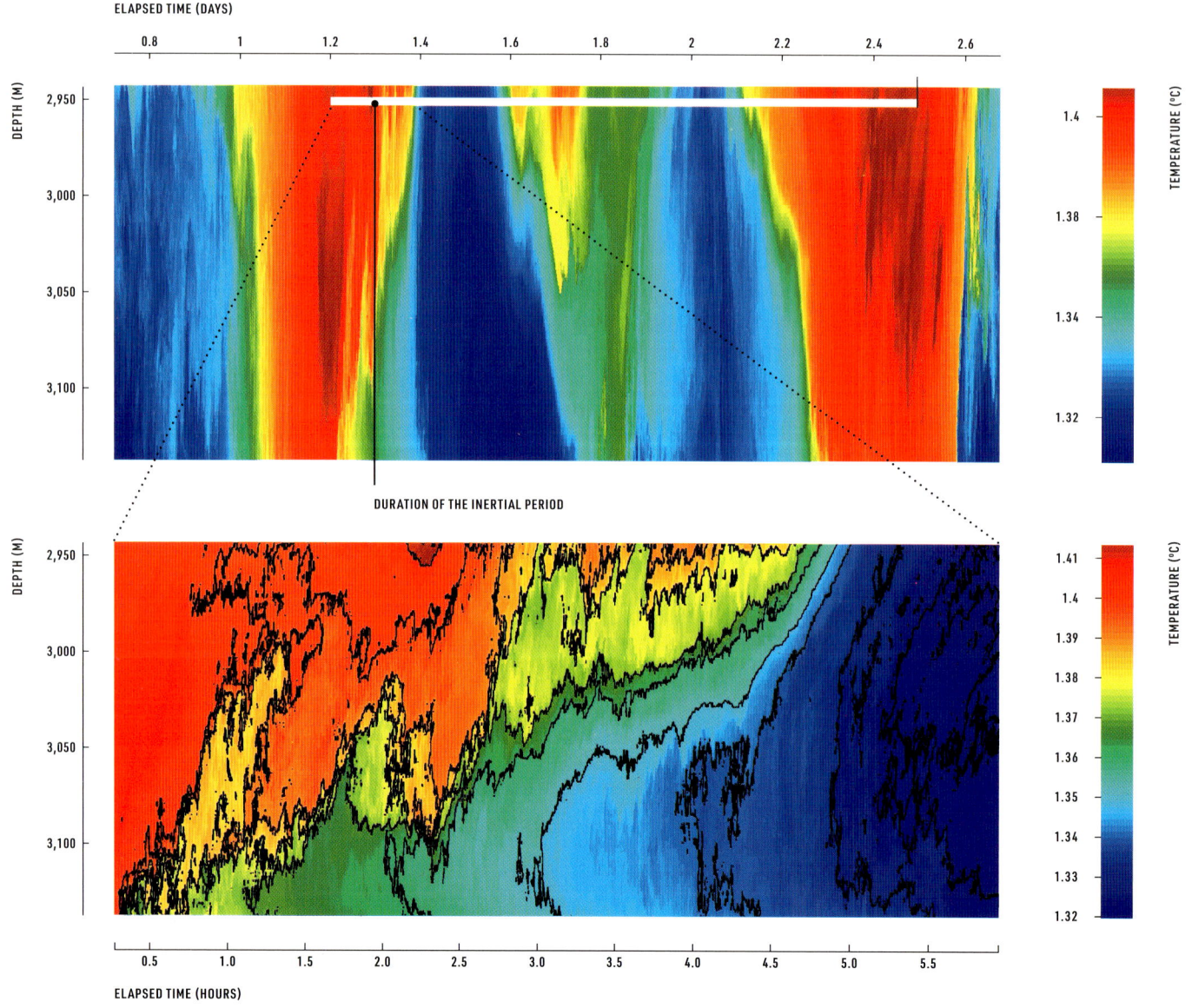

Backward-breaking internal wave

A large backward-breaking internal wave of an upslope moving solibore observed using temperature sensors near the top of Great Meteor Seamount, North Atlantic Ocean. The temperature ranges from 54.14 °F/12.3 °C (blue) to 57.38 °F/14.1 °C (red). The arrows roughly indicate the direction of motions. For example, around the strong front (change in temperature) the warm water is moving downward and the cold water upward at speeds of up to 6 in./s (15 cm/s).

ELAPSED TIME

14 MINUTES

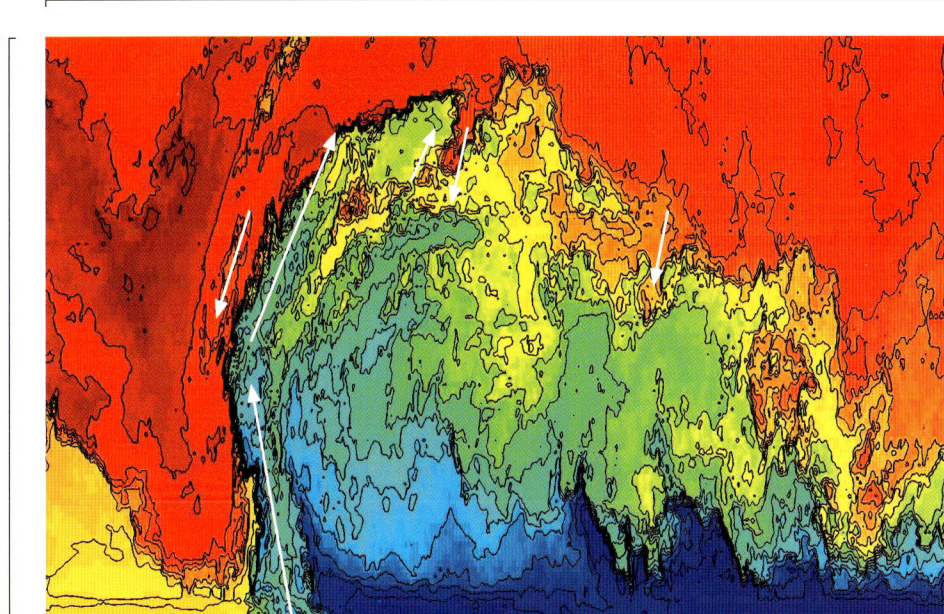

DEPTH (M)

500

550

AFRICA

Vigorous turbulence

Above topographic slopes, turbulent mixing and varying boundary-current flows alter the local stratification and thus the internal-wave slope at a particular frequency. As a result, internal-wave turbulent mixing also occurs episodically, mostly at topographic slopes that are just steeper (supercritical) than those of the prevailing internal waves. At times when conditions are right, the turbulence can be vigorous and spectacular, resulting in upslope-moving turbulent bores, or solibores—the steep leading edges of a traveling wave of water. Such bores of several tens of meters high have been observed in the deep sea. The impact of such vigorous events occurring a few times per (tidal) wave period on the seafloor fauna is large: sediments and nutrients are brought into suspension by vertical motions of more than 4 in./s (10 cm/s) up to 330 ft. (100 m) from the seafloor and are subsequently transported into the deep-sea interior. If one were at such a site, the event would feel like the passage of a Sahara dust storm, albeit underwater, in the pitch black of the deep sea.

Various observations have indicated that internal-wave breaking above topography is sufficiently strong in turbulent mixing to maintain vertical heat transport in the deep sea, thereby affecting the density stratification and thus the entire density-driven circulation of the ocean. An important property herein is that internal-wave breaking over sloping topography is very efficient. The back-and-forth sloshing of internal waves over a slope rapidly restratifies waters that are just mixed or homogenized. Homogeneous waters have a mixing efficiency equal to zero. Internal waves replace the mixed waters with stratified waters, which results in an efficient mixing property.

Turbulence over ridges and in trenches

Strong and efficient turbulent mixing in the deep sea can also be found when internal-wave motions pass over small ridges. Depending on wave phase, the flow may become critical when the particle tidal-flow speed exceeds the phase speed of the wave, and the internal-wave crest topples over its trough. Convective overturning into wave breaking results, which can be pictured as a deep-sea washing machine—one 330 ft. (100 m) in height. Such internal action—known as hydraulic-jump mixing—generally does not touch the seafloor, but merely affects the water column directly above the ridge. It is therefore important for further transport of heat and suspended matter into the deep-sea interior.

Thus, although internal waves of several tens of meters in amplitude move suspended matter and nutrients up and down regularly during their propagation, it is the turbulence they induce that actually mixes these suspensions, which eventually become transported by large-scale current flows along isopycnals. Transport along isopycnals is much easier than across them, also in the deep sea where density differences between water layers are slight and little force is required to cause turbulent mixing between them.

The deeper one gets into the sea, the weaker the vertical density stratification and thus the weaker the restoring force of internal waves (gravity). As a result, internal waves attain a more vertical path of propagation, and their amplitudes can grow larger. The weakest stratification is expected to be found in deep-sea trenches. Nevertheless, rare observations demonstrate that turbulence here is not negligible, being about twice as powerful as that in the surrounding deep ocean. This may explain why the carbon flux into trenches is approximately twice as large as that found outside of trenches. While stratification is weak, the nearness of topography, because the side walls of trenches are relatively nearby, may focus internal-wave breaking into puffs, shooting here and there into the deep trenches. Deep-trench turbulent puffs are observed to occur more vertically and more often than in the surrounding deep ocean. Turbulence in trenches is observed to be generated not as much by current shear as by more of a convective-type motion, with episodic vertical shots of denser and less dense waters into the relatively quiescent surroundings.

Hydraulic-jump mixing

A "washing machine" turbulent overturn over the Mid-Atlantic Ridge. The graphic shows data detected by high-resolution temperature sensors.

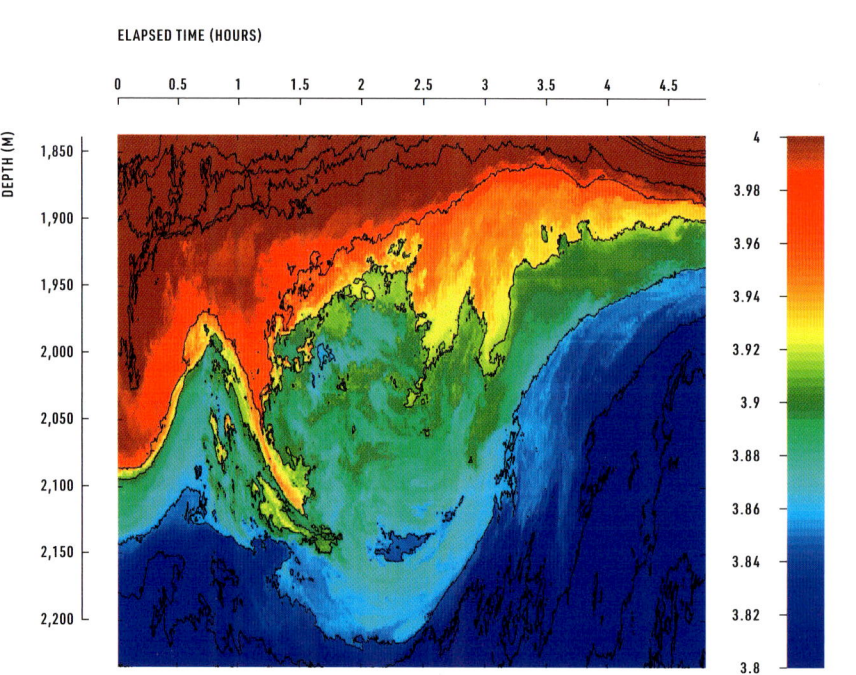

Mariana Trench turbulent jet

Occasional internal-wave breaking appears as puffs approaching the trench floor, as detected by temperature sensors. The white circles highlight relatively strong turbulent "puffs"; the white line indicates the direction of the turbulent puffs going down to the trench floor.

ELAPSED TIME (HOURS)

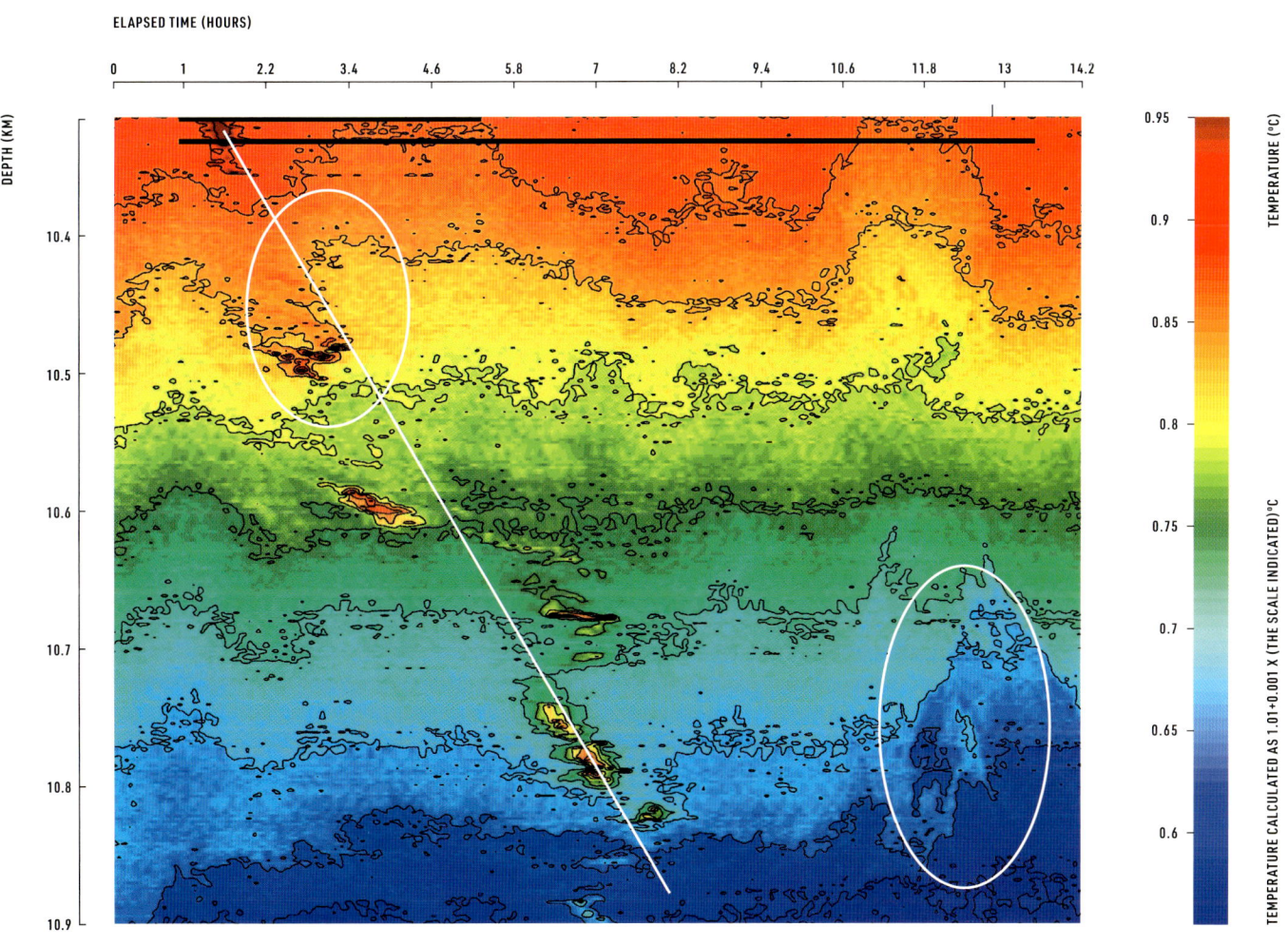

DEPTH (KM)

DEEP-SEA BASIN CHARACTERISTICS

The topographic basins that contain oceans and seas

Deep-sea basins are areas in the ocean that are partially, or entirely, closed off by topography from other areas. Often, a deep-sea basin has a relatively flat floor surrounded by higher underwater topography like seamount ridges, continental slopes, or continents. Or a basin may feature a stirring of waters around isolated seamounts or islands in an otherwise dull, flat deep seafloor. Basins include large ocean basins and small, marginal seas. These may differ in size and in particular topographic features, but their characteristics and most of the overall dynamics are similar. Some basins can have exchange with other nearby basins, via sea straits or over underwater ridges. Those that do not can have different circulation and physical characteristics because of the limited exchange with other basins.

Ocean basins

Ocean basins are the largest water basins on Earth. Meridional cross sections of all ocean basins demonstrate large oxygen content near the surface, modest to large oxygen content from the seafloor upward, and extended oxygen-minimum zones (OMZs) between 1,640 and 4,920 ft. (500 and 1,500 m), of which the intensity varies geographically. In all basins, turbulence in the interior is at a minimum around 3,280–6,560 ft. (1,000–2,000 m) depth, while larger wind-induced mixing occurs near the surface and internal-wave breaking induces more turbulence over sloping topography.

Major ocean basins

The five ocean basins include numerous subbasins that are separated, for example, via major underwater mountain ridges.

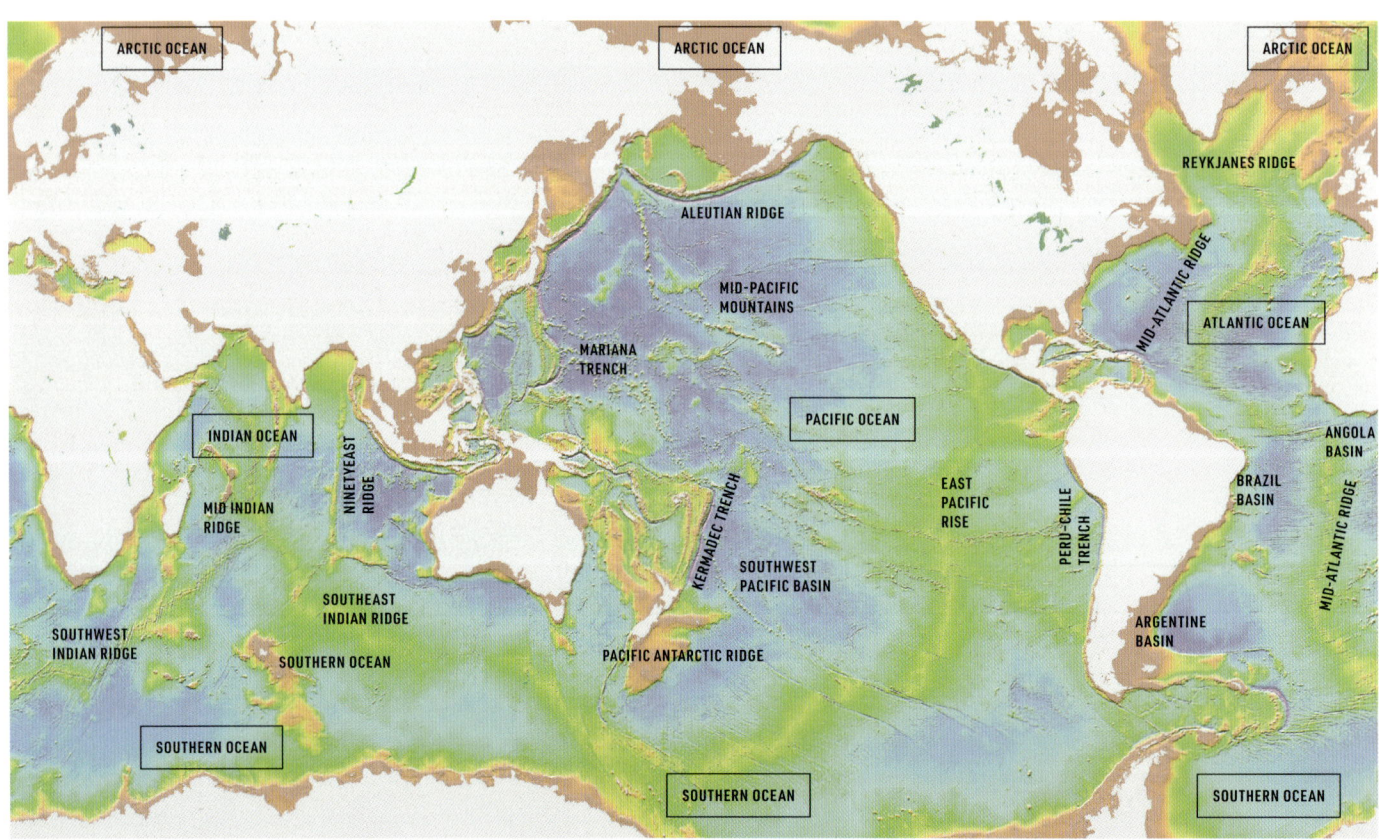

Atlantic Ocean basin
Large features are the drop
down from the continental
shelf to the abyssal plain
and the Mid-Atlantic Ridge.
Numerous small hills and
mounts are also visible.

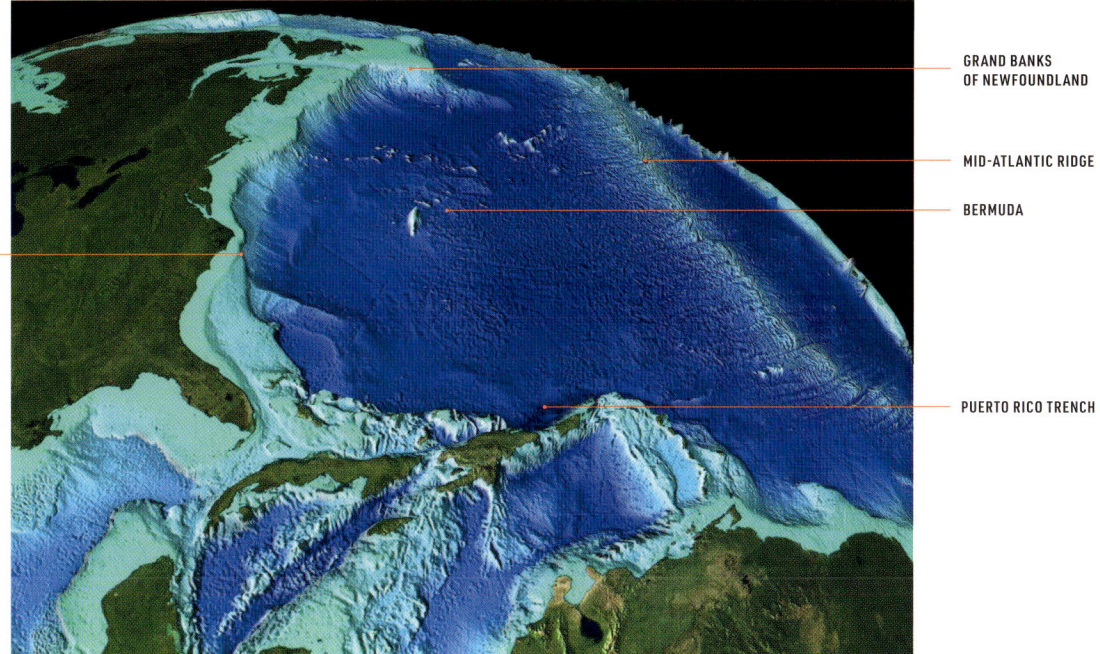

GRAND BANKS
OF NEWFOUNDLAND

MID-ATLANTIC RIDGE

BERMUDA

CONTINENTAL SHELF

PUERTO RICO TRENCH

Atlantic and Pacific Ocean basins

While ocean basins and smaller sea basins have many
similarities, a few characteristics unique to ocean basins
can be named. The Atlantic Ocean is the only ocean that
sees deep, dense water formation on both its southern
and northern boundaries. The Indian Ocean, and notably
the Pacific Ocean, see deep, dense water formation
flowing in only from the southern side.

In its northern reaches, the present-day Pacific Ocean is
almost blocked off from the Arctic Ocean by the Bering
Strait. It has limited openings in the Aleutian Ridge to
the Bering Sea, and its own surface circulation does not
allow for sufficient cooling and, especially, evaporation
to form deep, dense waters. The North Pacific
demonstrates a much stronger salinity contribution
to near-surface density stratification. Its near-surface
salinity is about 32.8 g/kg, while its salinity deeper down
is about 34.6 g/kg. For comparison, these values in the
North Atlantic are 34.7 g/kg and 34.9 g/kg, respectively.
Oxygen-rich dense waters are not uniformly found in the
North Pacific, and the OMZ is larger and stronger there
than in, for example, the North Atlantic.

While the Mid-Atlantic Ridge forms a seamount chain
bisecting the entire basin from south to north, the North
Pacific lacks a mid-ocean ridge. Deep flow, spreading
and mixing through fracture zones from eastern to
western basins in the Atlantic, hardly occurs in the North
Pacific. However, the equatorial flow system, of strong
east–west flows occurring between 10°S and 10°N, is
more or less similar for Atlantic and Pacific basins. This
flow system includes sharp contrast between westward
and eastward flows near the surface, and counter-flows
in various layers in the deep sea.

Indian and Southern Ocean basins

Because the Indian Ocean has limited extent north of the equator, its principal basin-wide gyre is found south of the equator. It thus has a very asymmetric appearance to its large-scale circulations, compared to those of the Atlantic and Pacific Oceans. Some of the world's warmest surface waters are found in its small northern part, which is blocked off by the South Asian subcontinent. A unique flow system with effects in the deep sea is the Agulhas Current, which is part of the large southern Indian Ocean gyre and transports warm waters near the surface and into the deep sea along the east African coast. It is the largest wind-driven flow after the Antarctic Circumpolar Current. The Agulhas Current splits off mesoscale eddies—eddies that have a diameter of about 60 mi. (100 km)—and collides heavily with Atlantic Ocean waters south of Africa, then turns to follow the Antarctic Circumpolar Current.

The Antarctic Circumpolar Current dominates large parts of the Southern Ocean. It is a unique flow, since it passes all longitudes (see also pages 52–53). Its strongest flow is near the surface, since it is wind-driven, but substantial magnitude still reaches into the deep sea. The relatively warm water prevents the formation of dense water. Instead, deep, dense Antarctic water is formed in austral winter in adjacent seas, such as the Weddell Sea and the Ross Sea, which have wide openings to the deeper basins to their north. As these seas average 1,640 ft. (500 m) deep, the formed dense waters exit the seas down the continental slopes into the deeper-sea basins. The topography is thus important for the transport of different water masses and their suspended matter and nutrients.

Marginal, nearly closed-off basins

Other deep-sea basins are marginal seas that are connected to oceans via relatively small openings. Some of these marginal seas are largely deep sea. For example, the Sea of Japan reaches down to 5,740 ft. (1,750 m), the almost closed-off Black Sea to 7,220 ft. (2,200 m), and the Mediterranean Sea, at its deepest point, to about 17,225 ft. (5,250 m) below sea level. Similar to the hadal trenches (see page 74), these deepest points are not widely spread, flat seafloors but merely localized deep spots or holes, like an isolated mountain summit.

Marginal seas have characteristic deep-sea density stratification, water-flow circulation, and redistribution of matter and marine life, as in the ocean interior. However, some specific characteristics occur only in these seas.

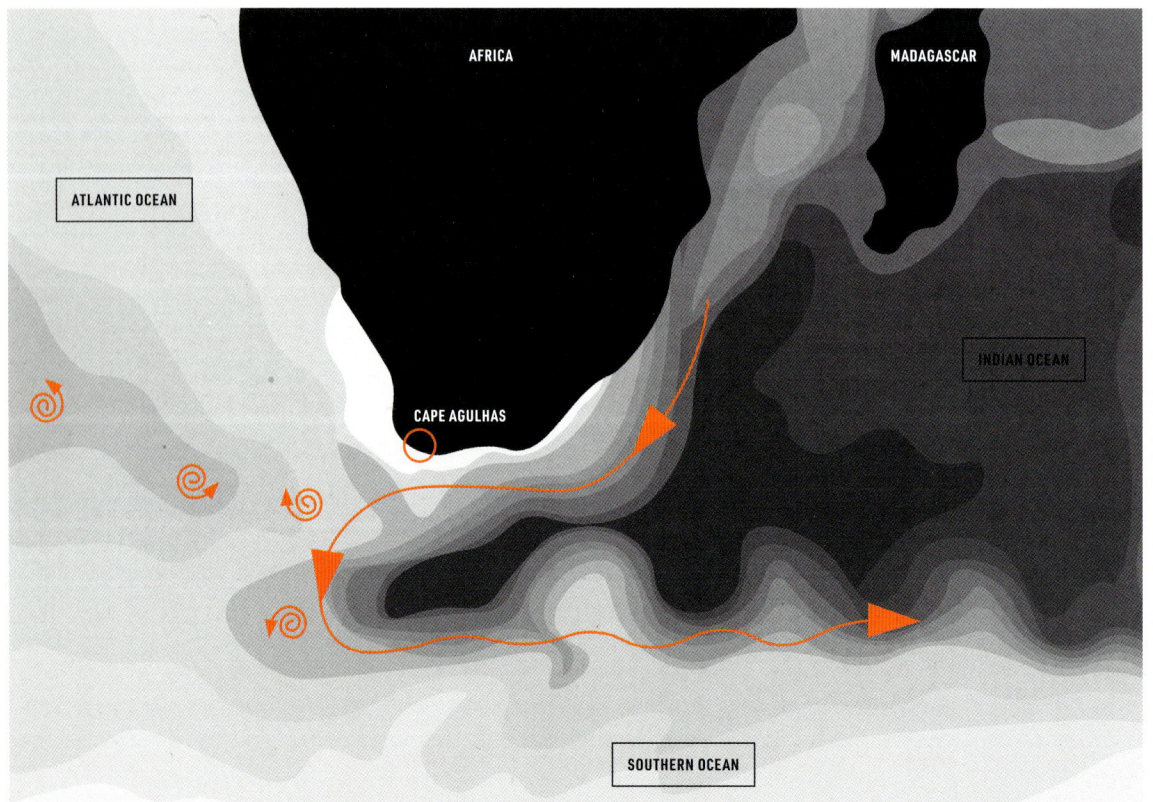

Sea of Japan

Top right: the Sea of Japan, also called the East Sea, is the marginal sea between the Japanese archipelago, Russian mainland, and the island of Sakhalin, the Korean Peninsula, and the mainland of the Russian Far East. Via sea straits between islands it is connected to the Pacific Ocean, the Sea of Okhotsk, and the East China Sea.

Agulhas Current, turning and eddies

Left and bottom right: Indian Ocean flow along the east African coast collides with Atlantic Ocean waters and diverges in the direction of the Antarctic Circumpolar flow. Many eddies spin off (left). This turbulence draws nutrients for phytoplankton up from the deep; the light blue swirl to the east of Cape Agulhas in the satellite image opposite shows a phytoplankton bloom in an area of cool, upwelling water.

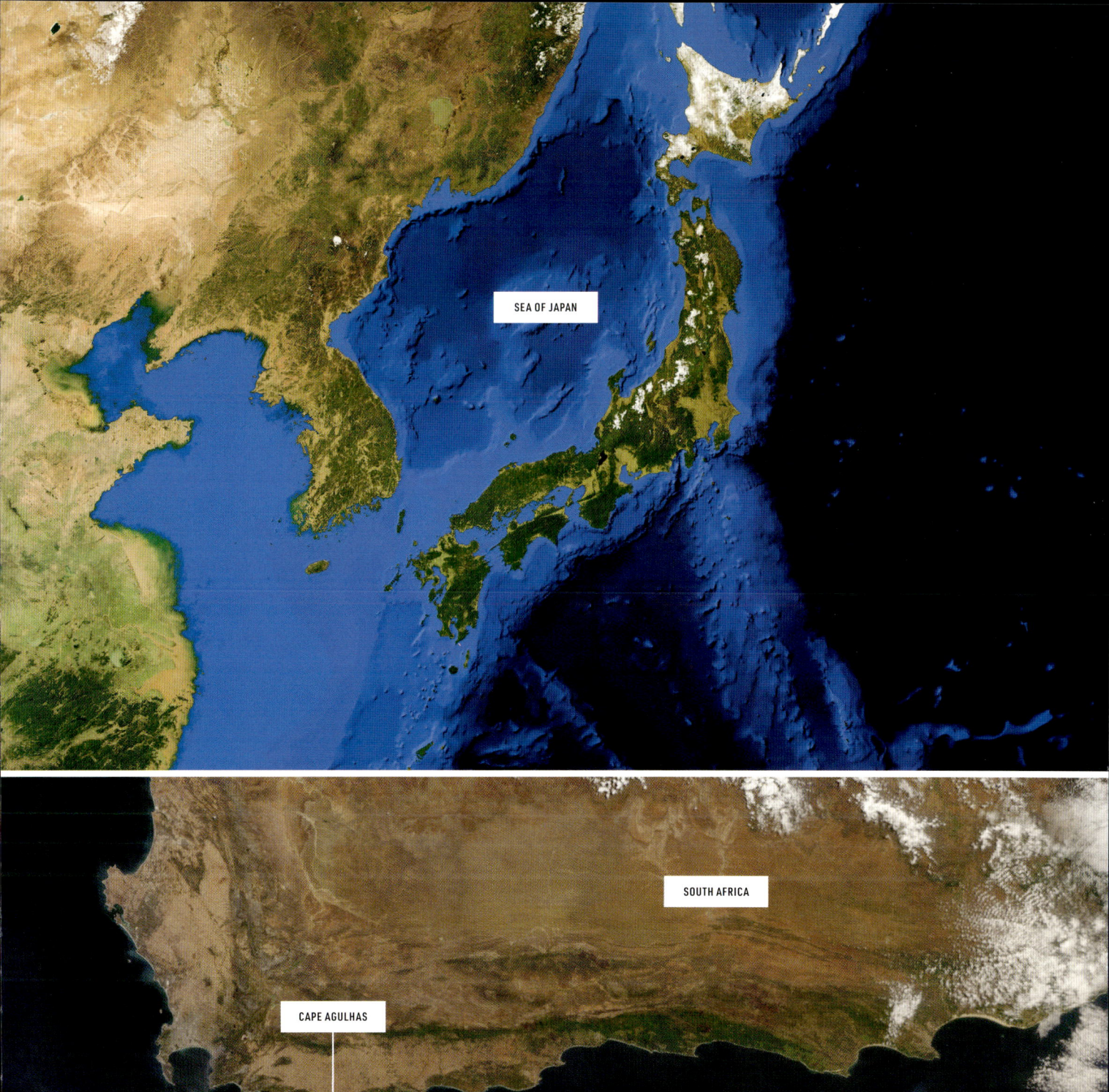

SEA OF JAPAN

SOUTH AFRICA

CAPE AGULHAS

Tides in marginal seas

As connections to oceans are small, tides are generally weak in marginal seas. This is because in such basins, tides can be directly forced only via Earth–moon (and sun) interaction and not driven to resonance—a reinforcing of a wave at a particular frequency determined by the size of a basin—via tidal current flows at their boundaries. The boundaries or openings to adjacent larger tide-containing basins are generally too small. If we compare the Bay of Fundy, along the eastern North American coast, where the world's largest tides are found, with the Mediterranean Sea, we notice that the Bay of Fundy has a wide opening along its entire southwestern boundary, whereas the Mediterranean has a relatively small opening in the Strait of Gibraltar. Inside the

Mediterranean near Gibraltar, tides are not negligible, but in the remainder of this marginal sea they are, with some exceptions—for example, near northern Italy, where tides are driven to resonance. Even in ocean basins, the tide-generating forces hardly directly drive tides. This is because the basin forms and scales do not fit the 6,200-mi. (10,000-km) scale of the tide. An exception is the Antarctic circumpolar basin, although even the potential direct tide-generation there is supported by some basin resonance. Nearly all ocean-surface tides of substantial amplitude are formed via resonance, in a generally complex way that is distantly reminiscent of the sound-producing resonance of organ pipes.

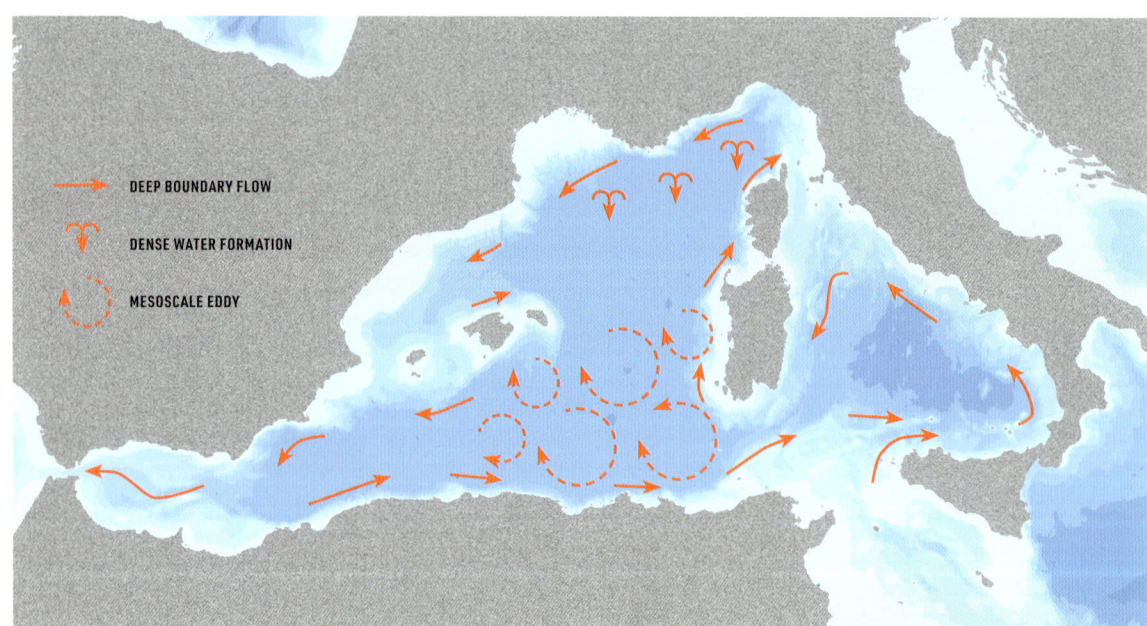

DEEP BOUNDARY FLOW

DENSE WATER FORMATION

MESOSCALE EDDY

Flow characteristics

Left: the Mediterranean Sea shows important flows that are also characteristic for the ocean: strong boundary flows, eddies, and deep, dense water formation, which is rarely found outside the polar regions.

Bay of Fundy and Mediterranean boundaries

Examples of boundary seas to the same scale: bottom far left, the semi-enclosed Bay of Fundy with the largest tides in the world and bottom left, the Strait of Gibraltar, which exchanges Atlantic and Mediterranean waters.

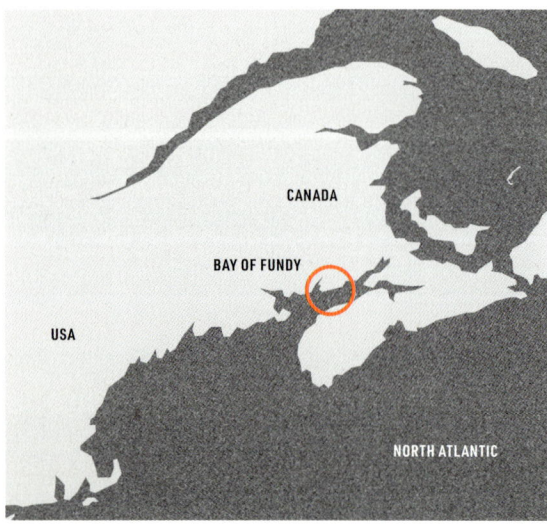

CANADA

BAY OF FUNDY

USA

NORTH ATLANTIC

STRAIT OF GIBRALTAR

NORTH ATLANTIC

MEDITERRANEAN

Mesoscale eddies
Top: strong mesoscale
eddies can reach all the way
to the deep seafloor.

Plankton in the Baltic
Bottom: swirls of plankton
blooms in the Baltic Sea.

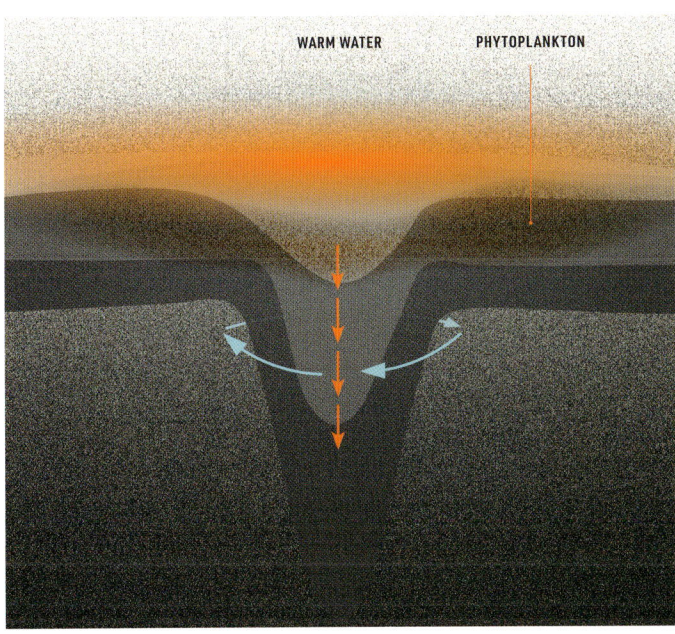

WARM WATER   PHYTOPLANKTON

Internal waves and flow in marginal seas

Marginal seas have internal waves and internal-wave turbulence supported by ample density stratification, despite the lack of tides, and hence internal tides. The dominant peak in a kinetic energy spectrum from any of these deep seas is found near the local inertial frequency, the lowest internal-wave frequency at which the largest shear is found. As in a tidally dominated ocean, internal waves at higher frequencies may thus break when deformed by the inertial-wave shear, or by shear generated by flow over topography, such as via boundary currents. Since the marginal sea basins are relatively small compared to ocean basins, continental slopes are closer. The land–sea exchange influences strong flows over the continental slopes. The irregular coastlines and canyons incising the continental slopes help destabilize such flows. The result is meandering of the boundary flow, including the splitting off of eddies, similar to the warm and cold rings found adjacent to strong ocean flows like the Gulf Stream and Kuroshio Current. Mesoscale eddies (diameter about 60 mi./100 km) and sub-mesoscale eddies (diameter 0.6–6.2 mi./1–10 km) can reach all the way to the deep marginal seafloors. They are important for the redistribution of suspended matter because they have relatively strong vertical motions of several centimeters per second, or several thousand meters per day, along their perimeter. This implies that fresh food, in the form of phytoplankton growing near the surface, may be transported to a seafloor 8,200 ft. (2,500 m) deep in about a day.

Black Sea
Surface flows marked by plankton (blue color differences) indicate various eddies and whirls. Deeper down in the Black Sea, the oxygen becomes depleted.

Density and deep sinking in marginal seas

The proximity of land and especially its mountain ranges creates specific physics in mid-latitude marginal seas—in particular the Mediterranean Sea, which shows where evaporation is high. The Mediterranean's waters, especially on its eastern side, become saltier over summertime. After further evaporating and cooling of near-surface waters over open sea during winter by strong cold and dry winds from mountain ranges such as the Alps and the Pyrenees, the relatively salty waters form a stark contrast with coastal waters influenced by river outflow at the end of winter. Then, they will reinforce the continental boundary flow, and they will sink. The sinking occurs over the deep basin but also in shallower waters over the continental shelf. The incised topography, the canyons of the shelf, plays an important role, here by guiding newly formed dense water into the deep (see pages 56–57).

Such deep sinking does not occur sufficiently in the deep Black Sea, which is characterized by rather rare anoxic (oxygen-depleted) waters below the wind-mixed upper 165–330 ft. (50–100 m) from the surface. Here oxygen, depleted by strong bacterial oxygen consumption, is insufficiently replenished by atmospheric oxygen transported from above, despite the inflow from the Mediterranean Sea (through the Bosporus strait) and abundant internal waves and associated turbulent mixing. The Black Sea is simply too strongly stratified, in both summer and winter, and lacks sufficient kinetic energy to carry the oxygen down. In addition, there are very minimal tides and thus no internal tides in the Black Sea. The inflow through the shallow (165 ft./50 m sill depth) and narrow (1.25 mi./2 km) Bosporus does mix some Black Sea waters with denser Mediterranean waters. This mixture sinks after entering the Black Sea but not in great quantities, nor to great depths, and does not provide sufficient mixing with interior Black Sea waters to overcome oxygen consumption.

Similar sinking of relatively dense Mediterranean Sea waters flowing into the Atlantic Ocean goes down to about 3,280–4,920 ft. (1,000–1,500 m), where these waters meet their density level. The sinking occurs in pulses that form intermediate lenses of relatively warm and salty waters in an environment of relatively cooler and fresher Atlantic waters of the same density. These Mediterranean lenses are transported through a substantial part of the North Atlantic Ocean, and traces are found as far south as the Cape Verde islands and farther north than Britain and Ireland. At these depths, such intermediate waters are blocked not by mid-ocean ridges but only by continental slopes.

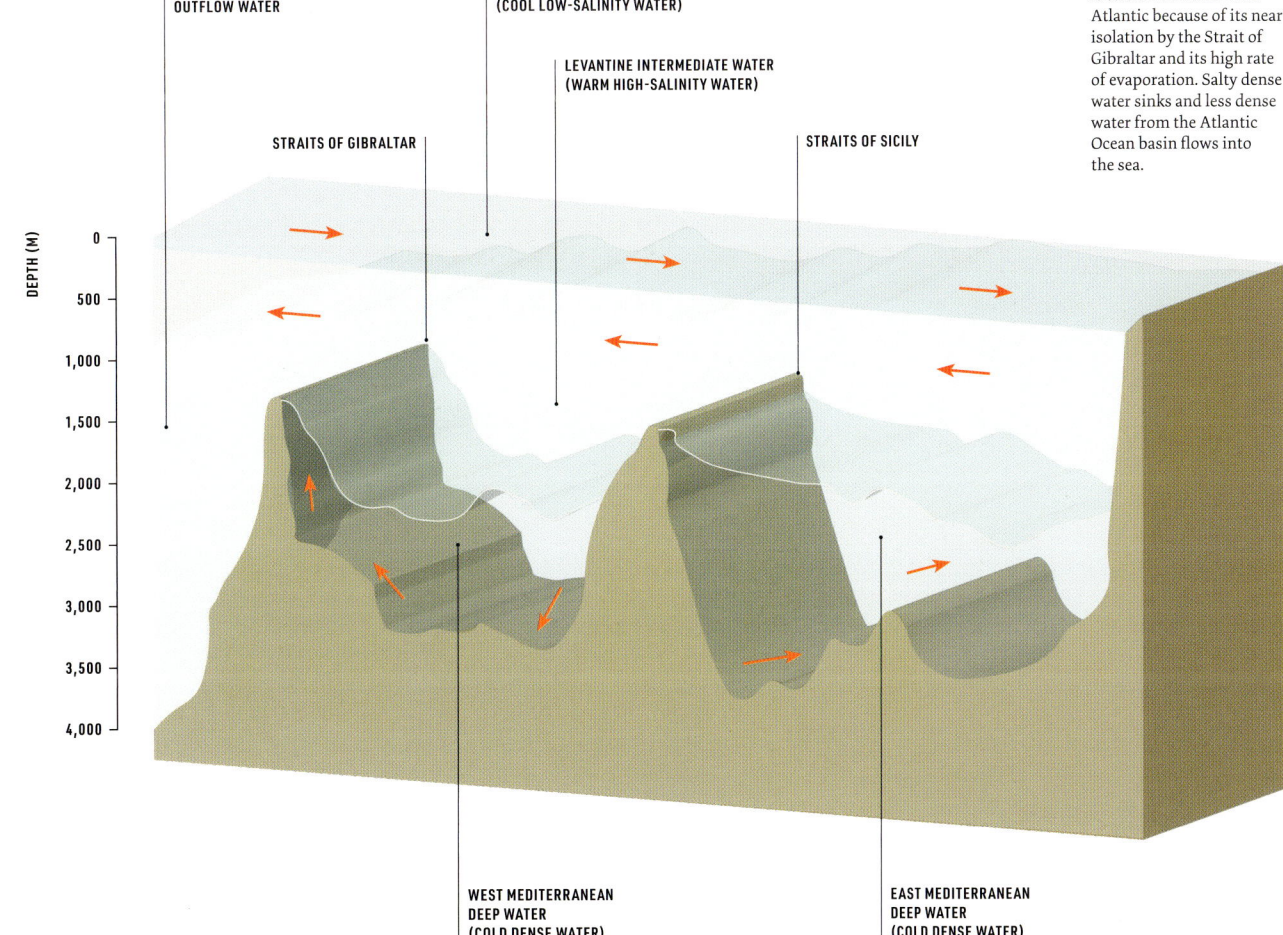

Mediterranean Sea water masses

A simplified graphic to show flow and deep sinking in the Mediterranean Sea. The Mediterranean is saltier than the North Atlantic because of its near isolation by the Strait of Gibraltar and its high rate of evaporation. Salty dense water sinks and less dense water from the Atlantic Ocean basin flows into the sea.

MEDITERRANEAN OUTFLOW WATER

MODIFIED ATLANTIC WATER (COOL LOW-SALINITY WATER)

LEVANTINE INTERMEDIATE WATER (WARM HIGH-SALINITY WATER)

STRAITS OF GIBRALTAR

STRAITS OF SICILY

DEPTH (M)

0
500
1,000
1,500
2,000
2,500
3,000
3,500
4,000

WEST MEDITERRANEAN DEEP WATER (COLD DENSE WATER)

EAST MEDITERRANEAN DEEP WATER (COLD DENSE WATER)

Basin connections

Subbasins are basins within the larger basins of oceans and marginal seas, separated by mid-ocean ridges or by sills in narrow straits. The individual subbasins are connected, not closed off, near the surface and are less distinct than marginal seas. Two subbasins may be connected in the deep sea via underwater fracture zones in mid-ocean ridges. The fracture zones are important conduits for the densest waters near the deep seafloors, in their propagation from one subbasin to the next. Due to acceleration of the flow near the seafloor when it moves into a fracture zone, these waters mix with overlying waters. Important examples are the near-equatorial Romanche and Vema Fracture Zones of the Mid-Atlantic Ridge. These guide, while modifying by turbulent mixing, the southwest Atlantic (originally Antarctic) deep waters to the northeast Atlantic subbasins, thereby crossing the equator. In the Mediterranean Sea, the eastern and western subbasins are connected via the Straits of Sicily (width 90 mi./145 km, maximum depth 1,050 ft./320 m) and Messina (width 3 mi./5 km, maximum depth 820 ft./250 m).

Small deep basins

Trenches are special hadal-zone deep-sea basins; they are deeper than the surrounding ocean floor, associated with subduction zones, where an oceanic plate of the earth's crust descends under a continental plate (see pages 46–47), and sloping topography is almost always nearby. The nearby topography may lead to turbulent events, such as breaking internal waves, which distribute resuspended matter into the trench interior. It may also lead to a unique water-flow circulation along the sidewalls. Observations of the conditions inside such trenches are rare because of the large ambient pressure and distance from the surface—specially designed and custom-made equipment is needed.

Sloping topography

The ROV *Deep Discoverer* climbs a steep slope along the wall of Guayanilla Canyon, south of Puerto Rico. Sloping topography such as this can give rise to internal waves and turbulent mixing.

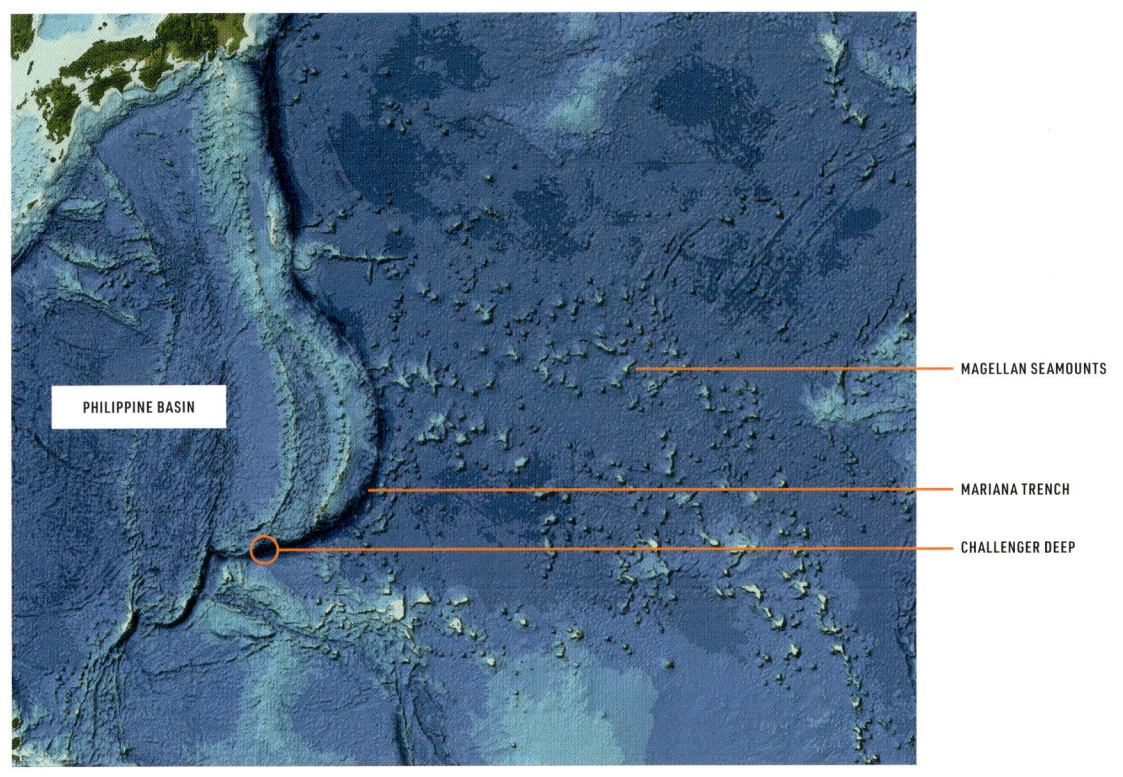

PHILIPPINE BASIN

MAGELLAN SEAMOUNTS

MARIANA TRENCH

CHALLENGER DEEP

VEMA RIDGE

ROMANCHE
FRACTURE ZONE

Mariana Trench

One of the deepest hadal-zone basins, the Mariana Trench in the western Pacific Ocean reaches its maximum depth of 35,840 ft. (10,925 m) at the Challenger Deep.

Fracture zones of the Mid-Atlantic Ridge

Numerous cracks and conduits intersect the Mid-Atlantic Ridge. These "fracture zones" guide deep water masses from one subbasin to another.

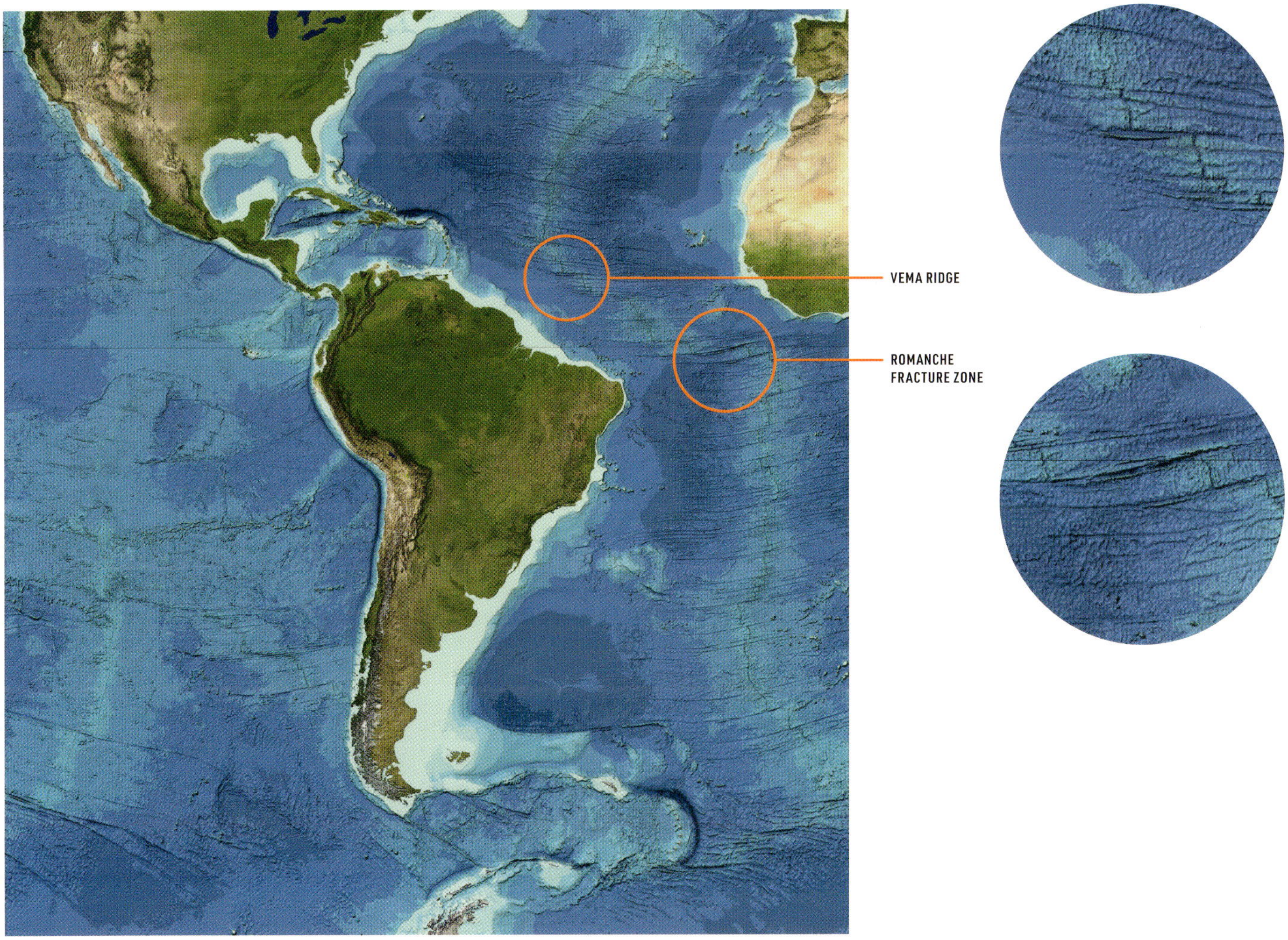

DEEP-SEA
ORGANISMS

BENTHIC INVERTEBRATES

Seafloor animals without backbones

Previous pages: **Red medusa** *Crossota sp.*
This small red medusa was encountered by an ROV as the jellyfish drifted near the bottom in the deep Arctic Ocean.

Animals without backbones—invertebrates—dominate the deep seafloor animal community known as the benthos (from Greek, meaning "depth of the sea"). In each location, the animals present are determined by seafloor substrate and by local oceanographic conditions that influence food availability. Animals may be adapted to live *in* or *on* the substrate; to anchor *in* soft sediment or *to* rocks; to filter feeding, deposit feeding, or carnivory. We see the same major animal groups as in shallow water but with different adaptations to help them thrive in this extreme environment. Of 35 animal phyla, all but one (the velvet worms, Onychophora) have some marine species, but just a few phyla encompass the most commonly encountered animals on the deep seafloor.

Venus Flytrap Anemone
Actinoscyphia aurelia

A Venus Flytrap Anemone enjoys an elevated position for suspension feeding on a deep-sea coral at 9,075 ft. (2,766 m) in the northwest Atlantic.

Some common benthic deep-sea invertebrates by phylum

Porifera

There are two major types of sponges in the deep sea: demosponges (class Demospongiae); and glass sponges or hexactinellid sponges (class Hexactinellida).

Cnidaria

This hugely diverse group includes hydrozoans (class Hydrozoa), tube anemones (subclass Ceriantharia), true anemones (order Actiniaria), zoantharians (order Zoantharia), and the different types of corals: stony corals (order Scleractinia); black corals (order Antipatharia); and octocorals (subclass Octocorallia), including mushroom corals (family Alcyoniidae), cauliflower corals (family Nephtheidae), bamboo corals (family Keratoisididae), and sea pens (order Pennatulacea). Cnidaria also includes various gelatinous groups found in pelagic waters (see pages 96–97).

Annelida

The most numerous group of annelids is the polychaete worms (class Polychaeta), including nereid polychaetes (family Nereididae) and tube worms (orders Sabellida and Terebellida, including the highly specialized family Siboglinidae, found in chemosynthetic ecosystems). Peanut worms (class Sipuncula) are also now classified as annelids (previously considered a separate phylum).

Mollusca

This group includes gastropods, or sea snails (class Gastropoda); bivalves (class Bivalvia), including clams, mussels, oysters, and relatives; and cephalopods (class Cephalopoda), which, in benthic environments, includes octopods and bobtails. (Other cephalopods, such as squids, nautiluses, and *Spirula*, occur in pelagic environments.)

Arthropoda

Arthropods of the deep seafloor comprise crustaceans (subphylum Crustacea) including crabs, shrimps, and lobsters (order Decapoda), acorn barnacles (order Balanomorpha), stalked barnacles (order Scalpellomorpha), isopods (order Isopoda), and amphipods (order Amphipoda); and sea spiders (subphylum Chelicerata, class Pycnogonida).

Echinodermata

Echinoderms are represented by five main groups: sea cucumbers (class Holothuroidea); sea stars (class Asteroidea), including brisingids (family Brisingidae); brittle stars (class Ophiuroidea); feather stars and sea lilies (class Crinoidea); and urchins (class Echinoidea), including heart urchins (order Spatangoida) and echinothuriids (family Echinothuriidae).

Porifera—deep-sea sponges

The most primitive invertebrates are the sponges, phylum Porifera. Deep-sea sponges fall mostly into two groups: the demosponges, which have a skeleton comprising fibers made of the protein spongin and/or siliceous spicules; and the glass (or hexactinellid) sponges, which have a skeleton of long siliceous spicules similar to glass fibers, including unique types of spicules never seen in demosponges. Sponges are nearly all filter feeders and use tiny beating flagella to draw water through pores, called ostia, in their surface, to trap tiny food particles. They have no mouth or stomach: cells extend and wrap themselves around the food particles and ingest them by phagocytosis, literally "cell eating." Sponges are prevalent where marine snow (aggregations of organic matter) rains from above, bringing an abundance of food—particularly at seamounts and submarine canyons, where currents and local oceanography entrain resources. Remarkably, a handful of demosponges are carnivorous; these are all closely related, in the family Cladorhizidae, suggesting that carnivory has evolved just once in Porifera. Carnivorous sponges lack ostia and instead have filaments covered with dense tiny spicules that act like Velcro to trap small swimming crustaceans such as copepods (subclass Copepoda). It is thought that this is an adaptation to the paucity of food suitable for filter feeders in many parts of the deep sea.

> *"Deep-sea sponges fall mostly into two groups: demosponges, which have a skeleton comprising fibers made of the protein spongin and/or siliceous spicules; and the glass (or hexactinellid) sponges, which have a skeleton of long siliceous spicules similar to glass fibers, including unique types of spicules never seen in demosponges."*

Carnivorous demosponge
Chondrocladia sp.

Top left: like all carnivorous demosponges, *Chondrocladia* feeds by trapping small invertebrate prey on its surface. The prey is then enveloped and digested by sponge cells that migrate to where the prey is trapped.

Lyre Sponge
Chondrocladia lyra

Center left: carnivorous demosponges (family Cladorhizidae) show remarkable morphological diversity: species usually have many appendages to increase the surface area to entrap prey. *Chondrocladia lyra* was named for its vertical appendages that resemble the strings of a lyre.

Carnivorous demosponge
Asbestopluma monticola

Bottom left: this species was first described from the Davidson Seamount off central California 4,200 ft. (1,280 m) deep. Researchers found partly digested crustaceans trapped in the sponge lattice.

Glass sponge
Family Euplectellidae

Top right: a euplectellid sponge with its intricate siliceous skeleton. Members of the genus *Euplectella* have been given the common name Venus's Flower Basket and are often home to a pair of shrimp that spend their whole life in the basket and breed therein.

Hexactinellid spicules

Center right: siliceous spicules unique to glass sponges include six-pointed ones, which put the "hex" into hexactinellid.

Hosting animals

Bottom right: sponges can provide substrate for other animals. Here, a glass sponge provides a home for brittle stars, which wind their long arms around the sponge, and a squat lobster hiding among the folds of the sponge skeleton.

Cnidaria

Some members of the phylum Cnidaria are suspension feeders, capturing suspended particles with their tentacles, but this highly diverse benthic assemblage, which encompasses anemones, corals, and zoantharians, includes many carnivores too. For example, true anemones mostly trap small animals with their stinging tentacles, pulling them into their gastrovascular cavity—a digestive sac with a single opening. Here, these prey animals are broken down into small parts and then phagocytosed by cells lining the sac. Anemones are often found attached to rocks, but there are burrowing species too. When threatened by a predator, anemones will retract their tentacles to form a tight ball. Their cousins the tube anemones (or ceriantharians) have, as their name suggests, a long tube extending into sediment, and they retract into this when they sense danger. Ceriantharians often occur in beautiful "fields" extending across large areas of continental slope where material is being channeled off the continental shelf and provides a source of food. Zoantharians, colonial anemone-like creatures, cover rock surfaces and other animals with their interconnected polyps. In one strange symbiotic relationship with a hermit crab, zoantharians have taken the place of a shell. Less conspicuous but very common are hydrozoans, most of which are tiny predatory cnidarians with feeding polyps resembling those of corals. But within Hydrozoa, we see one of the most spectacular examples of deep-sea gigantism (see page 29), with occasional huge solitary hydroids (consisting of a single polyp) growing to 20 in. (50 cm) across or more.

Corals are the dominant fauna in particular parts of the deep sea (see, for example, pages 128–129). In shallow waters, corals host symbiotic algae in their tissues, which photosynthesize. In the depths of the ocean, where light does not penetrate, corals do not have this adaptation, but some still build reefs. The most widely known reef builder is the stony coral *Desmophyllum pertusum* (formerly known as *Lophelia pertusa*), which builds large reefs on continental slopes and seamounts at depths between 660 and 2,950 ft. (200 and 900 m) and at shallower depths in fjords at higher latitudes. Although *Desmophyllum pertusum*, as the best-studied and most common and wide-ranging, is the most familiar species, deep-sea corals vary hugely in shape and form, since "coral" is a common name for any cnidarian whose polyps secrete a skeleton. Black corals have a wiry black skeleton, and each of their polyps has six nonretractile tentacles spread in pairs along the branches. Octocoral polyps have eight retractable tentacles. Some octocoral skeletons consist simply of sclerites—microscopic needles or plates of calcium carbonate—giving the coral a "soft" appearance, as seen in mushroom and cauliflower corals. Other octocorals have an axial skeleton as well as sclerites: in many sea fans (e.g., family Plexauridae) this axis is made of a stiff, wood-like protein that is flexible enough not to break in currents; in bamboo corals the axis comprises calcium carbonate sections with proteinaceous nodes between them that provide flexibility as well as a bamboo-like appearance; in yet others (order Scleraxonia), the axis itself resembles a layer of sclerites. Sea pens are also corals, and they tend to have a central axis with polyps and polyp leaves further supported by sclerites. Most corals require some hard substrate to attach to (although *Desmophyllum pertusum* reefs effectively create their own substrate once they have reached a certain size), but sea pens are usually found on soft substrate, and they are fast movers. When disturbed, they can rapidly withdraw into the sediment, although the speed at which they do this depends upon the species.

Despite the darkness of the habitat, the lights of ROVs have shown many deep-sea corals to be brightly colored. Color results from pigments in the tissues of the corals; the pigmented molecules may simply have been present in organisms ingested by the coral and accumulated in the tissues; the color serves no practical function.

Sea pen

Left: a sea pen from deep waters of the Indian Ocean. Here, the polyps are arranged on flat "leaves," which extend outward from the central axis or rachis. In other species, the polyps may be attached directly to the rachis along its length, or attached in a cluster at the top of the rachis, leading to quite different overall shapes.

Black coral

Top: black corals are so called because of their black stems, which can be seen clearly against the orange tissues of this *Leiopathes*. The polyps are mostly closely set, but, where they are more widely spaced, it is possible to count the six tentacles on each. Unlike octocorals the tentacles have no pinnules.

Octocoral

Center: a pale pink mushroom coral (subfamily Anthomastinae) at 7,350 ft. (2,240 m) depth. Note the eight-tentacled polyps with their tiny delicate pinnules that make octocorals so distinctive.

Zoantharians

Above: in the foreground are bright yellow zoanthids. Zoanthids often colonize dead octocoral skeletons as they have here (see the living white octocoral in the background), but they may also be found growing on sponges, on rock, and even providing a "shell" for hermit crabs.

Echinodermata

Sea urchins are part of the very diverse phylum Echinodermata. While some urchins feed on coral, others, such as the venomous echinothuriids and burrowing heart urchins, are deposit feeders, taking advantage of fresh detritus on the seafloor. Echinothuriids can be the unwitting protectors of juvenile cusk eels (Ophidiiformes; see pages 90–91), which seem to use the urchins as shelter while foraging on otherwise exposed, open seafloors. Another group of deposit feeders is the sea cucumbers, exemplified by the genus *Scotoplanes*, which are also known as sea pigs. Roaming the seafloor, sea pigs disturb the sediment as they search for nutrients within it. They forage with mucus-covered tentacles, and eat only the freshest food; studies using geochemical tracers have shown that they feed very selectively on particles that have been on the seafloor for less than three weeks. Sea pigs walk relatively fast on their water-filled parapodia (lateral outgrowths resembling feet), and they seem to aggregate on the highest-quality food deposits. They may be found in groups of more than 500 individuals, all bizarrely facing the same direction like a marching army but probably simply orienting themselves into the current to help maintain their position. It seems most deep-sea animals are quite selective of the sediments they consume. Deposit-feeding snails select food that has settled in the past three months but perhaps lack the adaptations of speed and the specialized feeding appendages of sea pigs to allow them regular access to higher-quality fresher resources.

Many other deep-sea echinoderms—for example, sea lilies, feather stars, brittle stars, and brisingid sea stars—are filter feeders. While they may live singly on rocks or sediment, better access to the resources raining from above can sometimes be obtained by living on another organism; brittle stars entwine themselves around corals, and feather stars are often found atop hexactinellid sponges. Competition for resources is a constant theme among deep-sea organisms.

> *"It seems most deep-sea animals are quite selective of the sediments they consume. Competition for resources is a constant theme among deep-sea organisms."*

Echinoidea (urchins)

Top: cidaroid urchins have big solid spines. Note the flexible attachments of the spines though: they can maneuver them to hold on to octocorals. All urchins have powerful jaws that allow them to feed on coral tissue.

Asteroids (sea stars)

Center left: Bright orange brisingid sea stars are very common in the deep sea. They are highly distinctive and are often seen sitting atop other organisms.

Crinoidea (sea lilies and feather stars)

Bottom left: sea lilies are a very distinctive deep-sea taxon. Their arms hinge at the central disk such that they can "close" like a flower. Crinoids without a stalk (feather stars) can use those hinged joints to swim.

Ophiuroidea (brittle stars)

Center right: a brittle star entwines its five long arms around the branches of a bubblegum coral at 2,477 ft. (755 m) on the continental slope southeast of the USA.

Scotoplanes (sea pigs)

Bottom right: a sea pig (*Scotoplanes* sp.) observed by MBARI's remotely operated vehicle (ROV) *Doc Ricketts* at a depth of approximately 4,600 ft. (1,400 m) at Sur Ridge, a rocky ridge off the coast of Big Sur, California. Sea pigs move slowly across the seafloor using their six pairs of enlarged tube feet. This bilateral symmetry is unusual in adult echinoderms, which are more frequently radially symmetrical.

Annelida, Arthropoda, and Mollusca

The annelids, arthropods, and mollusks of the deep sea tend to be highly mobile, although there is a minority of sessile (fixed in place) or slowly moving forms. Polychaete worms are adapted to a variety of lifestyles. They may live on, or in, the sediment, and they may be deposit feeders, filter feeders, or carnivorous. Some polychaetes live commensally (in close association) with particular coral species. Among the polychaete worms, there are delicate tube worms that filter feed, in contrast to the carnivorous or deposit-feeding majority. Evolution has even produced tube-worm specialists that use symbiotic microbes to live in chemosynthetic ecosystems (see pages 158–167). Among the arthropods, acorn barnacles measuring 1 in. (2.5 cm) across cement themselves to rocky outcrops, and their enclosing plates, discarded over millennia, occasionally form a coarse shell bed on the seafloor. Deep-water mollusks include bivalves such as deep-sea file clams (*Acesta* spp.) that adhere to vertical walls. Members of these three phyla, as is the case in shallow waters, are also some of the largest and most mobile invertebrate predators. Octopods have strong beaks to tear flesh. Nereid polychaetes grow to nearly 12 in. (30 cm) long and have pincerlike chitinous jaws. Crabs, lobsters, amphipods, and isopods have strong mandibles. In the deep sea, the amphipods and isopods can reach sizes not known among their kin in shallow waters. The supergiant amphipod *Alicella gigantea* grows to 12 in. (30 cm) long, and the giant isopod *Bathynomus* to 20 in. (50 cm). *Bathynomus* species have been known to chew through cables on scientific equipment. Sea spiders, which are distantly related to these other arthropods, can also grow very large. The leg span of the largest known species, *Colossendeis colossea*, can reach more than 2 ft. (61 cm) across.

Polychaete worm
Family Syllidae

Top left: this diverse group has an active mode of life foraging on the seafloor among seaweeds, corals, rocks, and on soft sediments. This one was caught on a fragment of deep-sea sponge in which it was living. Sponges provide substantial habitat for other animals.

Octopods

Bottom left: a warty octopus, *Graneledone verrucosa*, on the seafloor in the Gully Canyon in the northwest Atlantic. A female of a closely related deep-sea species in the Pacific, *Graneledone boreopacifica*, was seen brooding her eggs for more than four years.

Sea spiders

Top right: a sea spider feeds on zoantharians growing on a bamboo coral skeleton at 8,605 ft. (2,623 m) on a seamount in the northwest Atlantic. Cnidarians are the main prey of sea spiders.

Chitinous jaws

Bottom right: polychaetes have many different feeding strategies—they may filter feed, deposit feed, have chemosynthetic symbionts, or be predators. Many predatory polychaetes have huge chitinous jaws.

Lebensspuren—life signs

We often know more about very large organisms living on the seafloor than we do about the very small ones, such as burrowing polychaetes and surface deposit feeders, living in or on the sediment, even though these small animals dominate the abyssal plains that make up the vast majority of the deep seafloor. Many of these organisms leave traces known as *Lebensspuren* (life signs). Lebensspuren can be tracks, casts, or burrows. Tracks include the imprint of the foot of a sea snail or that made by an urchin as its hundreds of tiny water-filled tube feet glide it over the substrate. Casts are the egested waste from feeding, and these can sometimes be so distinct that scientists can determine the species that left them. For example, some species of acorn worm (phylum Hemichordata, class Enteropneusta) leave large spiral fecal casts, while many sea cucumbers produce tightly coiled fecal casts. With burrows, we often simply don't know what causes the lebensspur. Some might belong to large single-celled organisms called xenophyophores (phylum Foraminifera, superfamily Xenophyophoroidea), which are not classed as animals but are placed in one of the many "candidate kingdoms" that are used to classify organisms that are neither plants, fungi, animals, nor bacteria. Others may be made by peanut worms or other burrowing invertebrates, but the difficulties of sampling fragile animals from within the sediment mean that our knowledge of these infauna (sediment dwellers) is incomplete.

BENTHIC FISHES

An introduction to the fishes of the deep seafloor

The deep seafloor is populated by representatives of most of the major orders of marine fishes. These species appear similar to their shallow-water cousins, sharing many of the same adaptations. Their bodies tend to be elongated. The majority have eyes, though bright colors are absent (most are neutral shades of gray), and a few are bioluminescent. The full range of feeding modes is represented, including large-mouthed active and ambush predation, consumption of small swimming prey, bottom feeding, browsing, sediment burrowing, and scavenging.

Jawless and cartilaginous fishes

The "lower classes" of fishes, the jawless fishes (Agnatha) and the cartilaginous fishes (Chondrichthyes), are mainly confined to slope depths of less than 9,840 ft. (3,000 m). In the first group are about 60 species of deep-sea hagfishes, eel-like jawless fishes that burrow and scavenge on the seafloor, some appearing in swarms at large food falls such as whale carcasses. The Chondrichthyes—chimaeras, sharks, and skates and rays—are important predators on, and above, the deep seafloor. The 33 species of chimaeras (also known as rabbitfishes or ghost sharks, order Chimaeriformes) are relicts of a formerly widespread group of fishes that retreated into the refuge of the deep sea during the Jurassic, 150 million years ago. They grow up to 5 ft. (1.5 m) long, including a long tapering tail; they have seven tooth plates that give a rabbitlike grin to the face; and feed mainly on bottom-living prey. About 200 species of true sharks live in the deep sea; they vary in size from the Greenland Shark (*Somniosus microcephalus*, order Squaliformes), which grows to 18 ft. (5.5 m) long and may live longer than 300 years, to small catsharks (order Carcharhiniformes) less than 20 in. (50 cm) long. Among the primitive sixgill sharks (order Hexanchiformes) are frilled sharks, which specialize in feeding on squid; and cow sharks, which pursue a more varied diet that includes other sharks and bony fishes as well as squid. The dogfish sharks (Squaliformes)—including bramble, kitefin, rough, gulper, and sleeper sharks—are the main deep-sea group, with over 100 species; they have large, oil-rich livers for buoyancy and small fins adapted for slow swimming. Chondrichthyes also includes 100 species of deep-sea rays and skates (order Rajiformes), a few recorded at depths greater than 13,125 ft. (4,000 m). They differ little from their shallow-water counterparts and, like them, lay their eggs in special nursery areas.

Hagfish
Family Myxinidae
Top left: Black Hagfish (*Eptatretus deani*) scavenging on the bones of a whale at 5,500 ft. (1,675 m) depth off the coast of California. This image was taken 18 months after the whale sank to the seafloor. Top right: unidentified hagfish at 2,500 ft. (760 m) depth on coral rubble in the Atlantic Ocean off Florida.

Ghost shark or chimaera
Family Chimaeridae
Center left: deep-sea image from the Sulawesi Sea. The lines on the snout are a network of receptors enabling the fish to find prey in the dark by detecting electrical perturbations produced by living animals.

Frill Shark
Chlamydoselachus anguineus
Center right: a primitive sixgill shark usually found on slopes at 400–4,200 ft. (120–1,280 m) in all the oceans. Live young hatch from eggs inside the female.

Sleeper shark
Family Somniosidae
These are the deepest-living sharks, recorded down to 12,140 ft. (3,700 m). The body is swollen with an enormous liver for buoyancy and the fins are reduced.

Bony fishes

Among the bony fishes (class Osteichthyes), only the ray-finned fishes (class Actinopterygii) have significantly colonized the abyssal ocean floor at depths below 9,840 ft. (3,000 m), where their presence is the result of multiple invasions by numerous families. These can be divided into ancient deep-sea lineages that no longer have shallow-water representatives, and secondary deep-sea families that are recent colonizers and have species found at all depths.

Among the ancient deep-sea lineages are orders of eels. Halosaurs and spiny eels (order Notacanthiformes) live exclusively in the deep sea, down to depths of 13,125 ft. (4,000 m), where they specialize in feeding on invertebrates. Adults can grow to over 39 in. (1 m) long, and their remarkable pelagic larvae can be even longer. Eight bottom-living families of true eels (order Anguilliformes) occur in the deep sea. The most conspicuous are the cutthroat eels (family Synaphobranchidae), which scavenge in large numbers at food falls on continental slopes and mid-ocean ridges, while a few species are found on the abyssal plains at depths greater than 13,125 ft. (4,000 m).

More typical fishes are the slickheads or smoothheads (family Alepocephalidae), which are deep-sea relatives of salmons and smelts with large eyes and luminescent organs. They forage over continental slopes on gelatinous zooplankton as well as bottom-living animals, down to 15,750 ft. (4,800 m) or deeper, and spawn large eggs on the seafloor. The deep-sea tripod fishes (family Ipnopidae) balance motionless above the seafloor on the tips of elongate pelvic- and tail-fin rays, intercepting small crustaceans with extended pectoral fins, although not all species use the iconic tripod stance. Found at depths down to 19,700 ft. (6,000 m) worldwide, they have small or degenerate eyes in contrast to some other groups of deep-sea fishes in which the eyes have become enlarged to detect very low levels of light. The two species of deep-sea lizardfishes (family Bathysauridae) have a spectacular crocodile-like appearance. They occur worldwide, lying motionless on the seafloor and then ambushing fishes and crustaceans.

It's likely that the cods (order Gadiformes), with 11 extant deep-sea families, including commercially exploited hakes, morids, and cods, at upper- and mid-slope depths, originated in the deep sea. The grenadiers, rattails, and whiptails (family Macrouridae) are gadiforms with long tapering tails; this deep-sea specialist family comprises about 400 species. The macrourid genus *Coryphaenoides* is found throughout the abyssal plains of the world's oceans and down to maximum depths greater than 22,970 ft. (7,000 m) within hadal trenches. These are the dominant predatory bottom-living fishes

Cutthroat eel
Below: there are over 40 species in this family. Their enlarged nostrils and forebrain confer an acute sense of smell for detecting food sources in the deep.

Cusk eel
Family Ophidiidae
Below center: image from 7,220 ft. (2,200 m) in the Pacific Ocean. Cusk eels have somewhat elongated bodies but are very distantly related to true eels. There are about 250 species.

Deep-sea tripod fish
Bathypterois viridensis
Bottom: standing on the deep-sea floor on the tips of its pectoral and tail fins, the fish is oriented into the current to intercept drifting food items. The pectoral fins open like an umbrella to help capture prey.

Deep-sea lizardfish
Bathysaurus ferox

Right: an ambush
predator, usually found
at 3,280–8,200 ft. (1,000–
2,500 m), where it sits and
waits on the seafloor to
pounce on mobile prey.

Following pages:
Pelagic cusk eel larva
Brotulotaenia sp.

The larva of a pelagic cusk
eel—adults live in the
dark open ocean down
to 8,200 ft. (2,500 m)
depth. Larvae hatch from
gelatinous floating egg
masses and develop in
the upper layers of the
ocean before descending
to adult depths.

of the abyss, but they are also scavengers that rapidly intercept and consume food falls that reach the seafloor. The secondary deep-sea lineages of bony fishes are highly versatile, adapted to a wide variety of habitats. The cusk eels are found from inshore to deep waters, reaching depths over 19,700 ft. (6,000 m); some species occur at hydrothermal vents (see pages 162–163). The large Pudgy Cusk Eel (*Spectrunculus grandis*), which grows to over 39 in. (1 m) long, is common on lower slopes and abyssal plains of the Pacific and Atlantic, where it ingests small invertebrates as well as detritus. The order includes the blind cusk eels (family Aphyonidae), rare deep-sea fishes that give birth to live young, which have become small and degenerate, surviving in the abyss, down to 16,400 ft. (5,000 m) deep, with minimal activity and low energy turnover.

The snailfishes (family Liparidae) are small (up to 10 in./25 cm long) and equipped with a sucker on the belly to attach to rocks, where parents brood their eggs.

There are more than 380 species adapted to different habitats worldwide, from coastlines down to 26,250 ft. (8,000 m) in the Mariana Trench, where the endemic Mariana Snailfish (*Pseudoliparis swirei*) is probably the world's deepest-living fish. With about 200 deep-sea species, the eelpouts (family Zoarcidae) are another diverse group. Long-bodied and eel-like, they lay egg masses on the seafloor. Eelpouts occur from the Arctic to the Antarctic in a wide range of habitats to depths greater than 16,400 ft. (5,000 m), where 11 species are found at hydrothermal vents and others on remains of marine mammal skeletons on the abyssal plains.

The flatfishes (order Pleuronectiformes) may be found at depths down to 6,560 ft. (2,000 m). The biggest are the halibuts, which can grow longer than 13 ft. (4 m) and are important for deep-water fisheries in northern polar seas. Among the smallest, at 2.5 in. (6 cm) long, are *Symphurus* tonguefishes, found at sulfur-rich hydrothermal vents on underwater volcanoes in the west Pacific.

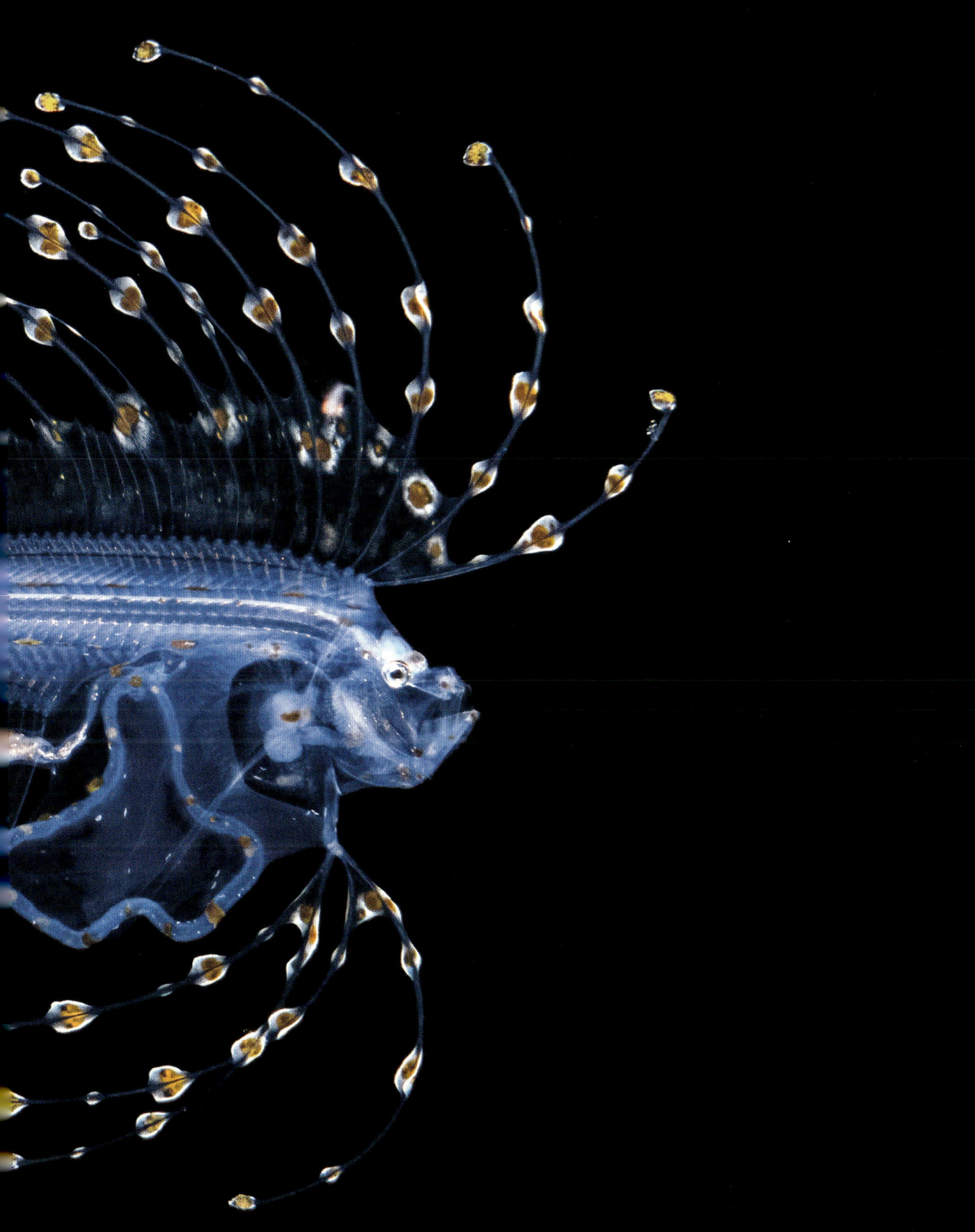

NONFISH PELAGIC ORGANISMS

Microbes, invertebrates, and nonfish vertebrates of deep open water

Pelagic waters encompass a huge midwater environment, away from the surface but above the bottom. Deep pelagic fishes are diverse and charismatic (see pages 104–109); here we cover the many other organisms found in the deep-water column, which range from invertebrates to microbes (both prokaryotes and eukaryotes) to, surprisingly, air-breathing vertebrates. Deep-pelagic invertebrates include representatives of almost every major evolutionary group (phylum), and a thorough presentation on pelagic invertebrates would require an invertebrate zoology textbook; we are limited here to a bare overview. The microbes include the classic groups of microscopic cells (e.g., bacteria, protozoans) living either freely or inside of animals, and also some free-living species (e.g., foraminifera and radiolarians) that are large enough to be seen without a microscope and which abundantly produce shells important in oceanic and planetary biogeochemistry. Nonfish vertebrates are mostly mammals diving to feed in mesopelagic (about 660–3,280 ft./200–1,000 m; see pages 170–171) and bathypelagic (generally deeper than 3,280 ft./1,000 m) waters, but some reptiles and even birds (penguins) enter the depths as well.

In an environment composed entirely of clear water and the animals inhabiting it, there is no place to hide. Those that live here also cannot sit down somewhere, other than on another animal, to rest and conserve energy. Therefore, we see among these organisms some broad themes, including buoyancy and various ways of looking like they are not there. These themes are explored in the section of this chapter on adaptations to living in the deep (see pages 110–121).

"In an environment composed entirely of clear water and the animals inhabiting it, there is no place to hide."

Salp chain
A colonial chain of salps. Looking somewhat like siphonophores, the jellyfish relative, the jet-propelled salps are in the phylum Chordata, which includes humans and all other vertebrates. Salp life histories are complex, with sexual reproduction, asexual budding, solitary individuals, and chains of clones.

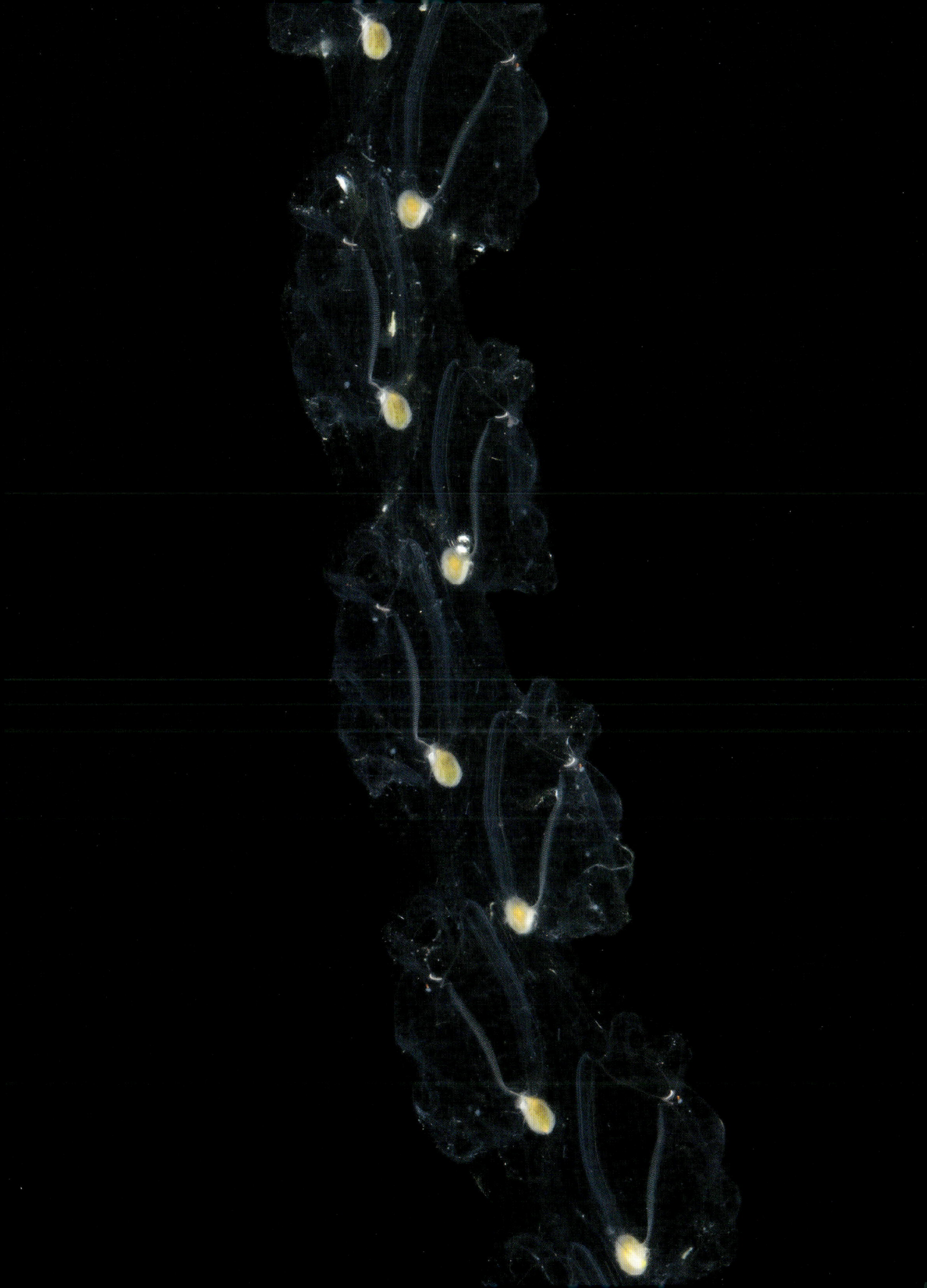

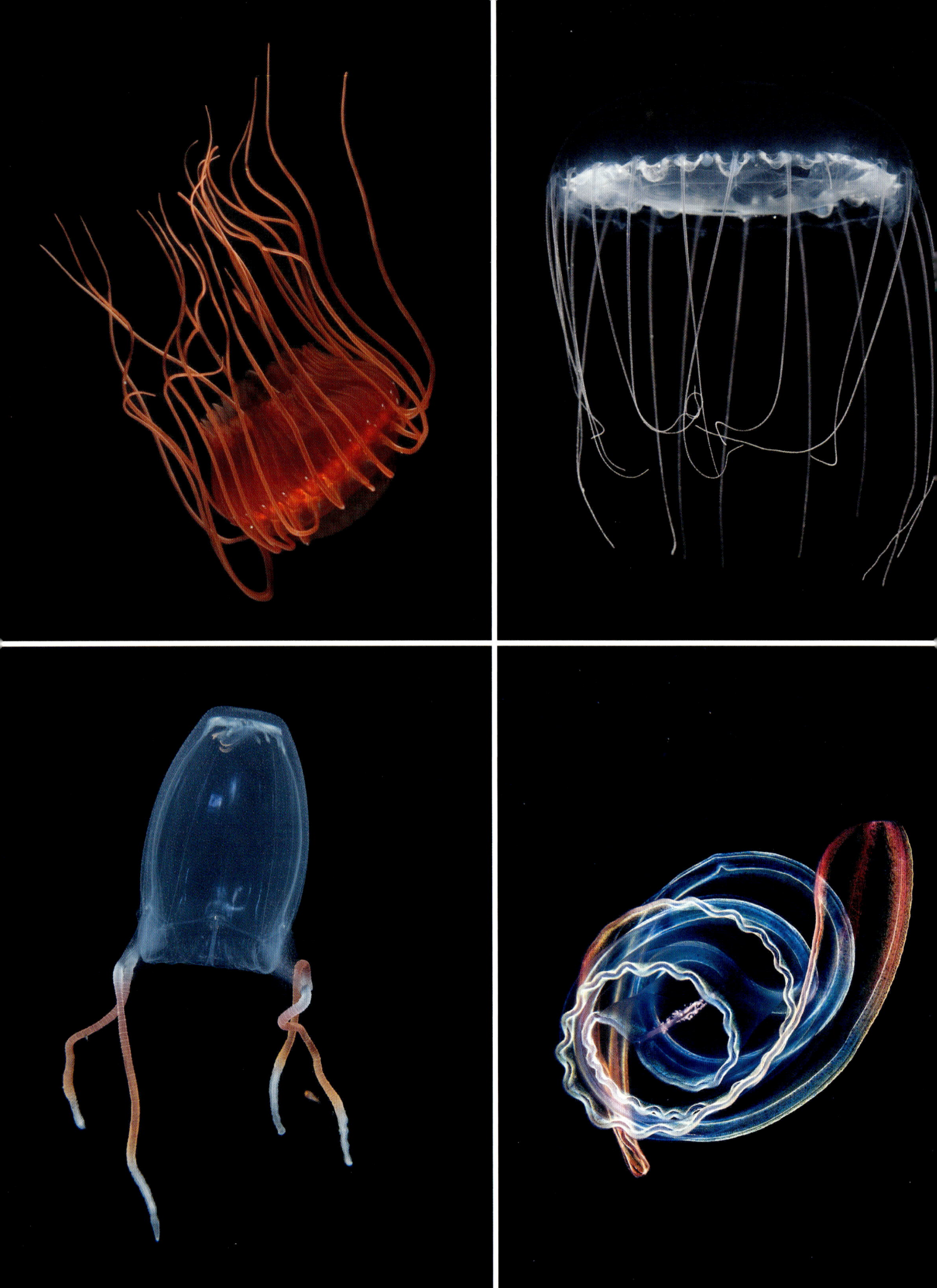

True jellyfish
Atolla sp.

Top far left: Atolla is a common deep-sea crown jellyfish. In addition to its bright-red color, common in midwater animals, it is capable of a characteristic bioluminescent display that rotates like a pinwheel. The display is known as a "burglar alarm" because it attracts large predators, probably to eat smaller predators that are attacking the Atolla.

Hydrozoan jellyfish
Solmissus sp.

Top left: the dinner-plate jellyfish, *Solmissus* spp., is often seen in the meso- and bathypelagic zones, looking somewhat like a flying saucer.

Box jellyfish
Alatina alata

Bottom far left: known as a Sea Wasp, this species of box jelly has long been known as a hazard to swimmers because of its very powerful stingers. We now know that Sea Wasps are also found in the deep sea because of observations from submersibles. Unlike other jellyfishes, box jellies have multiple image-forming eyes.

Comb jellyfish
Cestum veneris

Bottom left: the Venus' Girdle is a very unusual ctenophore. In this photo the flattened-out body is curled up in a defensive posture also often seen in other elongate transparent midwater animals like eel larvae.

Comb rows
Lampocteis cruentiventer

Right: most comb jellies (ctenophores) use rows of synchronized cilia for swimming.

Jellies, big and small

The most typical animals of the deep pelagic, and therefore of the earth because the deep-pelagic environment dwarfs all others, are various type of gelatinous organisms, often called jellies or jellyfishes. These include many species of cnidarians (phylum Cnidaria, characterized by stinging cells) and ctenophores or comb jellies (phylum Ctenophora, characterized by sticky cells), comparatively simple animals usually portrayed near the base of the conceptual tree of life. Remember, though, that these evolutionary lineages are very successful as well as ancient. Midwater forms include both individuals, swimming around by themselves, and colonies, some—such as siphonophores (class Hydrozoa)—with surprising specialization among the clonal individuals of the colony.

You may have learned in Introductory Biology that cnidarian jellies have a complex life cycle that alternates between benthic polyp (attached to the seafloor or another surface, with a ring of tentacles reaching up into the water) and pelagic medusa (swimming freely, more like what we think of as a jellyfish) stages. Species in the deep pelagic have no benthic substrates for polyps, and many have lost that stage, whereas others have solved the problem in unique ways, such as using the shells of planktonic swimming snails as a hard substrate on which their polyps can grow. Most people conjure up a medusoid (umbrella-like) mental image when they hear the word "jellyfish," and many cnidarian groups do have a medusa stage, including box jellies (class Cubozoa), true jellyfishes (class Scyphozoa), and hydrozoan jellies (class Hydrozoa). The last group includes colonial siphonophores made up of individuals specialized for swimming, catching prey, digesting, and reproducing. The maximum sizes of pelagic cnidarian species range from a fraction of an inch (a few millimeters) to tens of meters for free-swimming medusae to as much as 330 ft. (100 m) for some siphonophore colonies.

Because of a superficially similar appearance, comb jellies (ctenophores) are often confused with cnidarian jellyfishes, but they are not even closely related. Their tentacles, rather than having stinging cells like cnidarians, catch prey with "sticky" cells. Although most species swim using multiple "comb rows" of cilia, some species flap paired lobes of their bodies and look like the medusoid swimming jellyfishes.

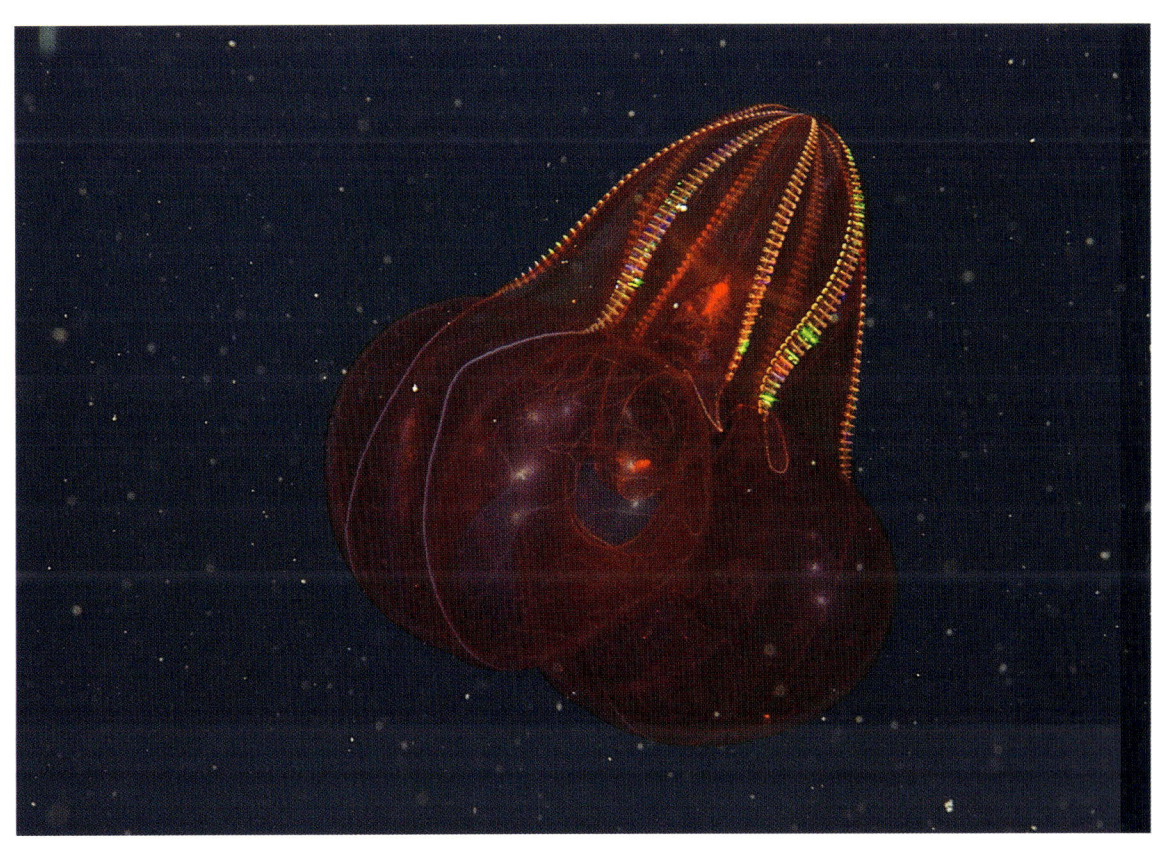

Tunicates

Another evolutionary group is often lumped into the category of "jellies," even though it is more closely related to humans than to cnidarians or ctenophores. This group is the filter-feeding tunicates (subphylum Tunicata). Like humans, these are chordates—characterized by a structure called the notochord, present in all members at some stage of development. Important pelagic tunicates include salps, which may be either colonies or free-living individuals, and pyrosomes, which grow as large dense colonies looking somewhat like swimming sponges (both in the class Thaliacea). The common name "pyrosome" refers to these animals' ability to produce bright luminescence. Another group is the strange, but sometimes abundant, larvaceans (class Appendicularia), which in adult form look like the "tadpole" larvae of other tunicates. When seen from a submersible, though, a larvacean usually will have secreted a complex feeding structure, called a "house," made of mucus. Larvacean houses, which are frequently abandoned, are major contributors to marine snow, and by sinking rapidly, abandoned houses deliver packaged detritus and its carbon to the bottom of the deep sea.

Larvacean house
A giant larvacean, *Bathochordaeus stygius*, in its "house." These pelagic tunicates create a complex inner filter (the many-lobed structure seen here) and a much larger gelatinous outer filter. The heart-shaped structure at the center is the head of the larvacean and its transparent tail can barely be seen extending upward from the head. Beating of the tail creates a current to force water through the filters, which catch food for the animal.

Marine microplankton

A mix of microplankton and mesozooplankton, ranging from microbes to many different types of zooplankton, including both holoplankton (that spend their entire lives in the plankton) and meroplankton (temporary residents of the plankton such as fish eggs and crab larvae).

100 µm

Zooplankton—the "drifters"

The jellies and tunicates described here are considered to be part of the community of "drifting" animals called zooplankton. "Drifting" is in quotation marks here because many zooplanktonic creatures actually are fairly good at swimming, though do not swim as fast or consistently as fishes, squids, and other aquatic animals. The large medusae and siphonophores are referred to as "gelatinous megaplankton." In addition to the groups described previously, other zooplankton, including sea butterflies and sea angels (class Gastropoda, order Pteropoda), arrow worms (phylum Chaetognatha), and other pelagic worms (class Polychaeta), are considered to be gelatinous because of their muscle consistency, but they neither look nor swim like a jellyfish.

The zooplankton assemblage is very diverse, including many species of crustaceans, some very abundant. Other diverse or abundant members of the zooplankton include snails (Gastropoda)—shelled and naked pteropods (order Pteropoda); sea elephants or heteropods (superfamily Pterotracheoidea), which are not closely related to pteropods; and unusual nudibranchs (order Nudibranchia)—and the aforementioned chaetognaths. Studies of zooplankton generally subdivide organisms by size because different study methods are used for different sizes: picoplankton (deep-sea examples include very small prokaryotic and eukaryotic

heterotrophs—i.e., nonphotosynthetic microbes) measure less than 2 µm (micrometers); nanoplankton (e.g., larger bacteria and protists), between 2 and 20 µm; microplankton (e.g., shell-forming foraminifera, important in forming sediments and in biogeochemical cycles), between 20 and 200 µm (0.2 mm); mesoplankton (e.g., many crustaceans, such as copepods; other examples include small pteropods and heteropods), between 0.2 and 20.0 mm; macroplankton (larger crustaceans, mollusks, and pelagic polychaetes), between 20 and 200 mm (0.75 and 8.0 in.); and the megaplankton, over 200 mm (0.75 in.) but not strong swimmers.

Another important division in zooplankton is based on whether the organisms spend their entire lives as part of the zooplankton. Those that do are called holoplankton, whereas temporary members are meroplankton. The latter category includes early life stages of many deep-sea benthic species. This, in turn, greatly increases the biological diversity of the midwater community at all taxonomic levels. For example, there are no holoplanktonic sponges, but sponge larvae disperse via deep currents and thus are meroplanktonic. The meroplankton also includes early stages of the category covered next, the strong swimmers of the nekton, including fishes.

Nekton—the active swimmers

The fishes, which are an abundant and diverse component of the assemblage of swimming pelagic animals known as the nekton, are covered elsewhere (see pages 104–109). Other major components of the nekton include such crustaceans as shrimps (order Decapoda), shrimplike krill (family Euphausiidae), and a poorly known group called lophogastrids (order Lophogastrida), possibly related to the so-called opossum shrimps (order Mysida).

Cephalopods (Cephalopoda) are also important in the deep-pelagic nekton. In addition to diverse families of oceanic squids (order Oegopsida), pelagic cephalopods include a few species of bobtails (order Sepiolida) and incirrate octopods (suborder Incirrata), finned octopods (suborder Cirrata), and a very unusual species without any close relatives, the Ram's Horn Squid (*Spirula spirula*), which has an internal coiled calcareous shell with gas-filled chambers. There is no hard line to define a species as nekton or plankton, as exemplified by some pelagic gastropod mollusks, such as the sea elephants. Active visual predators with well-developed eyes, sea elephants are generally considered as a group to be planktonic, but some large species, up to 20 in. (50 cm) long, are capable of very rapid, eel-like swimming in addition to their cruising mode of flapping a belly fin. If nektonic creatures are defined functionally by swimming ability, these large, strongly swimming snails seem to qualify.

Another open question about deep-sea nekton is whether small but actively swimming species of the major groups of nekton should be considered separately as micronekton. These organisms are generally sampled in small nets with small mesh. Such nets are easier to use than the very large gear needed to sample some of the larger nekton, so micronekton have been studied more often and more comprehensively than larger species or individuals.

A pelagic red shrimp

Top: red color is very common in the mesopelagic zone, especially among crustaceans. This is because the blue-green ambient light in this zone is completely absorbed by the red pigments. Because the light does not reflect, predators cannot see the shrimp except by looking up for a silhouette against the dimly lit surface overhead.

Antarctic Krill
Euphausia superba

Center: these large euphausiids form huge, dense swarms that are targeted by many predators. These swarms can be so dense that water appears reddish when the swarm is at the surface and the vertical migration of the swarm between the mesopelagic and epipelagic zones can physically stir the water column.

Ram's Horn Squid
Spirula spirula

Bottom: this species has a gas-filled chambered shell inside the end of its body near the fins. This positive buoyancy was thought to cause the animal to swim with its fins upward but, when finally seen in situ from an ROV, it actually balances and swims with the arms upward.

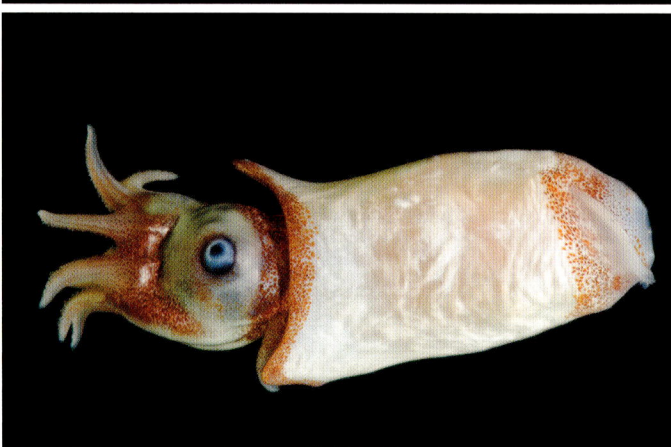

Emperor Penguin
Aptenodytes forsteri
This large bird can dive for over 20 minutes to as much as 1,640 ft. (500 m) depth in the mesopelagic to feed on krill, fishes, and squids.

Leatherback Sea Turtle
Dermochelys coriacea
The Leatherback Sea Turtle, the largest living turtle, is the most pelagic of the sea turtles. They sometimes dive to depths greater than 3,280 ft. (1,000 m) in search of their gelatinous megaplankton prey.

Air-breathing nonfish vertebrates

Speaking of large nekton, what about whales and other marine mammals? Some species indeed belong to the deep-sea nekton. Although baleen whales and most dolphins stay near the surface, many toothed whales dive deep to feed on other members of the nekton in the mesopelagic and bathypelagic zones. These include the Sperm Whale (*Physeter macrocephalus*), the largest toothed whale. Other important deep-feeding whales include beaked whales (family Ziphiidae), pilot whales (*Globicephala* spp.), Melon-headed Whale (*Peponocephala electra*), Dwarf Sperm Whale (*Kogia sima*), Pygmy Sperm Whale (*K. breviceps*), and several others. We think of these mammals as deep-*diving* whales, but some, such as the beaked whales, spend nearly all their time deeply submerged, surfacing for air only occasionally. They should probably be thought of as deep-*living* whales that have to surface periodically.

Elephant seals (*Mirounga* spp.) also dive deeply to forage for cephalopods and fishes. Some sea turtles, especially the Leatherback Sea Turtle (*Dermochelys coriacea*), do as well, in this case to forage on gelatinous megaplankton. There are even birds, the large King Penguin (*Aptenodytes patagonicus*) and Emperor Penguin (*A. forsteri*), that forage as deep as the mesopelagic zone for krill and other micronekton.

Swimming up instead of diving down

A final category of the pelagic fauna is the animals that swim up from the deep-sea bottom, sometimes very far up (hundreds of meters or more above the bottom). In addition to fishes, bottom-living cephalopods and decapod crustaceans are often seen swimming up above the bottom, as are such echinoderms (phylum Echinodermata) as mud-eating sea cucumbers (class Holothuroidea) and feather stars (class Crinoidea). The more we use submersibles for direct observations of animals in the deep, the more often we are surprised by what we see swimming; brittle stars (Echinodermata, class Ophiuroidea), sea spiders (phylum Arthropoda, class Pycnogonida), long-legged isopods (Arthropoda, family Munnopsidae), and acorn worms (phylum Hemichordata, class Enteropneusta) are among the unexpected swimmers and drifters from the benthic fauna.

Melon-headed Whale
Peponocephala electra

Melon-headed Whales are among many species of toothed whales that feed on fishes, squids, and shrimps deep in the water column. This small (ca. 9 ft./2.7 m maximum) species lives in large groups usually in deep waters throughout tropical areas of the world.

Swimmers and drifters

This sea cucumber (*Enypniastes eximia*), top image, and feather star (order Comatulida), bottom image, are two echinoderms among many benthic species in several phyla that often swim up, sometimes many feet, into the near-bottom layer of water either to escape predation or to reposition themselves on the bottom.

PELAGIC FISHES

An introduction to the fishes of deep open waters

Pelagic midwater fishes live in an unbounded three-dimensional environment with no solid surfaces, and the many species show an extraordinary variety of adaptations to life at different depths, from the dimly sunlit twilight of the upper ocean to the total darkness of the abyss. Most species employ bioluminescence, either for camouflage or communication. Upper-ocean species tend to have mirrorlike silvery skin, dark shading on the upper body, and light organs on the underside that obscure the silhouette of the fish in the down-welling light. Deeper-living species, found below 3,280 ft. (1,000 m), are dark-colored. Shapes vary from the elongate, exemplified by the Slender Snipe Eel (*Nemichthys scolopaceus*), which has over 750 vertebrae (more than any other vertebrate), to the spherical, represented by the Atlantic Footballfish (*Himantolophus groenlandicus*).

Sharks

Few sharks live in the deep pelagic zone. The Goblin Shark (*Mitsukurina owstoni*), which can grow over 20 ft. (6 m) long, floats motionless in midwater, feeding by sucking in prey such as other fishes, octopods, squids, and gelatinous plankton, using its remarkable protrusible jaws. Two families of small sharks, the kitefin sharks (family Dalatiidae) and the lantern sharks (family Etmopteridae), have evolved bioluminescence for camouflage, enabling them to stalk prey in the twilight zone of the mesopelagic zone. The Cookie-cutter Shark (*Isistius brasiliensis*) uses its bioluminescence to confuse sea mammals so it can take disk-shaped bites out of the skin and blubber of seals, porpoises, dolphins, and whales.

Goblin Shark
Mitsukurina owstoni

Right: this strange-looking fish has the fastest moving jaws of any shark. The jaws can shoot forward at 10 ft. (3.1 m) per second, without the rest of the body moving, taking prey by surprise.

Cookie-cutter Shark
Isistius brasiliensis

Bottom right: this shark emits light along its underside, which makes it invisible in the dim down-welling light. The conspicuous black band is unlit and remains visible to attract predators that perceive a small food item. The shark then takes a bite out of the would-be attacker. Below: the prey of a Cookie-cutter Shark can also include larger sea mammals such as dolphins—an unfortunate Spinner Dolphin (*Stenella longirostris*) shows the circular damage from the bite (close-up of teeth at bottom).

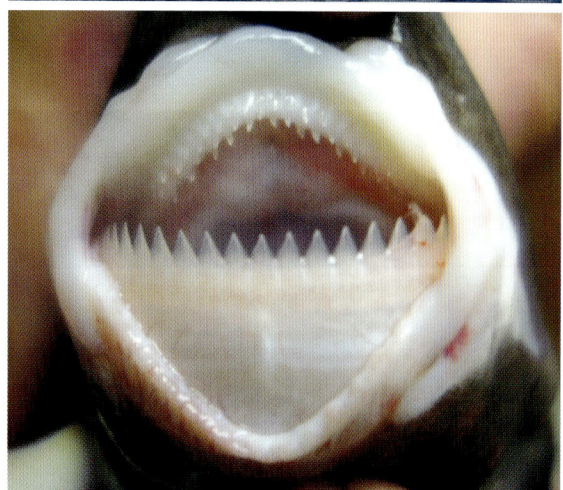

Forage fishes of the deep scattering layer

The twilight (or mesopelagic) zone of the upper deep sea is dominated by small fishes, predominantly less than 4 in. (10 cm) long, belonging to four families, the lanternfishes (family Myctophidae), bristlemouths (family Gonostomatidae), hatchetfishes (family Sternoptychidae), and lightfishes (family Phosichthyidae). Some of the most abundant vertebrates on the planet, they occur in vast quantities, possibly cumulatively weighing as much as 100 billion tons (100.10^9 tonnes) globally. All have light organs on the underside of the body. These fishes, together with invertebrates such as krill, form the deep-sea community of sound-scattering animals observed by echosounders during the day in several layers at 1,640–3,280 ft. (500–1,000 m) depth. At night, they move toward the surface to feed on plankton, then descend again at dawn. These forage fishes are consumed by a wide variety of predators, including mammals and birds diving from the surface as well as fishes and cephalopods attacking from below.

Two other families, the barreleyes and spookfishes (family Opisthoproctidae) and the deep-sea smelts (family Bathylagidae), occur in the pelagic zone in much lower abundance. Quite rare, barreleyes and spookfishes are small (6 in./15 cm), with upward-looking eyes specially adapted for spotting the transparent gelatinous plankton upon which they feed. One species, *Macropinna microstoma*, can rotate the eyes to keep track of the prey until it is ingested. The deep-sea smelts are dark-colored and are often caught with lanternfishes in tropical as well as polar seas.

Predatory fishes

An extraordinary variety of predatory fishes exists in pelagic waters between 660 and 6,560 ft. (200 and 2,000 m), stalking the creatures of the deep scattering layer (DSL) by either moving up and down, following the daily vertical migrations, or by sitting and waiting at depth. This group includes over 280 species of dragonfishes (family Stomiidae), elongate, dark or black fishes with large teeth and light organs along the belly. Most have a barbel attached to the lower jaw with a complex set of light organs at the tip that acts as a lure to attract prey. The viperfishes (*Chauliodus* spp.) are dragonfishes with no barbel, but they do have a lure on an extended dorsal fin ray. The lower jaw, with its extraordinarily long teeth, opens very wide, allowing ingestion of large prey. The stoplight loosejaws (*Malacosteus* spp.) have no floor to the mouth, also enabling the taking of large prey. These dragonfishes have unique light organs below the eyes that illuminate prey with red light that is invisible to their prey, which can only detect blue light. Collectively, the dragonfishes are able to consume most of the standing crop of small pelagic fish biomass every year.

There are three families of midwater predatory lizardfishes: the telescopefishes (family Giganturidae), pearleyes (family Scopelarchidae), and sabertooth fishes (family Evermannellidae). The last two have upward-looking tubular eyes for spotting prey from below. The telescopefishes have forward-looking eyes, so they must float vertically, head up, to look upward. A large mouth and distensible stomach enable large prey to be taken. The pearleyes have special windows around the base of their eyes, allowing them to peep downward to avoid dangers from below, while the main tubular eyes look upward, searching for food. The sabertooth fishes have large, fanglike teeth.

Lovely Hatchetfish
Argyropelecus aculeatus
The reflective silvery skin and blue bioluminescence shining out from windows on the underside of the belly and tail makes these fish almost invisible in the mesopelagic zone.

Bioluminescing deep-sea lanternfish
Myctophum spinosum
View of the underside of the fish with photophores emitting blue light—these obscure the silhouette in down-welling light.

Dragonfish
Stomias boa
Lights on the tip of the chin barbel help attract prey that are grabbed in the jaws with inward-pointing teeth. Light organs under the chin and belly aid camouflage.

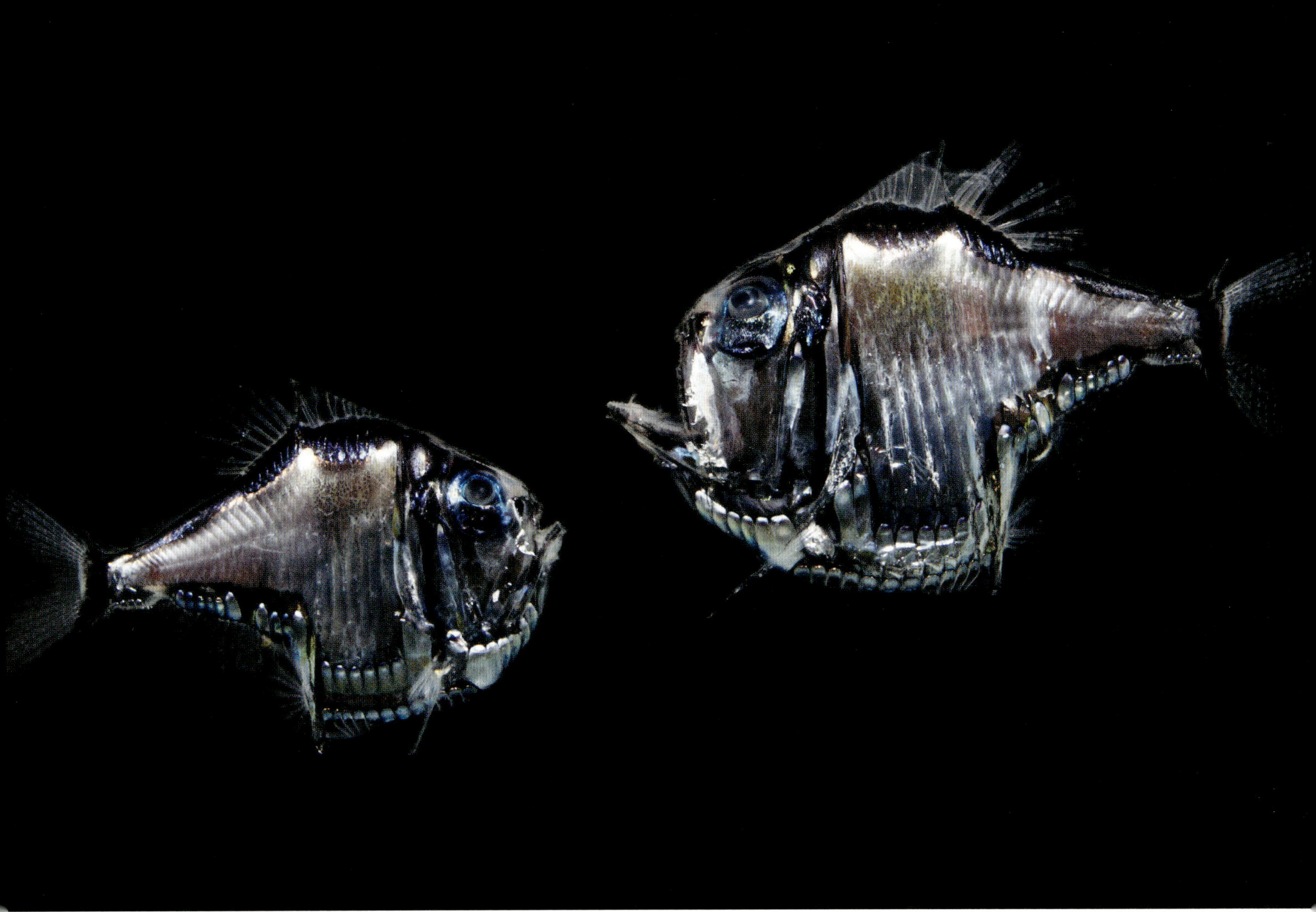
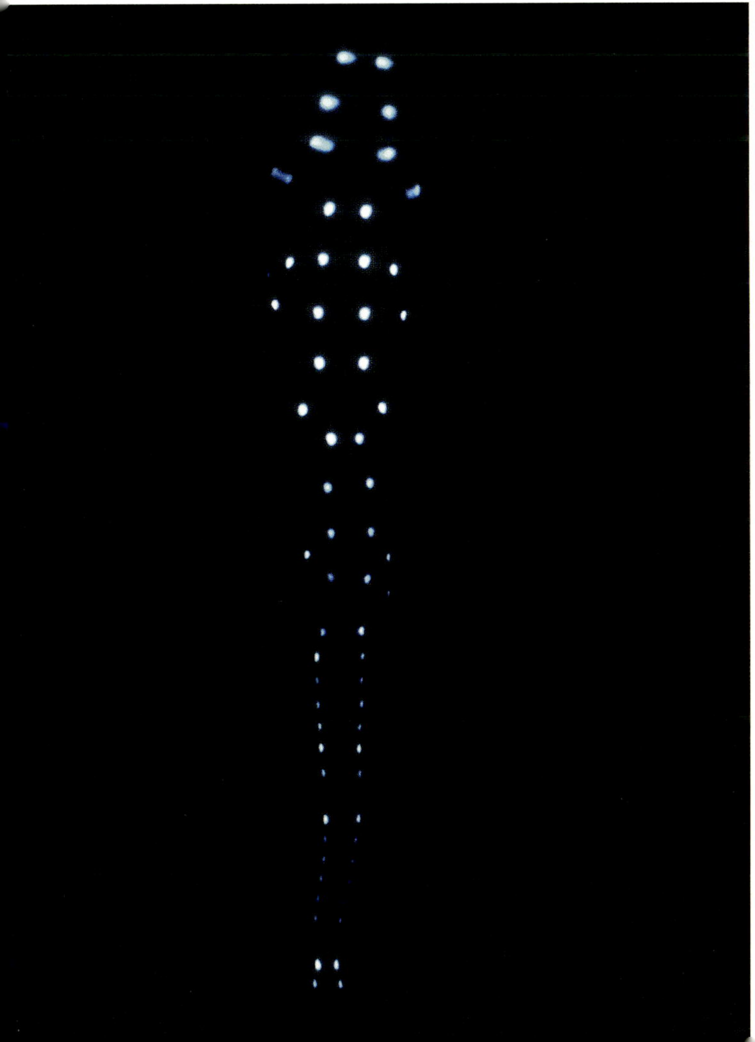
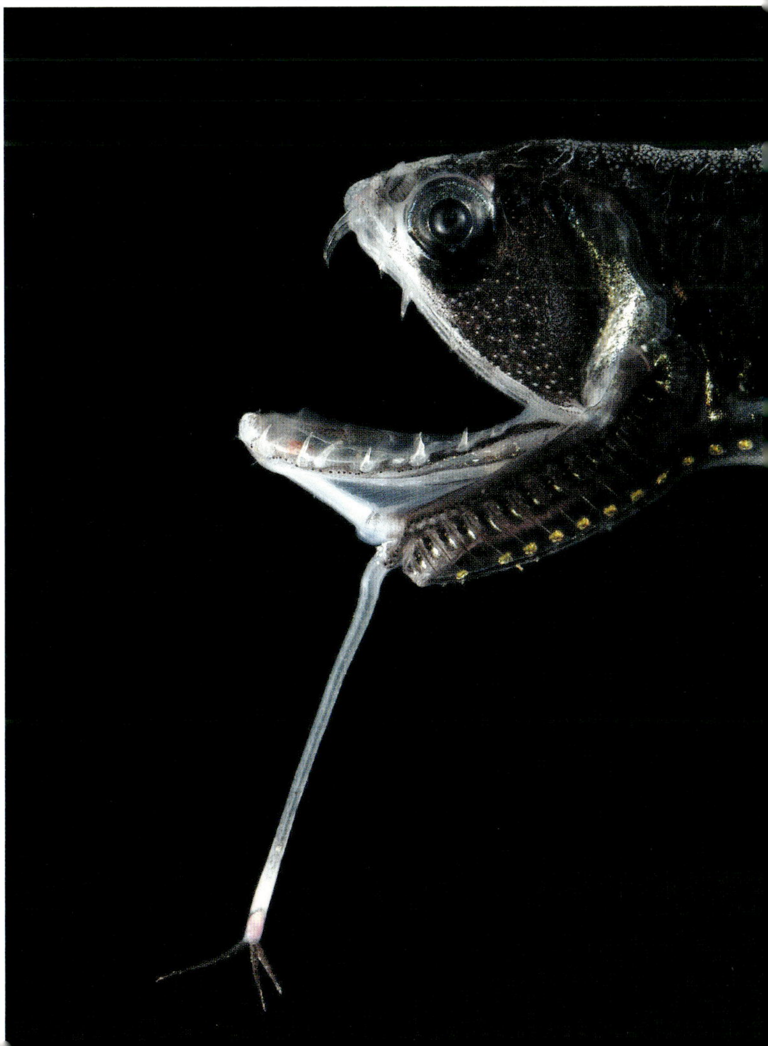

Two kinds of predatory eels, large-mouthed gulpers and slender-jawed species, are found in deep pelagic waters. The gulpers comprise two families, Eurypharyngidae, with a single species, the Pelican Eel (*Eurypharynx pelecanoides*); and Saccopharyngidae, the swallowers. Members of both are black, rendering them invisible at depth, and have very large mouths and distensible abdomens, enabling them to engulf very large prey, mostly fishes. The Pelican Eel is found worldwide, from temperate to tropical seas, usually at depths of 3,280–4,600 ft. (1,000–1,400 m). The 10 species of swallowers, which occur down to 9,840 ft. (3,000 m), have a luminous organ on the tip of the tail that may act as a lure. The slender-jawed species include the snipe eels (family Nemichthyidae), which are the most elongate vertebrates; they use their long, narrow jaws for catching crustaceans, mainly krill and shrimps. The sawtooth eels (family Serrivomeridae), less extreme in shape, feed on fishes and cephalopods as well as crustaceans.

Pelican Eel

Eurypharynx pelecanoides

Left: like its avian namesake, this fish has an enormous mouth cavity to engulf large volumes of water containing food. It expels the water through the gills retaining food items, which are then swallowed.

Redvelvet Whalefish

Barbourisia rufa

Bottom left: named for their whalelike shape rather than size, this species grows to only 15 in. (39 cm) long and occurs worldwide, usually associated with slopes, seamounts, and ridges.

Ceratioid Anglerfish

Melanocetus sp.

Bottom: female with a dwarf male attached below. Encounters with partners in the deep sea are so rare that when he succeeds, the male holds on until spawning. In some species, he becomes a permanently attached parasite for life.

Fishes of the great depths

The most abundant fishes of the great depths (the bathypelagic zone, generally deeper than 3,280 ft./ 1,000 m, are the whalefishes (order Cetomimiformes). They feed mainly on crustaceans, and some species are bright orange or red when freshly caught. The females grow to a full adult size of 16 in. (40 cm); the males are much smaller and may never feed during their adult life. The most specialized bathypelagic fishes are the deep-sea anglerfishes (family Ceratiidae). Many of the 170-odd species are rare, known from only a few specimens. As with the whalefishes, the males are much smaller than the females. The adult females float in midwater using a bioluminescent lure on the dorsal fin to attract prey. The males have large eyes and an excellent sense of smell, allowing them to actively pursue prey and mating opportunities. The dwarf male attaches to the female for mating, and in some species this attachment becomes permanent—the male loses its feeding, digestive, and swimming organs, becoming totally dependent on the female as an external parasite.

ADAPTIONS TO LIFE IN THE DEEP

Surviving the high pressure, the limited food, the cold, and the darkness

We generally think of the deep sea as a harsh environment where life is nearly impossible. But the conditions are really not harsh for species that have evolved to live there. Adaptations of these organisms to life in the cold, dark, high-pressure environment with limited food are many. We present here concepts behind these patterns of adaptation, focusing on four topics important to life in the abyss: energy management in a food-limited environment, common metabolic characteristics, sensory issues, and the production of light (bioluminescence).

Energy management

Except in relatively small chemosynthetic ecosystems, deep-sea life is entirely dependent on imports from the productive surface layers of the ocean. The greater the depth, the less food is available and, faced with a chronic shortage of vital resources, deep-sea animals respond with a combination of three strategies: maximize input (eat as much as you can), maximize assimilation (digest and store as much of what you eat as possible), and minimize expenditure of assimilated resources (don't use up all your fuel immediately).

Examples of adaptations to maximize feeding are seen in giant isopods (*Bathynomus* spp.) that efficiently follow scent trails to large food falls (such as a whale carcass), where they can pack their extendable digestive system; fishes (e.g., *Chiasmodon niger*) that subdue and swallow whole prey animals that are larger than themselves; and squat lobsters (family Galatheidae) that sit above the bottom on tall coral colonies and grab prey, drifting or swimming past, in their long claws. Efficient digestion often involves elongate intestines. In sea cucumbers, the long intestines are looped inside the body, whereas in larvae of some fishes the intestine actually hangs outside of the body. Some vertically migrating midwater animals slow their digestion by descending into colder, deeper waters during daytime. Minimizing expenditure also generally means moving as little and as slowly as possible. Some sea lilies and sponges move so gradually that we have recently been surprised to find that they move at all. Other benthic animals usually crawl slowly, although they can swim if threatened.

Keeping afloat is one of the challenges met by midwater animals, with no bottom on which they can sit. Muscles and skeletons are heavier than seawater, making the animal sink. To overcome this, some animals have a reduced skeleton, or have increased volume by adding fluid to the muscles. This is why many deep-pelagic animals have a gelatinous consistency similar to that of a jellyfish. Density can be decreased further by internal replacement of the heavier chloride salts from seawater with lighter substances such as ammonium. Even greater lift can be achieved by gas-filled buoyancy chambers, but these need to be carefully regulated to maintain constant buoyancy, and gas may be continuously lost to the surrounding water. These problems can be countered by replacing the gas with fats and oils that are relatively incompressible and can act as a food reserve. Some animals, such as amphipods and pelagic octopods, hitch a ride on buoyant gelatinous megaplankton, perhaps eating them as well.

Sinking can also be prevented by active swimming—this is like the difference in flight between a blimp and an airplane. Sharks and some squids create additional lift as water passes along the undersurfaces of their bodies and fins while they swim horizontally. Some animals, such as pelagic snails, energetically swim upward against gravity between periods of drifting and sinking.

Sea cucumber
Peniagone sp.
One way to make most efficient use of scarce food resources is to have a long intestine to process the useable components of what is eaten. In this swimming benthic sea cucumber that eats sediment, the intestine is long and forms a loop visible through the translucent body wall.

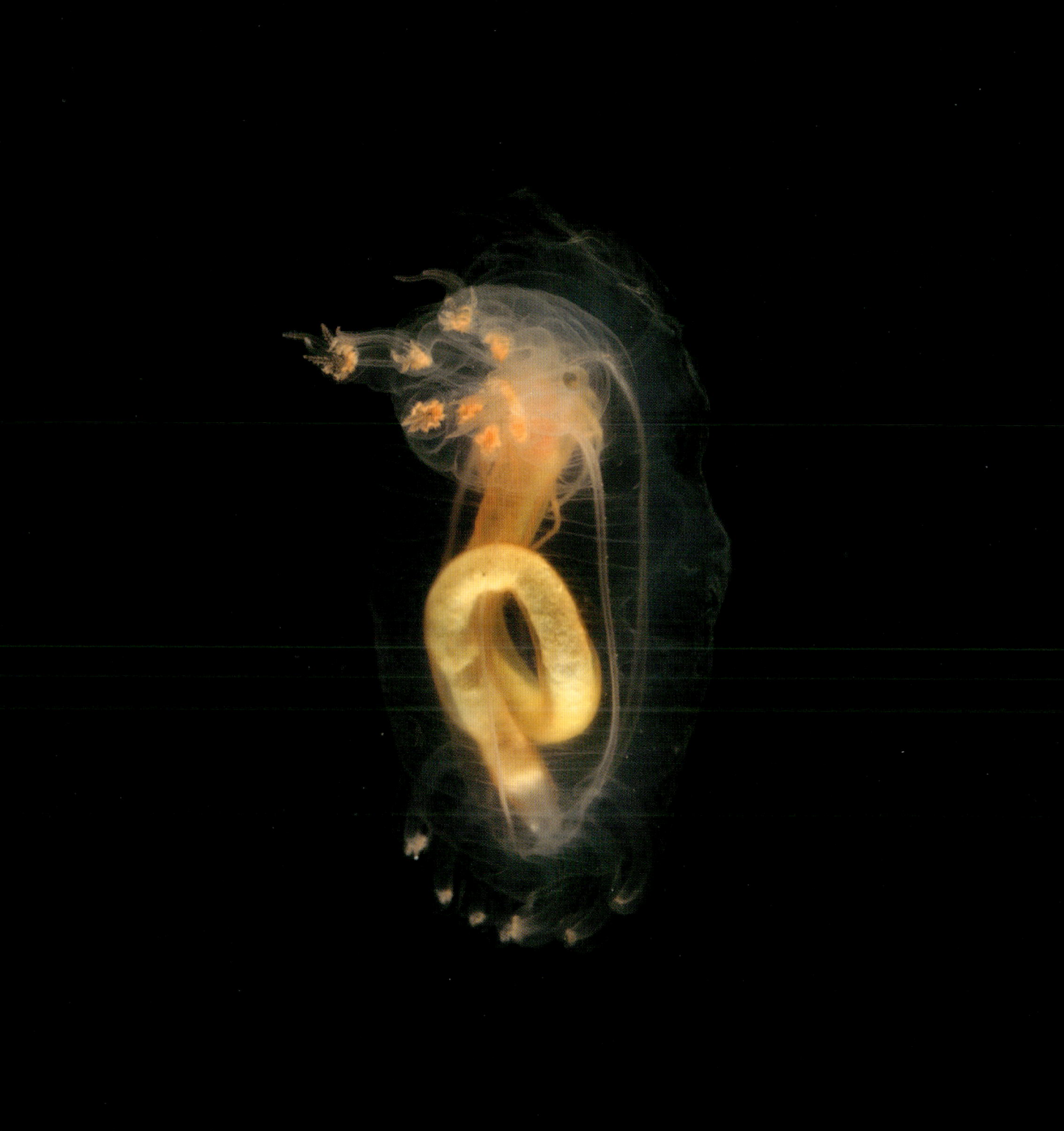

Metabolic issues

Except in oxygen-minimum zones (see pages 172–173), cold, deep waters contain adequate oxygen to support the metabolic needs of deep-sea animals. Also, oxygen is more soluble in colder water—Antarctic icefishes (family Channichthyidae) can even survive without red blood cells; oxygen is dissolved directly in their blood. Many invertebrates make use of copper-based hemocyanin rather than iron-based hemoglobin, and hemocyanin has a higher affinity for oxygen at lower temperatures.

The reduced muscle and tissue density mentioned above results in less metabolic activity for a given animal size. Also, low temperature and high pressure both tend to reduce rates of metabolism. Some deep-sea animals have been found to have enzymes modified in chemical structure, making them less sensitive to pressure changes but with the penalty that metabolic rate is decreased. The maximum depth of fishes and cephalopods is probably related to pressure sensitivity of their proteins. Cell membranes are also affected by low temperature and high pressure, becoming more rigid and less permeable. An evolutionary solution to this problem is to increase the amount of unsaturated fats in the membranes to maintain normal transfer processes into and out of the cells.

Sensory challenges

To animals evolved in the deep sea, the darkness there is neither complete nor uniform. The daily and seasonal changes in dim, down-welling sunlight in the mesopelagic and upper bathyal (benthic area of continental and island slopes at the same depths as the mesopelagic, 660–3,280 ft./200–1,000 m) environments are replaced below 3,280 ft. (1,000 m) by many types of bioluminescence (see pages 118–121). The darkness also increases dependence on senses other than vision, including detection of vibrations and pressure waves, chemical production and detection, and detection of electrical and magnetic patterns.

Often people expect deep-sea animals to be blind, like many cave creatures, but almost none are completely sightless. Even the "Blind" Octopod *Cirrothauma murrayi* can detect light. A very important reason for this is the widespread and common occurrence of bioluminescence. Color vision is unnecessary in the deep, since only blue-green light penetrates any distance through seawater. Most deep-sea animals can see only blue-green wavelengths, except for rarities such as the two species of Stoplight Loosejaw (*Malacosteus niger*), which generate their own red light that their prey cannot detect.

Stoplight Loosejaw
Malacosteus niger

Left: a dragonfish with unusual red lights below its eyes for stealthy illumination of prey. The loose jaw enables ingestion of enormous prey items. Lives at 1,640–12,800 ft. (500–3,900 m) depth.

"Blind" Octopod
Cirrothauma murrayi

Right: the "Blind" Octopod isn't actually blind. It has regressed eyes that are essentially small, cuplike structures lined with a retina of photoreceptor cells. Unlike most cephalopods, the eyes of this species lack lenses to focus an image on the retina.

Eye design and vision

Most complex eyes of deep-sea animals are variations of two basic designs: camera-type and crustacean-type. Camera-type eyes, with a single spherical lens focusing on a retina, are found in fishes and cephalopods but also in some unexpected organisms, like pelagic polychaete worms and box jellies. A common deep-sea variation is tubular eyes. Often mislabeled as "telescopic," they look somewhat like a telescope, but they provide no magnification. What the tubular form does is reduce the size, and therefore the silhouette, of the eye while retaining its full aperture to capture the maximum amount of light from a certain direction (usually looking upward).

Crustacean-type eyes come in several varieties. Most are compound (multifaceted), similar to the compound eyes of insects. Adaptations to the deep sea include size (some cover the entire head) and shape (bilobed to split attention laterally and above) as well as reduction and positioning of the sheets of photoreceptors. A strange crustacean-type version is the overdeveloped larval-type eyes of the large pelagic ostracods (seed shrimps) of the genus *Gigantocypris*. Their eyes are like parabolic mirrors focused on a few simple photoreceptor cells.

Some animals have light sensitivity with no visual acuity. For example, echinoderms have opsins (light-sensitive proteins) in their skin cells, which, in some species, are found in small, simple concentrated areas called eyespots. These can detect light patterns, such as the blocking of down-welling light by a passing animal, without focusing an image. Similarly, some fishes (*Ipnops*, a tripod fish genus) and crustaceans (*Rimicaris*, a hydrothermal-vent shrimp genus) have entirely sacrificed acuity for sensitivity, eliminating lenses and spreading photoreceptor cells over a larger area. Some species have retained lenses for focused images but have extra portions of retina to increase the field of detection beyond the edges of the main image. Some animals, especially cephalopods, have, in addition to their eyes, extraocular photoreceptors—vesicles to detect, but not focus, light.

Many other variations in eye design have been found in the deep. The eyes of histioteuthid squids (family Histioteuthidae), sometimes called "cockeyed squids," are dimorphic—two different designs in the same head; one is a normal eye looking horizontally and somewhat downward, while the other is a tubular design looking upward. The spookfishes (*Dolichopteryx* spp.) functionally have four eyes because each tubular eye has an extra opening with a mirror that focuses light on an accessory retina. Their relative *Macropinna microstoma* has tubular eyes set inside a transparent head so they can rotate depending on the fish's orientation. Many species have eyes set away from their heads on long stalks, especially during early development. Heteropod snails (superfamily Pterotracheoidea) have well-developed eyes with lenses but strangely shaped retinas. At least some species rapidly rotate their eyes up and down, presumably accumulating an image of their surroundings from one complete cycle. Some hydrothermal-vent shrimps supposedly can get very close to the scorching fluids without getting cooked because they can see faint chemical luminescence created by the fluids (see page 162).

Living in an environment with little light requires efficient capture and use of every photon possible. Because the light is essentially one wavelength, cone cells with color receptors are eliminated. Retinas packed with monochromatic rod cells or their invertebrate equivalents can capture more photons if pigment density is increased or if the cells are lengthened or multiple layers are stacked over each other. If a photon makes it through all that, the chance of absorbing it doubles with the addition of a reflector (called a tapetum, seen as the eyeshine reflecting from a cat at night) behind the retina, which sends the photons back for a second pass through the retina.

Barreleye
Macropinna microstoma

Top left: the green, upward-looking eyes inside a transparent dome on top of the head aid the search for planktonic food. When a prey item is identified, the eyes rotate forward to get a close-up binocular view for the final attack.

"Cockeyed" squid
Histioteuthis heteropsis

Top right: this and related species of the family Histioteuthidae have one large tubular eye aligned generally upward to see silhouettes against the dim glow of the surface, whereas the other eye is more normal in shape and looks horizontally and slightly downward, searching for animals around it and probably for bioluminescence. As the squid hovers obliquely in the mesopelagic zone, it rotates, visually sweeping the volume in its vicinity.

Giant ostracod
Gigantocypris muelleri

Bottom left: this pelagic giant ostracod (or seed shrimp) has very unusual eyes. Each eye has a parabolic reflector that concentrates the limited available light onto a small cluster of photoreceptor cells.

Spookfish
Dolichopteryx rostrata

Bottom right: view from above of the upward-looking tubular eyes used for foraging in the down-welling light of the mesopelagic. The eyes also have accessory lobes and retinas for sideways and downward-looking vision to avoid predators.

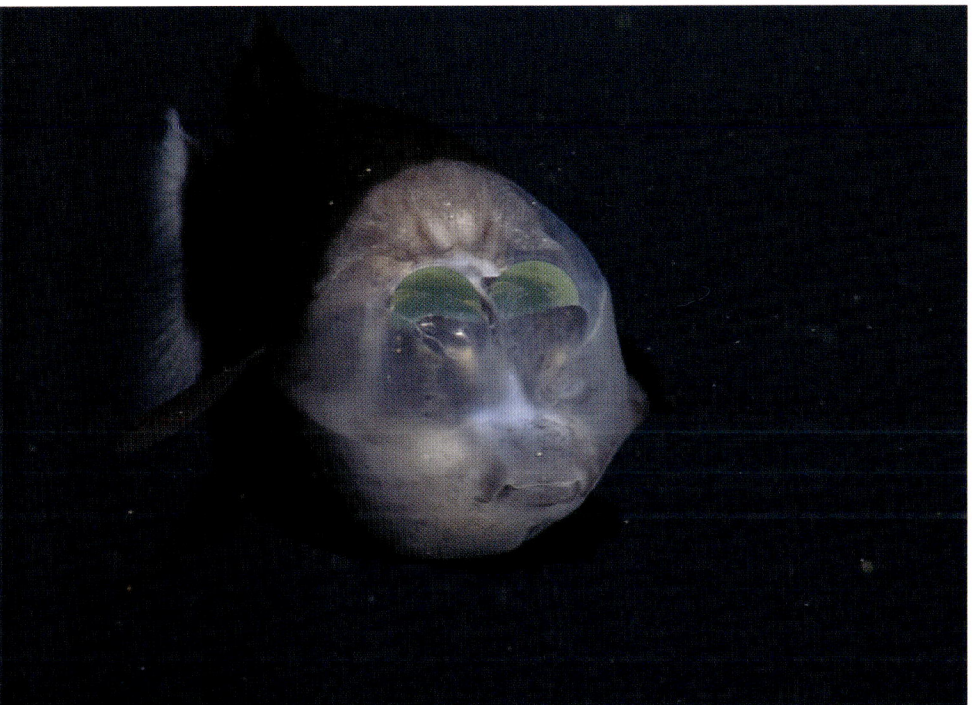

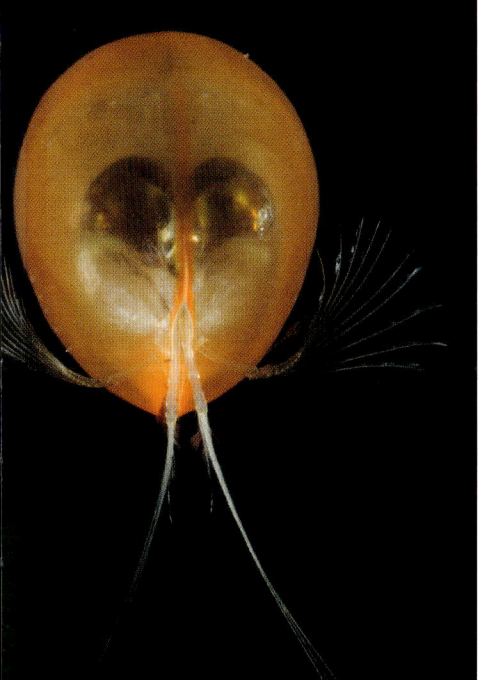

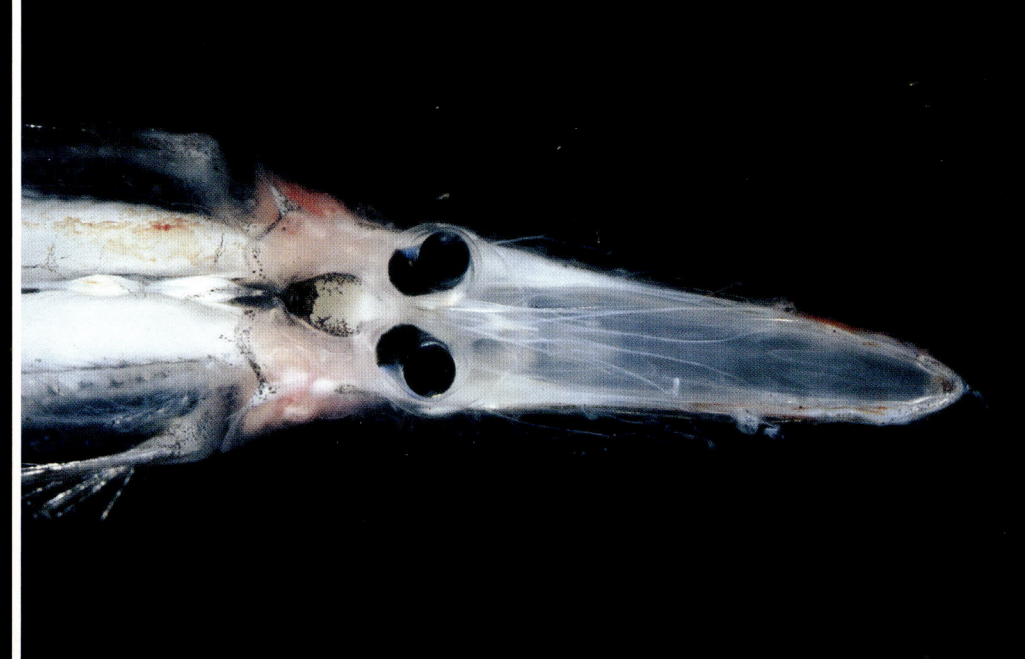

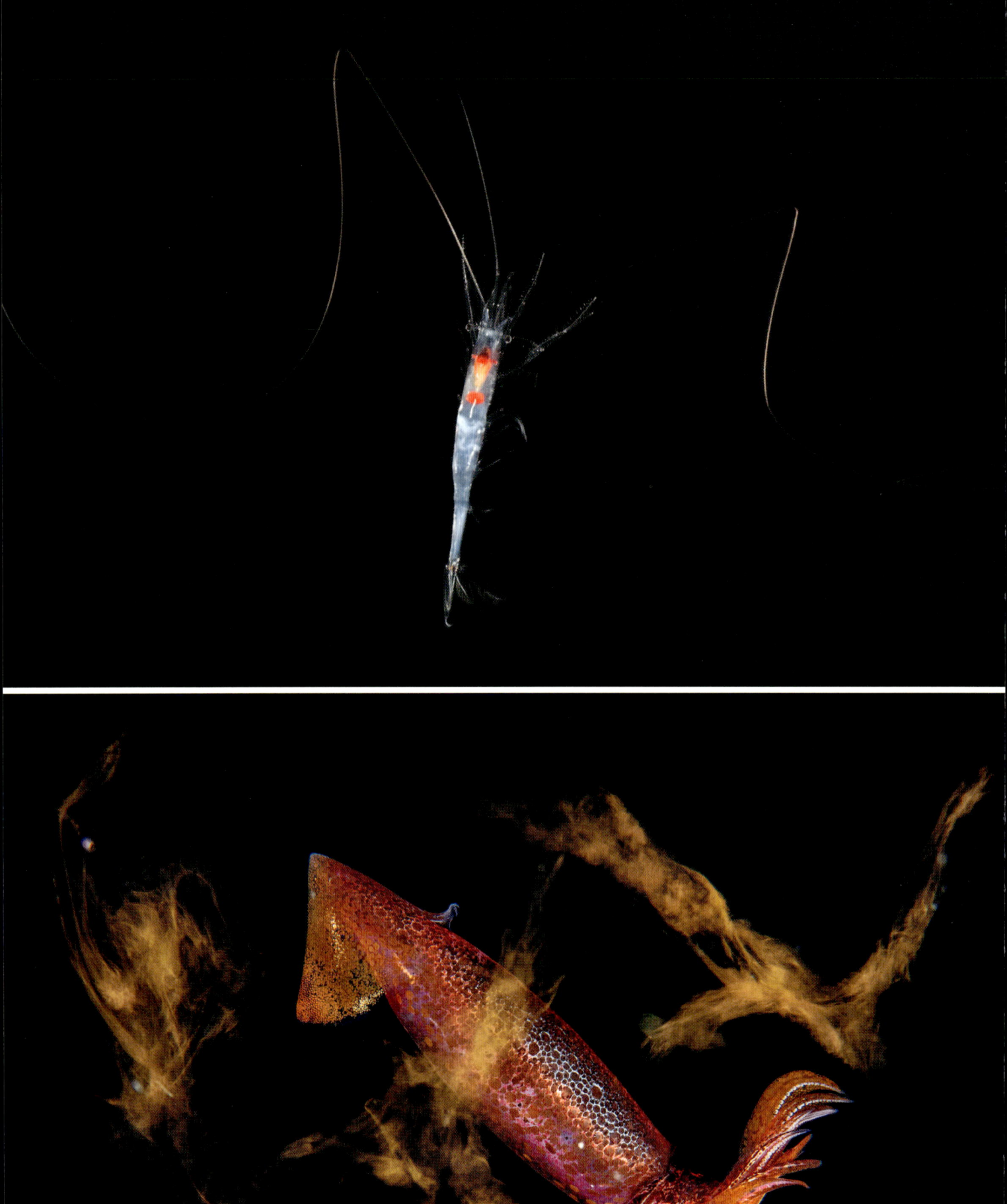

Hearing and feeling

With vision limited, other senses are important. Hearing (vibrations) and feeling (pressure waves) are similar phenomena, based on detection of disturbances in the water, but pressure and vibration differ in how the energy impacts the animal; this, in turn, depends on whether the source is relatively far away or nearby. These differences are important for the structure of the detection organs, or mechanoreceptors. External hairlike setae of arthropods and cilia of animals without exoskeletons, while quite different in evolution and development, are similar in appearance and function, and both can be used to detect hydrodynamic disturbances. Their efficiency increases if the cells to which they attach are organized into concentrated organs. Such organs may simply be scattered on the external surface of the animal, or they may be further organized in pits and canals; membranes over these structures allow resonance and increased sensitivity. In some animals, these mechanoreceptors are even extended away from the body on stalks for early warning.

Similar mechanoreceptors are also located internally. Their connections to tissues of different densities (e.g., bone, gas bladders) can affect their function because vibrations move at different speeds through different tissues. In addition to detection of vibrations and pressure waves, such internal systems are used to monitor linear and angular acceleration of the animal for swimming control. Animals may produce sound as a signal, such as the vibrations created by the swim bladders of some fishes, generally living near bottom in the upper bathyal environment. It can also be a tool, as in the sonar used by deep-diving toothed whales to locate prey, or as a weapon, as when Sperm Whale (*Physeter macrocephalus*) sonar is so loud it can incapacitate the prey.

Chemical and electrical receptors

Chemical receptors for olfaction (smelling) are very similar in biological structure to mechanoreceptors. As with both eyes and mechanoreceptors, chemoreceptors may be extended out on stalks. Chemicals to be detected include those secreted by the individual animal, its conspecifics (members of the same species), and other external sources, including other animals and physical phenomena. Physical processes such as water stratification by density or current transport can result in concentrations and trails of dissolved chemicals. The functions of secreted chemicals and their detection in deep-sea animals are numerous. Pheromones are important in a diffuse, light-limited environment, and many unrelated deep-water species have chemoreceptors adapted to finding mates in the dark. In addition to recognizing conspecifics, smelling has other uses in the dark, such as allowing animals to detect scent plumes of prey, predators, and substrate. The last is particularly useful for larvae searching for a good place to settle and metamorphose.

Given the importance of chemical detection in the deep ocean, use of chemicals for defense seems a reasonable strategy. An aggressive form of chemical defense is to produce chemicals that repel a predator. If an animal produces a defensive chemical and a member of the same species detects it and interprets it as an indication that a conspecific feels threatened, that chemical can function secondarily as a threat alarm. Some fishes produce such chemicals, known as *Schreckstoff* (fright substance). Similarly, Antarctic octopods in aquaria sit motionless for long periods; however, if a drop of ink from their species is added to the tank, the octopod suddenly changes appearance and swims madly about.

Electroreceptors are also similar, structurally, to chemical and mechanoreceptors. They are often based in clusters of cells in pits or canals, often located around the faces of fishes—especially cartilaginous fishes (Chondrichthyes, the sharks, rays, and kin) that are particularly known to use electroreception for the final approach to prey. A more mysterious ability, at least in large swimming animals of the abyss, is detection and orientation to magnetic fields. Basically, if an organ is present that can detect a magnetic field, then swimming through that field can create an electrical signal that can be picked up by electroreceptors. Although such organs have been identified in some deep-sea animals, many large species in which the organs have not been identified travel long distances and must navigate somehow. Magnetic navigation seems the most reasonable explanation for such long-distance migrations, but it is largely unproven in most animals.

Long antennae

Top left: the long antennae that parallel the bodies of swimming sergestid shrimps (family Sergestidae) have mechanoreceptor setae to detect potential predators and prey in the water around them, functionally similar to lateral lines of fishes but extended outward from the body.

Masking chemicals signals

Bottom left: the masking of a chemical signal can prevent an encounter with a predator. This is likely why squids such as this ommastrephid (family Ommastrephidae) and mastigoteuthids (family Mastigoteuthidae) release a cloud of ink and then hang out within the cloud.

Bioluminescence

Biological production of light, or bioluminescence, is extremely common in the dark of the deep, especially in pelagic animals, but it even exists in sessile benthic animals. Because of the huge volume of the deep sea, even widely dispersed organisms have very large total populations. Bioluminescence, therefore, is arguably the most common form of communication among animals. The colors sparkling on animals in videos from submersibles should not be confused with bioluminescence. In most videos, the camera lights are so bright that the natural luminescence is overwhelmed, and the colors seen are reflections and refractions of the video lights.

Bioluminescence requires two types of chemicals (elegantly called "luciferins" and "luciferases") in the presence of oxygen. Some animals produce their own chemicals; others get them by eating luminescent prey. In both of these categories, animals with this "intrinsic" bioluminescence tend to have lots of light-producing organs, or photophores. Other animals recruit and culture luminescent microbes in special chambers. These "bacterial photophores" tend to be fewer in number. A few animals, such as anglerfishes, have both bacterial and intrinsic photophores, but most have one kind or the other. The color of almost all bioluminescence is similar to that of sunlight down-welling from the surface, but in rare exceptions, animals can produce and detect yellows and even reds.

Bioluminescence has many uses for both defense and predation. Some predators use it in lures to attract prey, as was illustrated memorably by the anglerfish in the movie *Finding Nemo*. Visual predators such as dragonfishes, which have large photophores underneath their eyes, use it in "headlights" to illuminate prey. Red headlights, matched with retinal ability to detect red, found in a few dragonfish species, are particularly effective for preying on the many red crustaceans in the deep.

As a defense strategy, many mesopelagic animals produce downward-aimed light, creating counter-illumination, which breaks up the animal's silhouette against the surface light and makes it harder for upward-looking predators to detect them. Many animals, including various crustaceans, polychaetes, squids, cephalopods, and gelatinous plankton, use a cloud of bioluminescent fluid or particles to escape. For example, the Vampire Squid (*Vampyroteuthis infernalis*) releases glowing particles from photophores on its arm tips when threatened. Some organisms will even leave a part of themselves behind to effect a successful escape. The pelagic Green Bomber Worm (*Swima bombaviridis*) has branchiae (gills) modified into bioluminescent sacs, which it drops as a distraction when threatened by a predator. Other worms—for example, scale worms—can drop bioluminescent scales to distract predators. When escaping, the pelagic sea cucumber *Enypniastes* sloughs off its bioluminescent skin, which sticks to and lights up the predator, making the predator vulnerable to predation.

Light production is also important for inter- and intraspecific communications (see pages 228–229).

Reflection

Top right: when they are photographed, the reflection and refraction of these "comb rows" (as seen here in *Hormiphora californiensis*) in the camera's lights is often misinterpreted as bioluminescence. Although many ctenophores are bioluminescent, that light is overwhelmed by the bright lights needed for most photography.

Dragonfish lures

Top far right: a scaleless black dragonfish (*Melanostomias biseriatus*) from the tropical Atlantic Ocean bathypelagic, showing the luminescent bulbs and branches on the tip of the chin barbel.

Bioluminescent cloud

Right: the deep-sea shrimp, *Oplophorus gracilirostris*, reacts to predators by vomiting a bioluminescent cloud, functionally similar to glowing ink. This glowing cloud distracts the predator while the shrimp escapes.

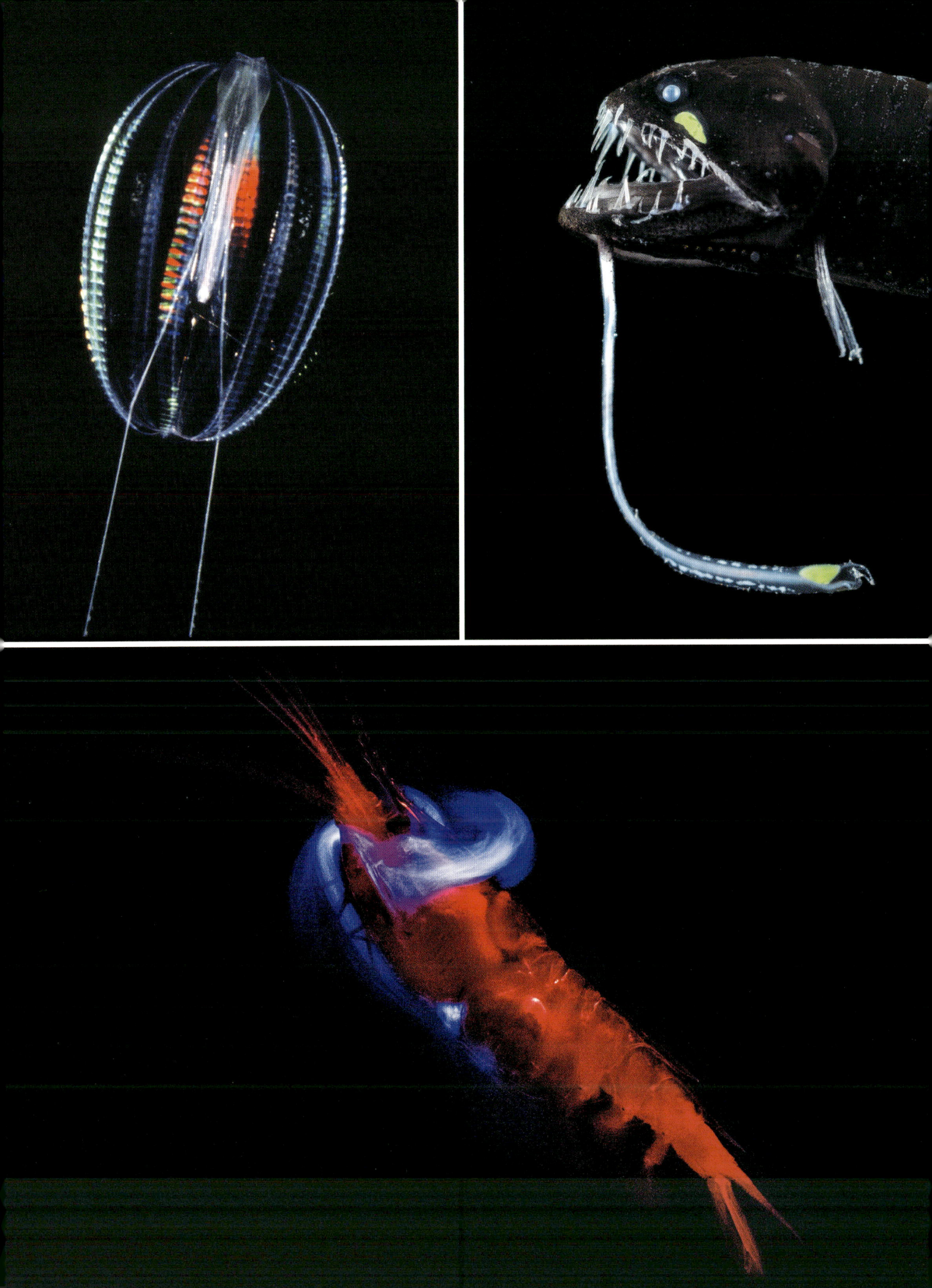

Glowing suckers

The slow-moving pelagic octopod, *Stauroteuthis syrtensis*, which traps water in the web between its arms and floats like a balloon, has glowing suckers that may lure prey toward it so that it has less active hunting to do.

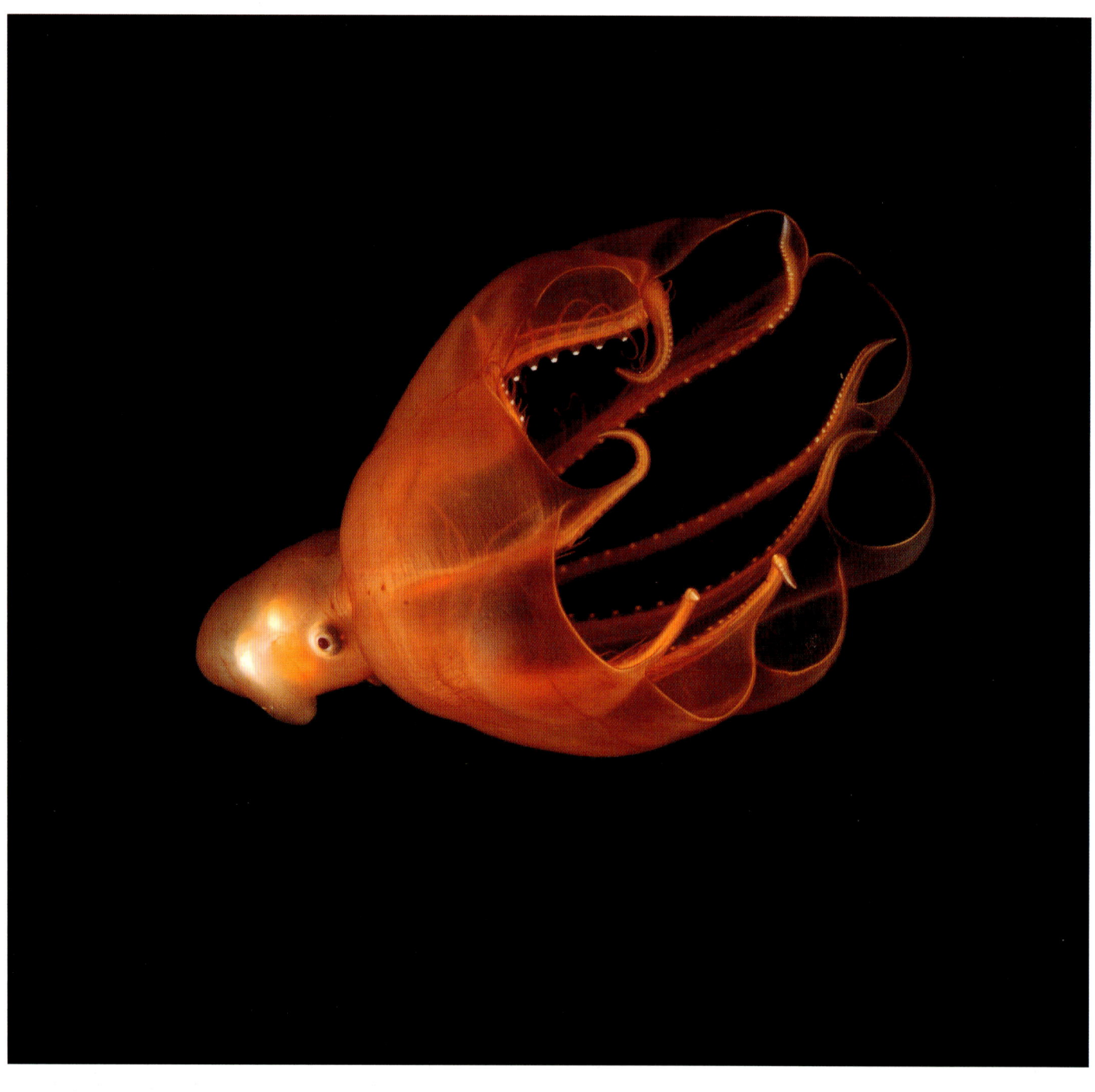

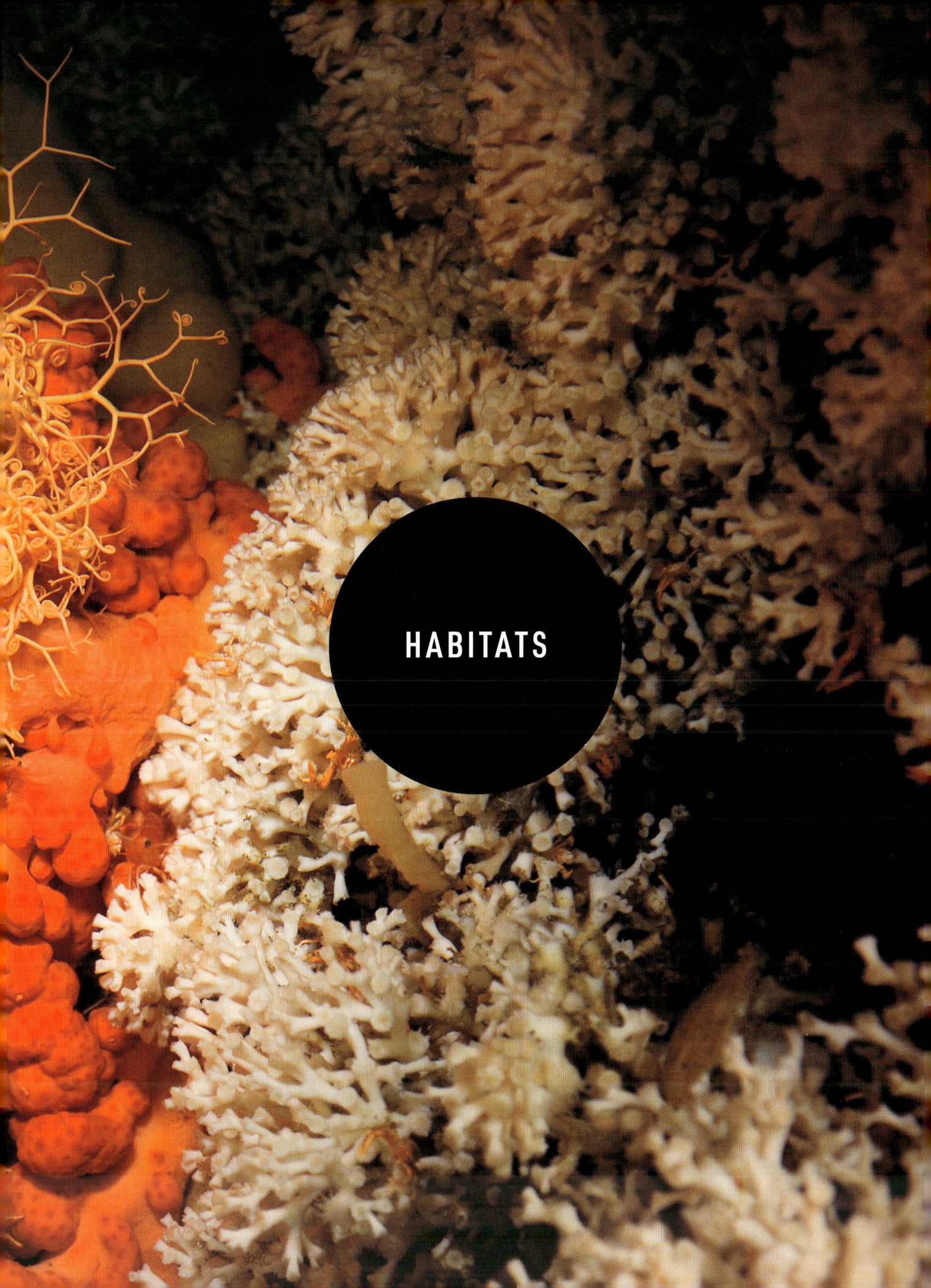

HABITATS

SLOPES AND SUBMARINE CANYONS

Topography and biodiversity of the continental slope

Previous pages: **Reef habitat**
Bubblegum coral (*Paragorgia arborea*) with basket star (*Gorgonocephalus caputmedusae*) within the *Desmophyllum pertusum* reef in Trondheim Fjord, North Atlantic Ocean, Norway.

Slopes

Continental slopes, extending from the outer edge of the continental shelf down to the deep seafloor, average about 25 mi. (40 km) wide, with depths from about 660–13,125 ft. (200–4,000 m). The slope is a zone of transition, supporting familiar shallow-water species and truly deep-water species of the abyss. The topographic incline is around 10%, about a third as steep as San Francisco's famous Filbert Street, and tends to be covered in soft sediments that are home to burrowers, deposit feeders, and organisms that don't need hard structures to anchor themselves. Many larger mobile species from the continental shelf can move down the slope into deep water: as adults, monkfishes (*Lophius* spp.) and halibuts (*Hippoglossus* spp.) move down to 3,280 and 6,560 ft. (1,000 and 2,000 m) depth, respectively; the Jonah Crab (*Cancer borealis*) migrates offshore to 2,460 ft. (750 m) depth in winter; and shortfin squids (*Illex* spp.) descend to 3,280 ft. (1,000 m) to sit on the bottom. On the slope, endemic specialist species are each confined to a particular depth zone. For example, among the fishes in the northeast Atlantic, Common Mora (*Mora moro*) is found at 1,640–4,265 ft. (500–1,300 m), Roundnose Grenadier (*Coryphaenoides rupestris*) at 2,300–6,230 ft. (700–1,900 m), and Günther's Grenadier (*Coryphaenoides guentheri*) at 3,940–9,430 ft. (1,200–2,875 m); while among sea stars, *Psilaster andromeda* lives at 1,640–4,130 ft. (500–1,260 m) and *Hymenaster pellucidus* at 4,790–10,170 ft. (1,460–3,100 m). Similar depth specializations occur in all animal groups, so from the upper slope down to the abyss, there are changes in the fauna analogous to the elevational succession of vegetation on a mountain slope from forests to snowfields.

Generally, the amount of food available decreases with depth, roughly tenfold for every 1.25 mi. (2 km), so that at the foot of the continental slope (13,125 ft./4,000 m), the food availability is 1% of that on the continental shelf. However, biomass, abundance, and numbers of species usually peak at mid-slope depths, partly due to overlap among shallow-water, slope, and abyssal species. Animals of the oceanic deep scattering layer (DSL; see pages 170–171) also impinge on the slope at these depths, providing an abundance of food for mid-slope creatures. Oceanic species of the DSL never encounter a solid surface during their normal lives, so they become disorientated by an encounter with the seafloor and are easy prey for slope-dwelling predators. However, off Hawaii and probably elsewhere, a "mesopelagic boundary community" is permanently associated with these bottom depths and migrates obliquely up and down the slope, rather than vertically as with most members of the oceanic DSL.

Continental-slope species live in a narrow strip of seabed subdivided by underwater canyons. Such patches of habitat can be quite small, with populations cut off from one another. Different groups of species may be found on slopes on opposite sides of an ocean, and the fauna also changes with latitude. Slope species tend to be regional rather than global.

Sea star
Psilaster andromeda

Top left: a predator on upper slopes of the northeast Atlantic Ocean where it feeds mainly on mollusks, small clams and snails, crustaceans, and worms.

Sea star
Hymenaster pellucidus

Center left: mid-slope species from the North Atlantic Ocean, the Arctic, and Antarctica. The short stubby arms are joined by an umbrella-like membrane and it moves over the seafloor on its tube feet.

Angler
Lophius piscatorius

Top right: the largest adults populate the upper slopes down to 3,280 ft. (1,000 m) depth. They lie half buried in the sediment leaping up to grab passing prey in the enormous mouth.

Common Mora
Mora moro

Center right: there are over 100 species of deep-sea cods (family Moridae) found on upper slopes around the world. *Mora moro* feeds on fishes and invertebrates.

Shortfin squids
Illex spp.

Bottom right: shortfin squids, in the family Ommastrephidae, are very important commercial resources as well as prey for many fishes and marine mammals. Sitting on the bottom during daytime at depths to greater than 3,280 ft. (1,000 m) and swimming upward at night as far as the surface, they form dense schools sometimes in very large numbers and can be the locally dominant large swimmers on the continental slopes.

Incising submarine canyon

The continental slope is interrupted by deep canyons eroded by violent turbidity currents loaded with abrasive rock particles that descend the slope and spread debris across the continental rise and abyssal plain. This creates very steep slopes and special semi-enclosed deep-sea habitats.

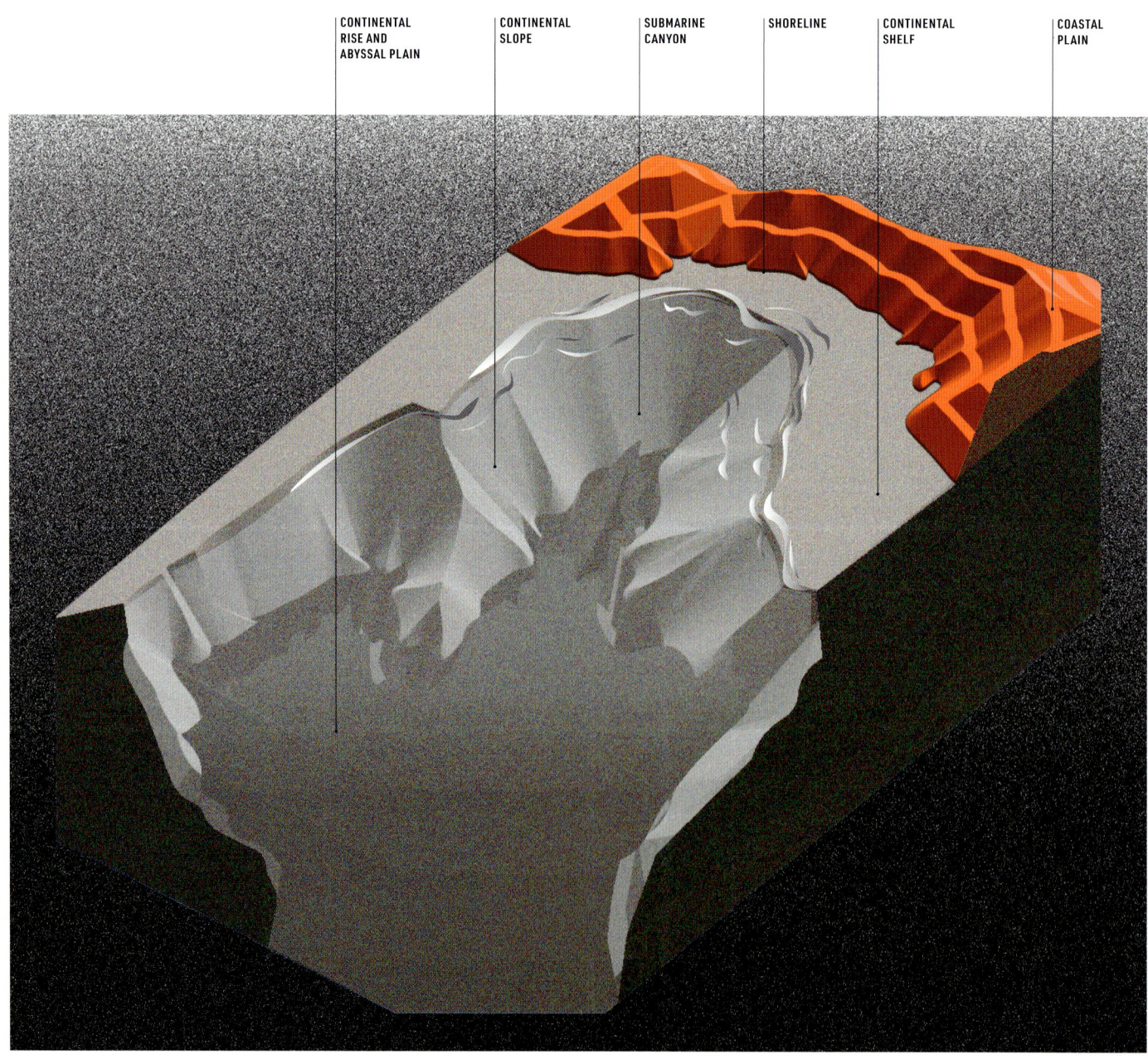

CONTINENTAL
RISE AND
ABYSSAL PLAIN

CONTINENTAL
SLOPE

SUBMARINE
CANYON

SHORELINE

CONTINENTAL
SHELF

COASTAL
PLAIN

Submarine canyons

In places, steep canyons are incised into the slope, sometimes with vertical walls plunging hundreds of meters. These underwater features can be twice the depth of the Grand Canyon. The steep-walled channels deflect currents and accelerate the movement of sediments and food downward. Internal waves (see pages 60–65), impacting and reflecting from the canyon walls, can help to distribute nutrients. Food is abundant in submarine canyons: phytoplankton and detritus from zooplankton form clumps of marine snow as they sink from the waters directly above them, and additional food arrives in cold, dense, winter waters that cascade off the continental shelf. The high food availability and patchy hard substrate allow the diverse habitats to be richly populated with sessile organisms such as corals (phylum Cnidaria) and sponges (phylum Porifera). These structure-forming assemblages cover the seafloor in places, forming a rich, three-dimensional matrix of animals sometimes compared with terrestrial forests.

"Animal forests" vary from coral and sponge gardens just 3.3–6.5 ft. (1–2 m) tall to carbonate mounds and coral reefs built over centuries and reaching tens to hundreds of meters high. These aggregations alter local current flows, trap nutrients, and provide substrate for other organisms to live on, crevices for them to live in, and places for them to lay eggs and rear young. Areas of animal forest are thus much richer in biodiversity than the open slope. Because of the fragile nature of the sessile animals that form them, some of these communities are recognized by the United Nations as Vulnerable Marine Ecosystems (VMEs) and therefore are important targets for conservation.

Coral and sponge grounds are not the only spectacular seafloor habitats associated with the continental slope. Cold seeps are also common, where leaking oil and methane gas support chemosynthetic communities—explored in more detail in Chemosynthetic Ecosystems, pages 158–167. In other places, where an oxygen-minimum zone (OMZ) (see pages 172–173) reaches the seafloor, bacterial mats can dominate, and the animal community is limited to small organisms that can cope with reduced dissolved oxygen. But on the boundaries of OMZs, mobile crustaceans (subphylum Crustacea) and echinoderms (phylum Echinodermata) often aggregate in large numbers, taking advantage of the food supply generated in the core of the OMZ.

Cold seep

Methane bubbles flow in small streams out of the sediment on an area of seafloor offshore Virginia north of Washington Canyon. Quill worms, anemones, and patches of microbial mat can be seen in and along the periphery of the seepage area.

Deep coral reefs and mounds

Hundreds of coral species lack photosynthetic algae (see page 82), but only a handful of these build reefs. In fact, just five species of stony corals (order Scleractinia) are responsible for most of the deep-sea coral reefs worldwide (see table below). The best studied of these is *Desmophyllum pertusum*, which occurs on continental slopes worldwide. In the Atlantic, particularly large areas of *D. pertusum* extend from the southern Barents Sea to West Africa and from Nova Scotia to Florida. A 2018 expedition off South Carolina explored *Desmophyllum* reefs that extended for 85 mi. (137 km).

In other areas, cold-water corals form mounds that develop over thousands of years, with framework coral (i.e., coral providing rigid structure) growing during interglacial periods and becoming packed with sediment during glacial periods. Such mounds reach heights up to 1,150 ft. (350 m) above the seafloor. While changes in oceanographic conditions can render mounds "retired," with no live coral, many still thrive today. Notable examples occur in the Porcupine Seabight off Ireland, where rich stands of *D. pertusum* cover the tops of numerous mounds, extending for miles and enriching the biodiversity of the area.

Deep-water coral reefs are recognized as Vulnerable Marine Ecosystems (VMEs) owing to their high probably of damage from contact with mobile fishing gears. The diversity of organisms found on *Desmophyllum* reefs is three times that of surrounding areas. The reefs provide hard substrate for other sessile filter feeders, food for grazers, and a home for scavenging and predatory mollusks, crustaceans, and echinoderms, and are presumed to be important breeding and nursery grounds for fishes. In 2018, a large shark nursery was found associated with an extensive *D. pertusum* reef on a carbonate mound off Ireland. Harboring an enormous density of Blackmouth Catsharks (*Galeus melastomus*), the seafloor was littered with shark eggs among coral fragments. ROV (remotely operated vehicle) surveys have covered only a fraction of these vast habitats, and this discovery shows how much we still have to learn about many of them.

Diversity on *Desmophyllum*

A *Desmophyllum pertusum* reef provides habitat for a myriad of invertebrate species. Top left: a squat lobster (*Eumunida picta*) on a reef off Florida. Bottom: a basket star (*Gorgonocephalus caputmedusae*) on a reef in Trondheim Fjord, Norway.

South Carolina reef

Top right: a deep-water *Desmophyllum* reef off South Carolina, as visible by multibeam bathymetry. The linear reefs can be seen as a series of mounds and ridges in red and yellow against the green and blue background. Depths range from approximately 2,300 ft./700 m (red) to 2,950 ft./900 m (blue).

Deep-water reef-building corals

Species	Depth	Geographic distribution	Notes
Desmophyllum pertusum	Commonly 165–3,280 ft. (50–1,000 m) but can be found deeper	Cosmopolitan	The most common and best-studied species; previously known as *Lophelia pertusa*.
Madrepora oculata	165–6,560 ft. (50–2,000 m)	Cosmopolitan	More fragile; often found together with *Desmophyllum* and *Goniocorella*
Solenosmilia variabilis	660–6,560 ft. (200–2,000 m)	Mostly cosmopolitan but absent from Antarctica and north and east Pacific.	
Enallopsammia profunda	1,310–5,740 ft. (400–1,750 m)	West Atlantic	Often co-occurs with other Atlantic framework-building corals
Goniocorella dumosa	330–4,920 ft. (100–1,500 m)	Around New Zealand and other Southern Hemisphere and Pacific areas	

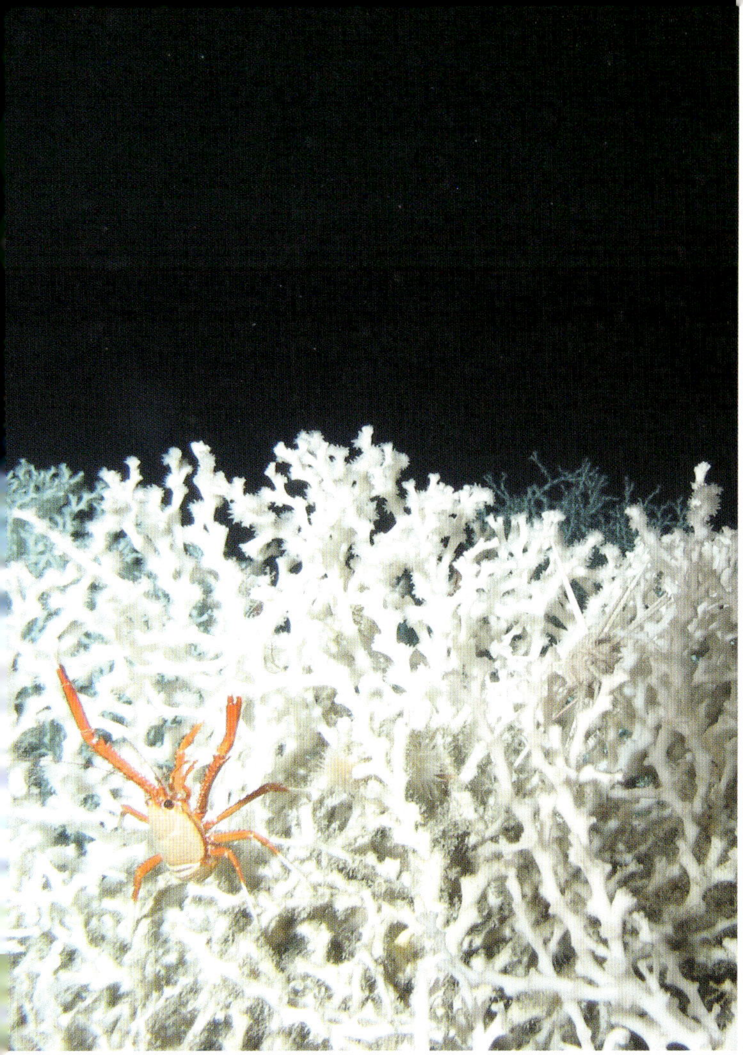

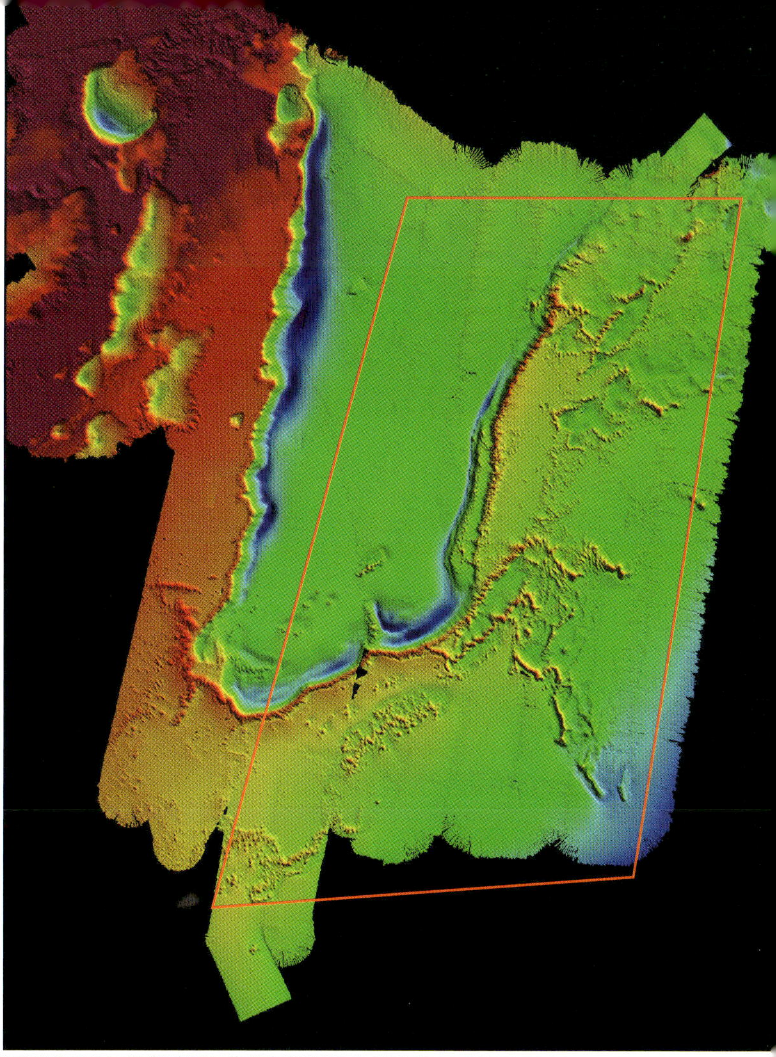

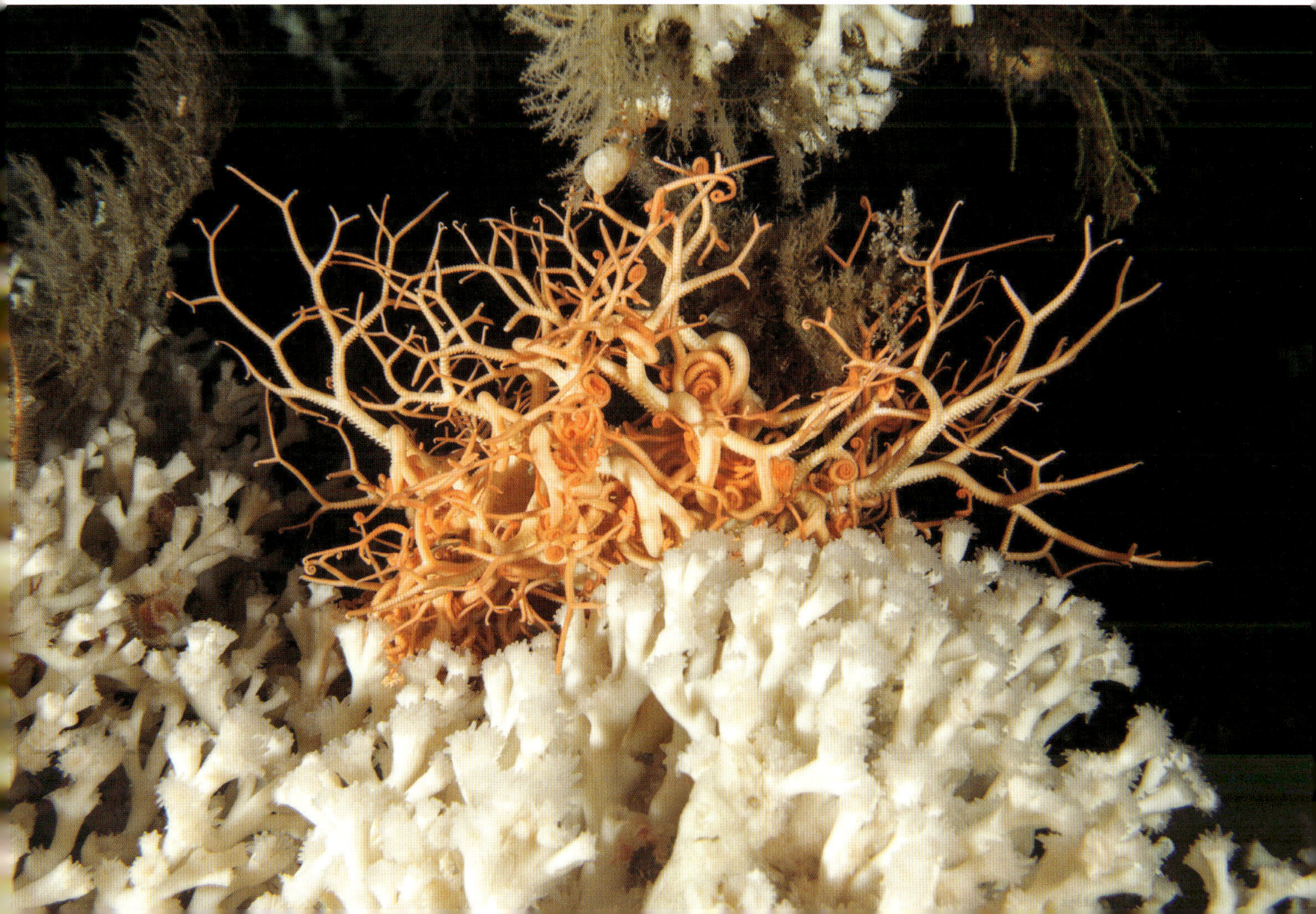

Coral gardens
Coral gardens may be dominated by one species, such as in Oceanographer Canyon on Georges Bank off northeast USA where a species of *Paramuricea* dominates (top), or contain a mix of species as in the Gully Canyon farther north (bottom), where gardens comprise bamboo corals, mushroom corals, and zoantharia among others.

Vulnerable slope and canyon ecosystems

Many areas on the continental slope have no stony, reef-building corals; instead there are coral gardens, which vary hugely, depending on the dominant coral species. Based on whether the seafloor is rocky or soft and local current flows, coral gardens may be dominated by stony cup corals; octocorals (subclass Octocorallia), including soft corals (suborder Alcyoniina); or black corals (order Antipatharia). Sea pens (specialized octocorals of the order Pennatulacea; see page 82) also often occur in large "fields" or "meadows" that may extend for many miles. Like cold-water coral reefs, these coral gardens form habitat for other species and are always areas of enhanced biodiversity. Sponges may also aggregate and form habitat (see box).

The classification of such ecosystems as Vulnerable Marine Ecosystems (VMEs) that warrant special protection is often based on how dense the component corals and sponges are. Yet there is no doubt that coral and sponge aggregations are highly threatened by fishing. They are particularly vulnerable to bottom trawling, where heavy gear and cables are dragged across the seafloor, but they may also be damaged by longline fishing, which trails baited hooks. Many of the species are not only fragile, but also extremely slow growing, meaning they may take decades or centuries to recover from damage. In recognition of this threat, many VMEs are now protected by conservation zones restricting bottom fishing. However, we have only limited knowledge of where these VMEs occur, and further mapping, monitoring, and exploration with remotely operated vehicles are essential to conservation efforts.

Because the corals have calcium carbonate skeletons, they are further threatened by ocean acidification (see pages 266–271). Rising atmospheric carbon dioxide concentrations caused by the burning of fossil fuels lead to more carbon dioxide dissolved in the ocean. This makes the oceans more acidic, which tends to dissolve calcium carbonate, making it harder for organisms to produce and maintain their skeletons. Climate change may also affect local hydrodynamic processes, reducing food supply to areas that are currently well supplied. These factors could impact coral gardens in the future if there is inadequate action to combat the climate crisis.

Sponge aggregations

Sponge aggregations can also form habitat. Where sponges aggregate, their spicules form dense mats, providing habitat for small organisms. The sponge species that form aggregations vary by depth, and aggregations may comprise just one sponge species or a mix of several. Here, at Davidson Seamount, 75 mi. (120 km) southwest of Monterey, at around 4,920 ft. (1,500 m) depth, habitat is formed by a species of *Staurocalyptus*, a hexactinellid sponge sometimes called the Picasso Sponge. In boreal waters, some types of sponge aggregations are referred to as "ostur." This word, Icelandic for "cheese," came from fishermen who used it for various sponges brought up in their trawls because of their distinctive smell and texture.

Recent Vulnerable Marine Ecosystem discoveries in the North Atlantic

Only a small proportion of the seafloor has been surveyed by cameras, and so we do not know the true extent of deep-sea coral reefs and gardens, sea pen fields, and sponge aggregations. Species indicative of their presence are sometimes recovered from trawls conducted for fisheries purposes, but the presence of a single individual does not guarantee the presence of an extensive, or dense, marine animal forest. However, knowledge of indicator-species distribution, combined with high-resolution sonar data indicating bathymetry (the depth of the ocean floor) and substrate (see multibeam map of South Carolina reefs, page 129), can be used to guide visual surveys, leading to spectacular finds.

Whittard Canyon

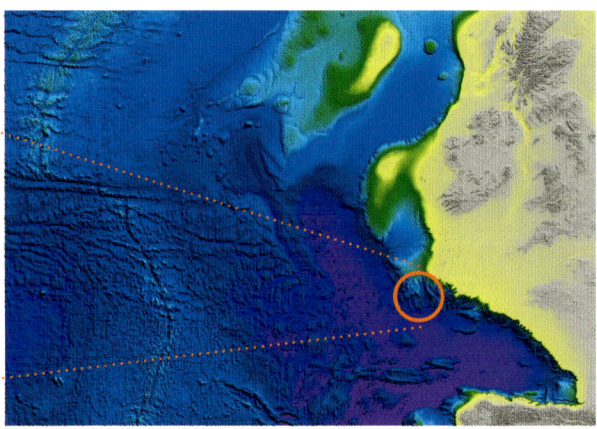

Whittard Canyon, southwest of Ireland, is a complex system with multiple canyon arms extending from the continental shelf. The canyon heads are characterized by vertical walls extending to around 1,970–2,625 ft. (600–800 m) depth. In 2012 and 2013, still-camera and remotely operated vehicle (ROV) footage showed these walls to be covered in a dense assemblage of deep-water oysters (*Neopycnodonte zibrowii*), clams (*Acesta excavata*), the framework-building coral *Madrepora oculata*, and large solitary polyps of the stony cup coral *Desmophyllum dianthus* (top image). The cryptic habitat provides refuge for many small mobile invertebrates, including shrimps and crabs (order Decapoda), mollusks (phylum Mollusca), and echinoderms (phylum Echinodermata), and hard structures for many sessile ones such as anemones and hydrozoans (phylum Cnidaria). Fishes are also abundant, from small North Atlantic codlings (family Gadidae) to large conger eels (family Congridae). The habitat varies slightly in different canyon arms, with oyster prevalence varying, but it is always dominated by filter-feeding bivalves providing attachment surfaces for other organisms.

Following pages:
Whittard Canyon

View from the foot of the slope, southwest of Ireland in the Celtic Sea, looking north toward the deep-water basin Porcupine Seabight and Ireland.

Baltimore Canyon

In 2019, an expedition to Baltimore Canyon, in a large deep-sea coral protection area off the US east coast, yielded exciting finds. Parts of Baltimore Canyon were already known to be coral-rich, but researchers exploring a new area discovered a wall of bright red, pink, and white bubblegum corals (*Paragorgia* sp.) at around 1,640 ft. (500 m) depth (top image). These distinctive octocorals were both densely spaced and very large, some extending more than 6.5 ft. (2 m) in width. Given the likely slow growth rate—probably just 0.4–0.8 in. (1–2 cm) per year—the corals are likely also very old and provide a stable habitat for many other species. Here they were interspersed with other octocoral species, and home to sea stars (class Asteroidea), brittle stars (class Ophiuroidea), shrimps, and other invertebrates.

Greenland coral gardens

In 2018 and 2019, towed camera sleds on the continental slope west of Greenland revealed a variety of soft-coral gardens (top image), including some dominated by cauliflower and mushroom corals (families Nephtheidae and Alcyoniidae). The habitats extend over nearly 200 sq. mi. (500 km²) and are adjacent to productive fishing grounds for halibut and prawns in the nearby shallower waters. These soft corals, which lack an axial skeleton, provide a lower-relief habitat than some other octocorals, but their effect on local biodiversity is similar.

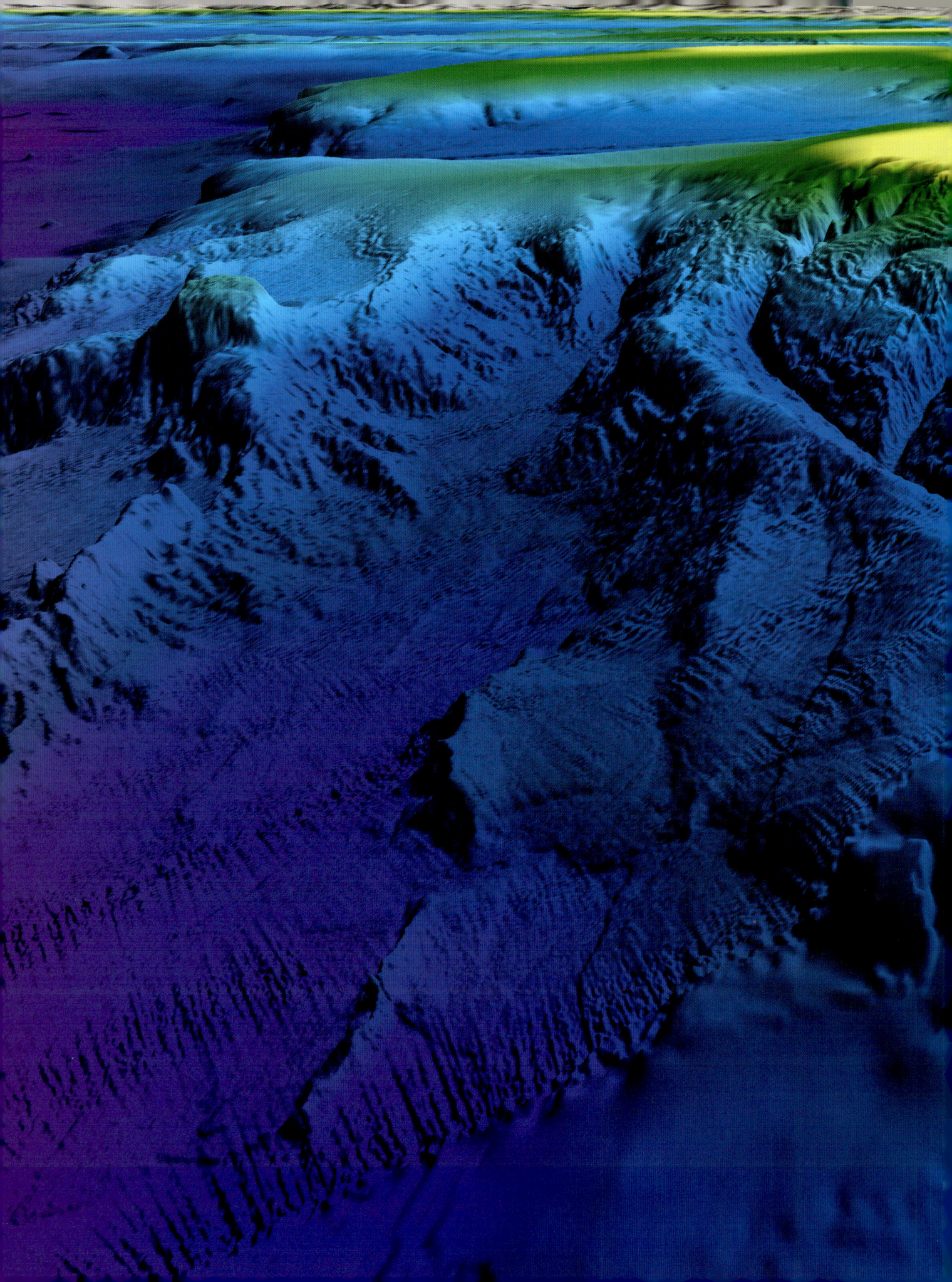

Pelagic habitats of submarine canyons

Nutrients cascading off the shelf and upwelling systems that bring cold, nutrient-rich water toward the surface combine to make submarine canyons ideal habitats for pelagic species too. Food availability is enhanced in midwater, leading to dense accumulations of krill (planktonic crustaceans; see page 100), squids (class Cephalopoda), and mesopelagic fishes. Submarine canyon shape further enhances food supply. Internal waves (see pages 60–65) can enhance mixing that in turn increases nutrient levels, while small particles rich in organic nutrients are resuspended from the bottom and accumulate in layers between water masses of different densities. These "nepheloid layers" (see page 176) can act as food-rich nurseries for some juvenile deep-sea shrimps and deep-sea fishes, while the animal forests on the canyon floor can provide nursery areas with refuges for vulnerable juveniles of certain other fishes.

The enhanced diversity and abundance of potential prey attracts higher predators. Large predatory fishes such as tunas, swordfishes, and sharks are often present in high numbers, and may also use the canyons as spawning grounds. Because some submarine canyons are used predictably by oceanic wildlife, including turtles, seabirds, and whales as well as fishes, designating these canyons as marine reserves could afford protection to threatened pelagic species.

Toothed whales may prefer to feed in submarine canyons, taking advantage of both the enhanced midwater prey and the benthic prey supported by the coral and sponge communities. In Kaikōura Canyon off New Zealand, Sperm Whales (*Physeter macrocephalus*) feed between 660 and 1,970 ft. (200 and 600 m) depth in the deep scattering layer (DSL; see pages 106 and 170–171), which is rich in mesopelagic fishes and shrimps, but spend more of their time foraging much deeper in the benthic boundary layer (the water directly above the seafloor) at the bottom of the canyon. Here, they probably prey on demersal (bottom-associated) fishes such as the New Zealand Ling (*Genypterus blacodes*, a type of cusk eel), whose own preferred prey of grenadiers (small long-lived benthic fishes of the family Macrouridae) is more abundant because of the presence of rich three-dimensional invertebrate ecosystems on the Kaikōura Canyon seafloor.

Off Nova Scotia, Northern Bottlenose Whales (*Hyperoodon ampullatus*) reside and forage in the Gully, the largest submarine canyon in the northwest Atlantic. The whale's preferred prey is the deep-water squid *Gonatus fabricii*. These squid feed on crustaceans early in life. As they grow, they move into deeper waters, and their diet changes to mainly fishes. The Gully is much richer in pelagic crustaceans such as shrimp and krill than the nearby slope regions, and its high diversity of deep-water corals supports a near-bottom ecosystem with enhanced abundance of demersal fishes, such that the Gully likely provides food for all the life stages of the whales' squid prey.

Swordfish
Xiphias gladius
This swordfish feeds down to at least 2,300 ft. (700 m) on deep-sea prey including squids, lanternfishes, other fishes, and crustaceans.

Atlantic Puffin
Fratercula arctica
Although they nest on cliffs, puffins spend much of their life at sea. Those that nest on islands in the Gulf of Maine exploit the enhanced resources associated with the canyons of Georges Bank, east of Cape Cod.

Sperm Whale
Physeter macrocephalus
Adult male Sperm Whales have been recorded diving and feeding in the food-rich Kaikōura Canyon off New Zealand spending about 15 minutes on the surface between dives that can last over one hour.

Atlantic Bluefin Tuna
Thunnus thynnus
Unfortunately now quite a rare species owing to overfishing, but a voracious predator on small schooling fishes, squids, and red crabs. Grows to over 1,000 lb. (450 kg) weight.

Northern Bottlenose Whale
Hyperoodon ampullatus
A male spy hops in the Gully, off eastern Canada, where this species feeds on or near the seafloor of the canyon in deep waters.

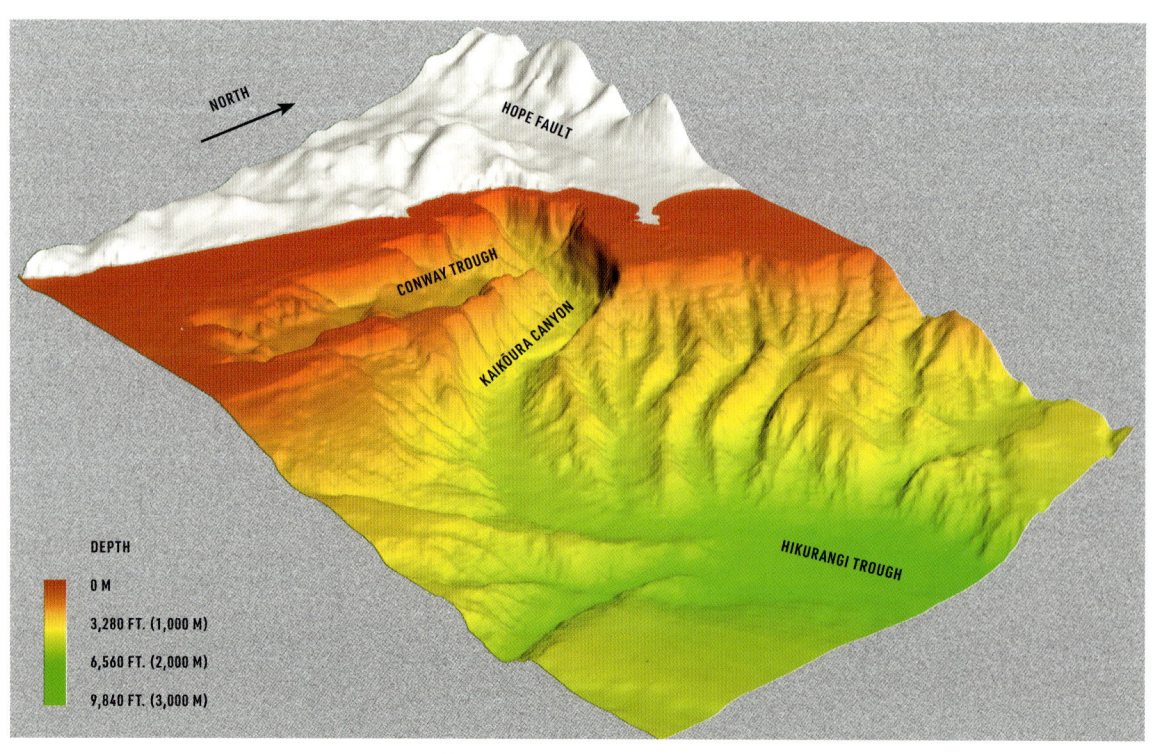

DEPTH
0 M
3,280 FT. (1,000 M)
6,560 FT. (2,000 M)
9,840 FT. (3,000 M)

NORTH

HOPE FAULT

CONWAY TROUGH

KAIKŌURA CANYON

HIKURANGI TROUGH

Kaikōura Canyon
The steeply incised canyon off New Zealand channels nutrients and resources to provide a rich feeding ground for Sperm Whales.

RIDGES, SEAMOUNTS, AND PLATEAUS

Hot spots of exceptional biodiversity in the deep sea

The global ocean floor is dominated by vast areas of flat, sediment-draped plains extending from 9,840 to 19,700 ft. (3,000–6,000 m) depth (see The Abyssal Plains, pages 146–153). However, multibeam sonar and satellite remote-sensing surveys reveal the presence of numerous underwater ridges, seamounts, and plateaus where the seafloor rises far above the surrounding abyssal plains. These features are colonized and exploited by a rich variety of life-forms and have a major influence on offshore deep-sea biodiversity.

Size, shape, and location

Ridges, seamounts, and plateaus vary in size and shape from small summits a mile or so in diameter to large structures thousands of miles in length. Ridges are part of the mid-ocean ridge system that extends 40,000 mi. (65,000 km) around the planet, often described as the earth's longest mountain range. The average depth of the top of the ridge is 8,200 ft. (2,500 m) below the sea surface, but some parts are much taller, in some cases forming islands, such as the Azores, Iceland, and the Galápagos. In contrast, seamounts are isolated conical remnants of underwater volcanoes, sometimes occurring in groups, such as the Emperor Seamount chain in the North Pacific and the New England Seamount chain in the North Atlantic. More than 33,000 seamounts are known in the world's oceans. Plateaus—steep-sided flat-topped areas—rise as much as 1.2–1.8 mi. (2–3 km) above the abyssal plains and constitute over 5% of the ocean area. Some, for example the Challenger Plateau off New Zealand and the Rockall Plateau between Iceland and the British Isles, are rocky remnants of the ancient continent of Pangaea, whereas others are of more recent geological origin.

For most creatures, whether the habitat is a ridge, seamount, or plateau does not make much difference; however, the habitat's size, shape, and location do influence the ability of animals to colonize and sustain populations in these remote areas. On the mid-ocean ridges, inhabitants can migrate and disperse through a globally interconnected habitat that is interrupted only by small gaps at fracture zones. Many seamounts, by contrast, are far offshore with no direct connection to similar habitats on other seamounts or continental slopes. Scientists once thought that this isolation might give rise to unique endemic life forms, as found on oceanic islands such as the Galápagos. New species have been discovered on seamounts, but subsequent research has usually found them elsewhere at similar depths.

Life on Pacific Ocean seamounts

Top: anemone near the summit of the Mozart seamount 560 mi. (900 km) northwest of Honolulu. Lower: *Circeaster pullus* starfish feeding on a coral colony *Victogorgia* sp. at 8,200 ft. (2,500 m) depth. The starfish everts its stomach to enclose and digest coral branches.

Blackbelly Rosefish
Helicolenus dactylopterus

A rockfish (family Sebastidae) known off the edge of continental shelves but also occurs on offshore shallows such as summits of seamounts and ridges.

Leafscale Gulper Shark
Centrophorus squamosus

Deep-sea, bottom-living sharks with their maximum depth limit of 9,840 ft. (3,000 m) can find suitable mid-ocean habitat on ridges, seamounts, and plateaus.

Seamount topography

The summit on the right, around 985 ft. (300 m) below the sea surface, is the Pao Pao Seamount in the South Pacific Ocean. The flat-topped peak 15 mi. (25 km) to the left is an unnamed guyot.

An abundance of life

Deep-sea, bottom-living sharks are found on the slopes of ridges, seamounts, and plateaus. They prefer mid-slope depths and cannot survive at depths greater than 9,840 ft. (3,000 m). This means they cannot move between seamounts by swimming along the seafloor. The Leafscale Gulper Shark (*Centrophorus squamosus*) can take shortcuts between suitable habitat patches by swimming in midwater thousands of feet above the abyssal plain until it encounters somewhere suitable to settle. Some species of rays and sharks lay their eggs on seamounts and plateaus, and these species are adept at utilizing widely dispersed patches of suitable habitat. Even sessile marine species such as corals and sponges have a remarkable ability to disperse their eggs and larvae over long distances, so the populations of seamounts generally resemble those of the nearest continental slopes. Indeed, the Mid-Atlantic Ridge harbors species known from both the eastern and western Atlantic continental slopes. Plateaus are sufficiently large to sustain relatively stable resident populations of animals. The considerable interchange of slope-dwelling fauna across the oceans is probably aided by the use of seamounts, ridges, and plateaus as stepping-stones, which are colonized by successive generations, enabling species to disperse over distances far greater than can be achieved by the movements of single individuals.

When these offshore geological features first began to be mapped in the 1960s and 1970s, researchers were astounded by the richness of life. This was partly because, unlike continental shelves and slopes, seamounts and ridges had not been disturbed by human activity; fishes occurred in pristine abundance among rich coral gardens. Unfortunately, once commercial fishers discovered this offshore wealth, it was quickly depleted (see pages 234–235). On the mid-ocean ridges, the biodiversity and abundance of life is enhanced by the presence of chemosynthetic ecosystems at hydrothermal vents (see pages 162–163). However, the specialized fauna at vent sites makes only a small contribution to the total abundance of life; most organisms on mid-ocean ridges thrive by exploiting material transported from the productive sunlit surface layers of the ocean.

Dynamics of the summit

The most biologically productive ridges, seamounts, and plateaus are those with summits less than 4,920 ft. (1,500 m) below the sea surface. The presence of the summit can create upwelling currents, bringing to the surface nutrient-rich deep water that fertilizes localized blooms of phytoplankton (plankton consisting of plant life; see page 12). In turn, the zooplankton (plankton consisting of animal life) feeding on the phytoplankton may be trapped by currents circulating around the summits, increasing the food available to organisms higher up the food chain. Dead plankton and their excrement contribute locally to the downward rain of particulate organic matter known as marine snow, and because of the short falling distance over the summits, this material arrives fresher and in much higher quantity on these raised seabeds than at greater depths. Detritus feeders on the slopes around the summits therefore live in an environment with a greatly enriched food supply.

Summits near the sea surface also intersect the depths of the deep scattering layer (DSL; see pages 170–171) and the mesopelagic zone with their great abundance of life. Forage fishes, krill, and other animals of the DSL undergo their 24-hour cycle of migration, ascending to the surface at nightfall and descending at daybreak (see page 106). During the morning descent, these creatures may become trapped on summits, unable to escape during the day from mobile predatory fishes, crustaceans, and cephalopods (such as octopods)

living on the slopes. The Roundnose Grenadier (*Coryphaenoides rupestris*) and slickheads (family Alepocephalidae), abundant demersal (bottom-dwelling) fishes found around summits, have large eyes and a highly adapted brain specialized for visual hunting of bioluminescent creatures of the DSL at the twilight depths of the lower mesopelagic zone.

The summits of seamounts and ridges have large areas of exposed hard rock, which is otherwise very rare in the deep sea. The rocks provide a firm foundation for attachment of a rich variety of sessile life, including corals, anemones, sponges, sea lilies, and feather stars (see pages 78–79). On seamounts off Australia and New Zealand, the biomass of the branching coral *Solenosmilia variabilis* is 29 times greater than in equivalent habitats on continental slopes of the mainland. Filter-feeders (e.g., foraminifera and sponges) and filter-feeder predators (colonial and solitary corals) make up a large proportion of the resident fauna on these seamounts, exploiting the rich material carried on ocean currents around the summits. Corals are efficient at capturing organic-matter particles and zooplankton. The deep-water coral *Desmophyllum pertusum* has been observed to capture live zooplankton, including arrow worms (phylum Chaetognatha) and krill and other small crustaceans. Coral and sponge beds act as forests (see page 127), providing shelter for animals that live in the interstices, places to make nests and lay eggs, and hiding places into which mobile species can retreat.

Seamount ecosystem

Seamounts provide rare patches of hard rock in the middle of the ocean for attachment of corals, sponges, and other bottom-living species. The surrounding waters attract shoals of fishes, squids, krill, and a variety of predators, including fishing boats.

Life on a seamount

The summit of a coral seamount on the Southwest Indian Ocean Ridge 575 ft. (175 m).
Above: a squat lobster (family Galatheidae) on a fan coral.
Center right: a scorpionfish (family Sebastidae) next to a mushroom coral at 1,500 ft. (460 m) depth on one of the Musicians Seamounts in the Pacific Ocean north of Hawaii.
Bottom right: a brinsingid starfish at 3,660 ft. (1,115 m) depth on a rich cold-water coral bed in the Huon Marine Park, south of Tasmania.

Thriving on the summit

Multiple factors combine to increase the abundance and diversity of life on, and around, mounts and ridges. One important effect is simply the varying elevation of the seafloor. This results in a range of niches within which species with different depth preferences can thrive; the greater the range of depths, the more species can survive, from abyssal species on the lower slopes to shallower-depth species around the summits.

A further general enrichment occurs through aggregation of fishes around summits and slopes. Highly migratory pelagic fishes such as sharks, billfishes, and tunas tend to congregate at seamounts, and several deep-sea species use seamounts as spawning areas. Slender Armorhead (*Pentaceros wheeleri*) is an oceanic bass that roams throughout the North Pacific feeding and growing during the juvenile stages of its life cycle. When these fish attain sexual maturity, at about two and half years, they migrate to seamounts of the Emperor Chain northwest of Hawaii, forming vast shoals over the summits. This effectively concentrates food reserves the armorheads have accumulated from thousands of square miles of the Pacific Ocean into small areas around the seamount summits. Their concentrated abundance on seamounts vastly exceeds the local food resources available—yet they stay on the seamounts for the remaining three to five years of their life, gradually losing weight as they spawn each December and January, until they become emaciated, moribund, and finally die. During their adult life on the summits, Slender Armorheads continue to feed on a variety of pelagic prey, including large amounts of gelatinous plankton, but this is insufficient to compensate for the loss of energy during repeated spawning.

The slower-growing Orange Roughy (*Hoplostethus atlanticus*) has a similar life cycle, with juveniles widely dispersed in the ocean until they reach the age of reproductive maturity, at around 30 years, when they begin to aggregate each winter for spawning on slopes and summits. In the southern Indian Ocean and the southeast Pacific, they spawn during July and August each year, at depths of 2,300–4,600 ft. (700–1,400 m). The roughy's life span extends to over 140 years, so an adult may continue spawning for more than a century. However, it is almost certain individuals do not spawn every year. Unlike the armorheads, roughies do not remain on the summit, but migrate after spawning into the open ocean to replenish their food reserves, returning to the spawning grounds when they have regained enough energy.

Slender Armorhead
Pentaceros wheeleri

On a seamount in the extreme northwest of the Hawaiian Island Archipelago 1,400 mi. (2,260 km) from Honolulu, Hawaiian monk seals dive to over 1,475 ft. (450 m) depth to catch these fish.

Emperor Chain

The Emperor Seamounts and Hawaiian islands comprise a chain of over 80 undersea volcanoes stretching 3,850 mi. (6,200 km) across the Pacific Ocean.

Orange Roughy
Hoplostethus atlanticus

An aggregation of adult fishes at about 2,590 ft. (790 m), near the summit of the Pedra Seamount, south of Tasmania, Australia.

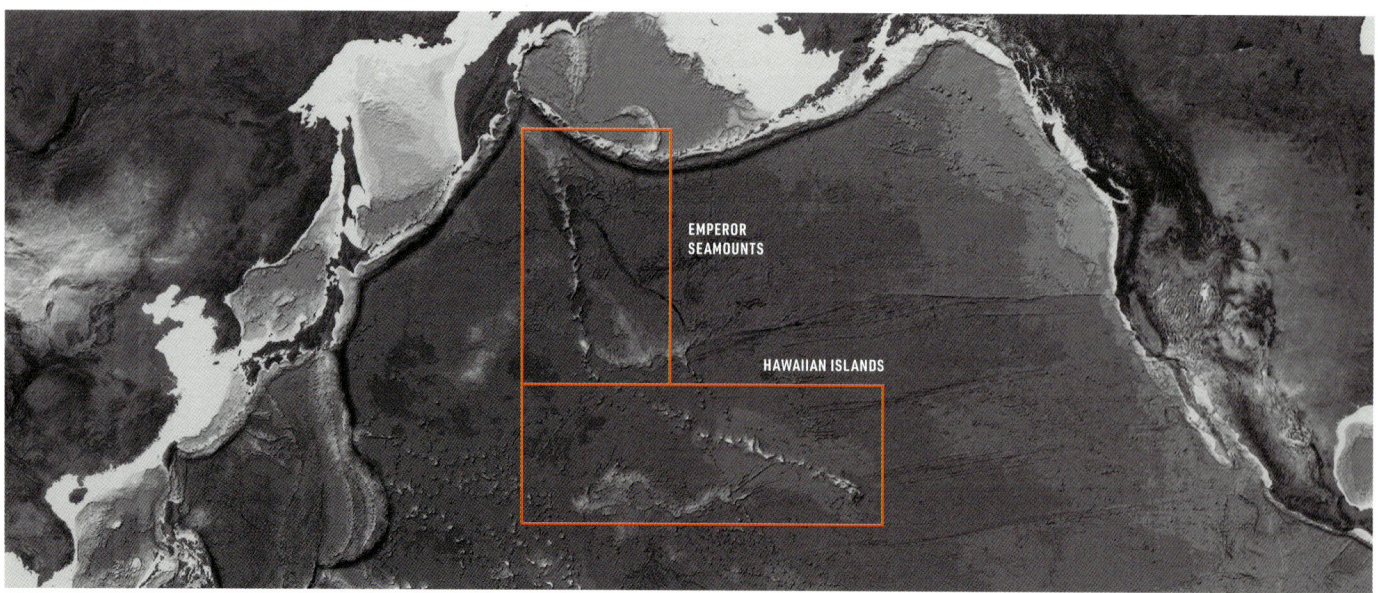

EMPEROR
SEAMOUNTS

HAWAIIAN ISLANDS

Predators and foragers

The rich food resources of seamounts, ridges, and plateaus also attract toothed whales. The largest of these is the Sperm Whale (*Physeter macrocephalus*), which grows to over 50 ft. (16 m) long and attains an adult weight of 40 tons. Sperm Whales dive to a maximum depth of 3,280 ft. (1,000 m), hunting for mesopelagic squids that become concentrated on the slopes of summits. They consume a wide variety of squids, including slow-moving gelatinous histioteuthids (family Histioteuthidae) and fast-swimming muscular ommastrephids (family Ommastrephidae). An adult whale requires about 3% of its body weight per day, which is equivalent to 1,000 medium-size squids; female and smaller male Sperm Whales feed on smaller prey, and large adult males eat the largest prey, including the Giant Squid (*Architeuthis dux*), which can grow over 33 ft. (10 m) long. In some areas, especially at high latitudes, Sperm Whales also consume quantities of fishes. In the northwest Atlantic, there is evidence of seasonal migrations, with Sperm Whales visiting the New England Seamount chain predominantly in the spring.

Sperm Whales tend to dive down as a group, minimizing the effort of descending by regulating their buoyancy, while using highly sensitive sonar in the bulbous head to detect prey. They feed deep and then ascend to the surface, where they digest and empty their bowels, further fertilizing the productivity of the ocean above the summit, and thereby reversing the "biological pump" of mesopelagic diel vertical migrators that actively transport nutrients into the deep. The deepest-diving whale is Cuvier's Beaked Whale (*Ziphius cavirostris*), which has been recorded at a depth of 9,816 ft. (2,992 m) and routinely forages down to 3,280 ft. (1,000 m), consuming deep-sea squids, fishes, and crustaceans. Along the mid-ocean ridge in the North Atlantic, Long-finned Pilot Whales (*Globicephala melas*) and three species of dolphins consume mainly mesopelagic fishes and squids, each of these mammal species specializing on different prey. Common Dolphin (*Delphinus delphis*) feeds on Glacier Lanternfish (*Benthosema glaciale*) and dragonfishes (family Stomiidae). White-sided Dolphin (*Lagenorhynchus acutus*) feeds on the lanternfish Rakery Beaconlamp (*Lampanyctus macdonaldi*). Striped Dolphin (*Stenella coeruleoalba*) consumes the Cockatoo Squid (*Teuthowenia megalops*). Pilot Whales feed on Glacier Lanternfish and gonate squids (*Gonatus* spp.).

Bird foragers

In addition to sea mammals and fishes, birds are also attracted to offshore shallows. Northern Fulmars (*Fulmarus glacialis*) make foraging trips of up to 18 days duration from their nesting sites in Britain and Ireland, flying as far as the Mid-Atlantic Ridge, where they feed on surface fishes, krill, and squids. Black-footed Albatross (*Diomedea nigripes*) and Cook's Petrel (*Pterodroma cookii*) aggregate at seamounts in the northeast Pacific and exploit mesopelagic fishes and squids. It is not yet clear how these shallow-diving birds access these deep-sea prey species. It has been suggested that the birds feed at night when the prey is near the surface, but no one has observed these birds feeding nocturnally. During the day they may be able to opportunistically take moribund or injured individuals that float to the surface. Most seamounts, ridges, and plateaus attract far-ranging seabirds, including shearwaters and petrels as well as the albatrosses.

Northern Fulmar
Fulmarus glacialis

These birds range far offshore for feeding. Leaving its partner to incubate eggs, one individual traveled over 3,850 mi. (6,200 km) during 15 days to feed on the Mid-Atlantic Ridge before returning to take its turn at the nest.

Squid prey
Histioteuthis sp.

These squid are found in quantity along the Mid-Atlantic Ridge, where they are preyed upon by deep-diving whales. Histioteuthidae make up one third of the diet of Sperm Whales around the Azores islands.

Sperm Whale
Physeter macrocephalus

A group of Sperm Whales descending at the start of a feeding dive off the island of Pico in the Azores archipelago on the Mid-Atlantic Ridge. They obtain 75% of their diet by intercepting luminous shoals of slow-moving squid floating in midwater at great depth.

Cuvier's Beaked Whale
Ziphius cavirostris

The deepest-diving whale, here seen breaching the surface of the Atlantic Ocean. They feed almost exclusively (over 90%) on cephalopods, plus a few crustaceans.

THE ABYSSAL PLAINS

Living in the abyss

The abyssal plains occupy 75% of the ocean's area, equivalent to over half of the surface of planet Earth. They are covered almost entirely with a layer of fine sediment, averaging 1,475 ft. (450 m) thick, which obscures any topography of the solid crust beneath, creating a relatively smooth, soft surface that slopes imperceptibly from about 9,840 ft. (3,000 m) at the continental margins to a maximum depth of 19,700 ft. (6,000 m) at the margins of hadal trenches.

Life in the sediment

The deep-seafloor sediment includes fine clay particles, originating from land as wind-blown dust or from rivers, mixed with ooze made up of skeletons and shells of microscopic planktonic organisms that have sunk to the bottom from the upper layers of the ocean. When a sample of this sediment is brought to the surface, it is pale-colored, smells clean, and is pleasant to handle, unlike the stinky black sediment often found in fresh water or estuaries. The reason is that the seawater above the abyssal plains is well oxygenated, supporting a vast army of creatures that burrow in, scrape, and browse on the sediment, continuously cleaning the seafloor so that very little organic debris accumulates.

Most of the food supply to the abyssal plain comes from a continuous rain of small organic debris falling from the surface layers of the ocean, sometimes dense enough to appear as marine snow. The amount is quite variable, depending on productivity in overlying waters. In the weeks following a spring phytoplankton bloom in the upper ocean, a massive accumulation of fresh, green chlorophyll-rich debris may completely cover the seafloor.

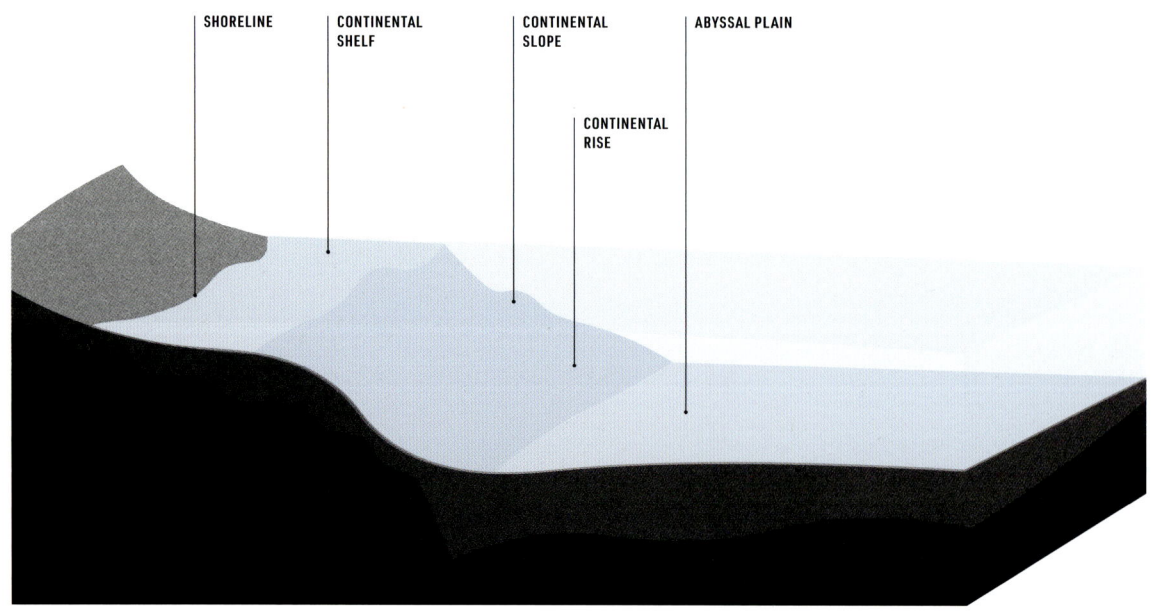

SHORELINE CONTINENTAL SHELF CONTINENTAL SLOPE ABYSSAL PLAIN CONTINENTAL RISE

Abyssal plain
An example of a passive continental margin without a trench, such as in the Atlantic Ocean, where the seafloor descends from a flat continental shelf to a steep continental slope, followed by a gentle slope of the continental rise down to the flat vastness of the abyssal plains that amount to about half of the solid surface of the planet.

Occurring every year, or once every few years, this rich bonanza from the surface stimulates a massive feeding frenzy among deposit-feeding creatures on the abyssal plain, which clean the seafloor within a few weeks. The rain of small particles is supplemented by food falls of various sizes, including carcasses of small fishes, gelatinous zooplankton, sharks and other large fishes, and even whales. These are consumed by voracious scavenging species equipped with sharp appendages and teeth that quickly remove all the soft flesh.

Animal life within the abyssal sediment is restricted to an extraordinarily thin layer, no more than 4 in. (10 cm) thick, and most life is found within the top 1.2 in. (3 cm). The feeding tentacles of some animals may reach up to 12 in. (30 cm) above the seafloor, and a few corals and sponges may stand over 3.3 ft. (1 m) high. In addition, there are mobile creatures above the sediment, floating or swimming in the bottom 330 ft. (100 m) of the water column—the so-called benthic boundary layer.

Marine snow

This is made up small particles of dead animals, skeletons, and organic detritus that fall into the deep sea from surface layers of the oceans. In the lights of a submersible, it looks like falling snow in the prevailing darkness of the deep sea (top right). Flurries reach the seafloor (top left), which becomes covered with drifting particles stirred up by mobile animals.

Abyssal Grenadier
Coryphaenoides armatus

Above left: this species occurs close to the seafloor over all of the abyssal regions of the world's oceans, except for some of the deeper parts of the Pacific Ocean. It is a mobile hunter and scavenger; each individual may travel thousands of miles per year.

Gummy Squirrel
Psychropotes longicauda

Above right: a sea cucumber about 23.5 in. (60 cm) long browsing over manganese nodules covering the abyssal sediments of the Tropical North Pacific Ocean. It feeds on marine snow detritus. The squirrel-like tail may aid orientation or swimming in the bottom currents.

Meiofauna and macrofauna

The lifestyle of abyssal creatures is to a large extent defined by their body size. The meiofauna comprises tiny animals, typically a fraction of a millimeter long, small enough to live in the spaces between grains of sediment. The most common members of the meiofauna are nematode worms (related to the threadworms that infect cats and dogs), which occur in an amazing variety of species including deposit feeders, carnivores, and consumers of bacteria. Tiny crustaceans, known as harpacticoid copepods (subclass Copepoda, order Harpacticoida), have diverse shapes adapted for burrowing or crawling through the sediment. Seed shrimps, or ostracods (class Ostracoda), are also abundant. Eight-legged tardigrades (phylum Tardigrada), often called "water bears," are relatively rare but spectacular when viewed close up under a microscope. In addition to these multicellular animals, single-celled organisms are an important part of this deep-sea Lilliputian fauna. The most important are the foraminifera, or "forams" (phylum Foraminifera), amoeba-like protozoans with chalky external shells. Found in an extraordinary variety of shapes, forams contribute greatly to the biodiversity of the deep sea.

The next largest size category is the macrofauna, animals that are visible to the naked eye, ranging in size from 1 mm to several centimeters. Up to three-quarters of all macrofaunal animals are bristle worms (class Polychaeta). Abyssal polychaete species tend to have shorter bodies with fewer segments than the familiar ragworms (family Nereididae) found on seashores. The abyssal polychaete macrofauna includes burrowing forms, deposit feeders, carnivorous species, and surface-mobile species. Hundreds of species can be found living together in some regions. The next most abundant creatures are crustaceans, including amphipods (order Amphipoda), related to the sandhoppers found on beaches; hooded shrimps (order Cumacea), with a large head-body and a segmented tail; tanaids (order Tanaidacea), which are small, elongate bottom-living relatives of crabs and lobsters; and isopods (order Isopoda), related to the terrestrial wood louse (or roly-poly). These crustacean groups have evolved in the deep sea, and numerous unique species are continually being discovered. Mollusks make up about 10% of the abyssal macrofauna by number. The most numerous in the deep sea are clams (class Bivalvia) that burrow or crawl on the surface of the sediment and have well-developed digestive systems for processing the detrital food available in the deep. Gastropods, or snails (class Gastropoda), occur everywhere, including species with coiled shells that are probably carnivorous, and simple limpet-like scavengers.

Marine worm
Laetmonice sp.

Top: a polychaete or bristle worm of the sea mouse family (Aphroditidae) discovered on the abyssal plain of the central Pacific Ocean at 10,1057 ft. (3,096 m) depth, where it was observed crawling on the seabed. The body is caked with fine pale sediment picked up from the seafloor.

2 CM

Abyssal isopod
Paropsurus giganteus

Bottom: a deep-sea "daddy long legs" or munnopsid isopod, walking on the abyssal seafloor. The red-light spots are 11.4 in. (29 cm) apart to indicate the size of the animal, which can also swim. Sense organs of the long legs and antennae aid detection of food.

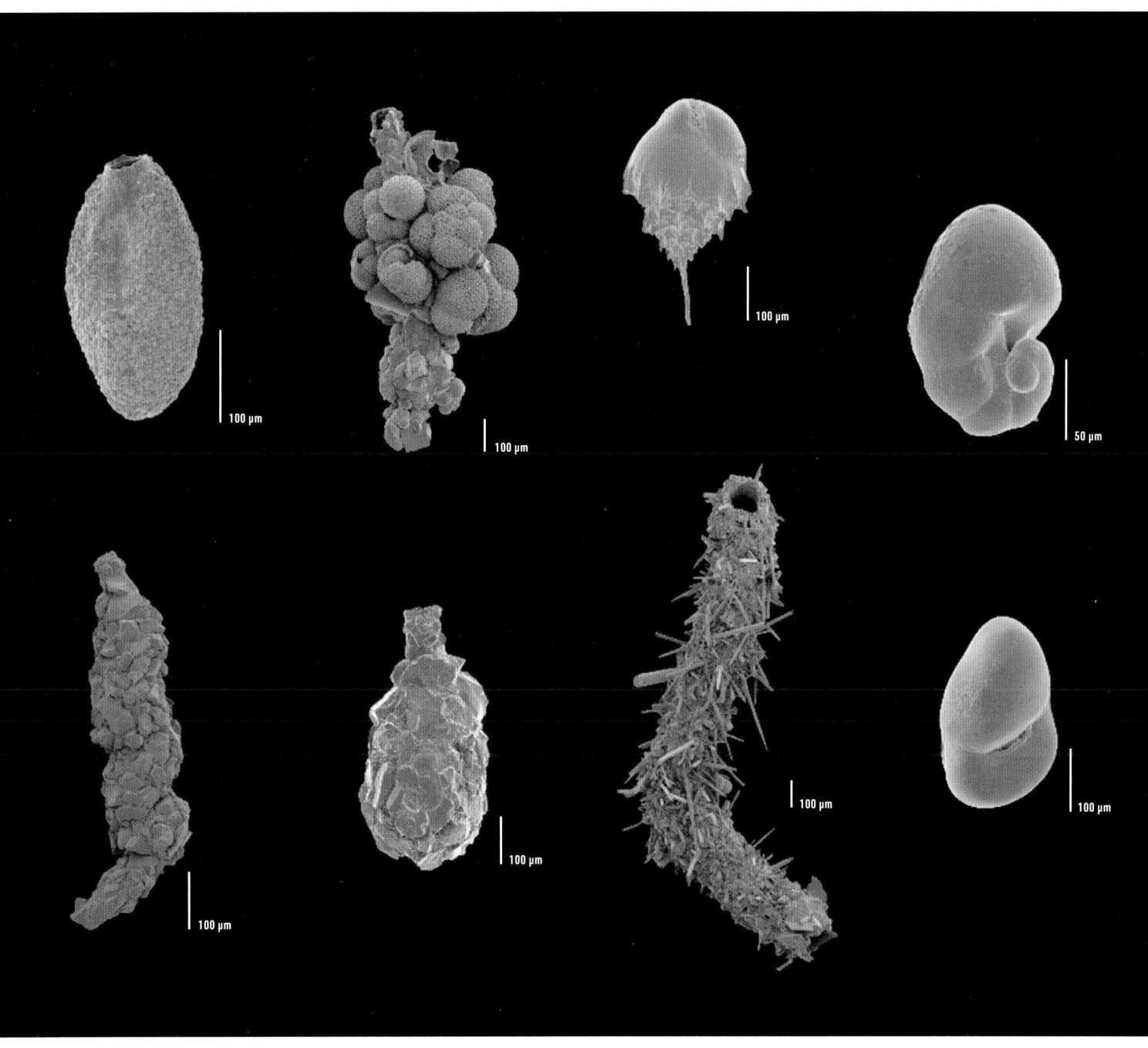

Deep-water benthic foraminifera
Tiny single-celled organisms that live on, or in, seabed sediments. Examples are from the Rockall Trough in the northeast Atlantic. They come in three forms: soft-shelled; agglutinated, which make a test or shell from material such as silt and sand grains; and calcareous with a hard chalky shell. Top row from left to right: Saccamminid, soft-shelled; *Reophax agglutinatus*, agglutinated; *Bulimina striata*, and *Nonionella iridea*, both calcareous forams. Bottom row from left to right: *Lagenammina* sp., agglutinated; *Saccorhiza ramosa*, calcareous; *Reophax scorpiurus*, agglutinated; *Gyroidina* sp., calcareous.

Megafauna

The largest size category of abyssal animals is the megafauna, which includes everything larger than a few centimeters, up to, and including, fishes over 3.3 ft. (1 m) long. The abyss is too deep for sharks, so the largest fishes are the ubiquitous cusk eels (order Ophidiiformes) and grenadiers (family Macrouridae) (see pages 90–91). Surprisingly, one conspicuous group of animals that fits within the megafaunal size category is the giant protozoans known as xenophyophores ("bearer of foreign bodies"; superfamily Xenophyophoroidea). These are single-celled organisms, and one species, *Stannophyllum zonarium*, can grow to over 10 in. (25 cm) in diameter, although it is only 1 mm thick. A more typical xenophyophore is *Reticulammina* sp., which resembles a rough, rounded pebble, about 2 in. (5 cm) in diameter, scattered on the seafloor. The body is hard, made up of an organic tubular network, to which particles are stuck, including grains of sediment and fragments of shells of other animals, to make solid armor (known as "test"). The living protoplasm amounts to less that 1% of the total weight of the animal. Multiple nuclei are found throughout the giant single cell, which feeds by extruding amoeba-like pseudopodia and gathering material off the seafloor or capturing food particles carried on the bottom current. There are about 75 abyssal species of xenophyophores adapted to different feeding conditions, some on the surface of the sediment and others with a root extending a few centimeters into the sediment. The body of a xenophyophore is also colonized by bacteria and small attached animals, so the creature becomes a small, self-contained deep-sea community.

The most conspicuous megafaunal members of the abyss are sea cucumbers (class Holothuroidea), swarms of which often comprise most of the biomass present on the abyssal plain. Typically up to 12 in. (30 cm) long, these animals feed mainly on detritus by crawling across the seafloor, ingesting the surface sediment, digesting the contained organic matter, and leaving a trail of feces. Recent studies have shown that species specialize in utilizing differing parts of the available food; some process fresh chlorophyll-rich deposits, whereas others can digest older material. Most crawl using the rows of feet on the underside of the body or by undulating movements of the body. Several species have gelatinous bodies that are close to neutral buoyancy, enabling them to swim above the seafloor, drift on the bottom current, and land at a new feeding location (e.g., *Enypniastes* spp.). Species of *Psychropotes* have a mysterious upward-floating tail that young individuals use for swimming, but adults appear unable to swim.

Sea cucumber
Amperima sp.

Sea cucumber browsing on the seafloor at 21,325 ft. (6,500 m) depth along the edge of the Java Trench in the eastern Indian Ocean.

Giant deep-sea shrimp
Cerataspis monstrosus

A giant deep-sea shrimp on the Porcupine Abyssal Plain at 15,750 ft. (4,800 m) depth in the North Atlantic. Antennae much longer than the body of the animal detect odors and movements of potential prey.

Xenophyophores

Giant, stone-like unicellular organisms are abundant on abyssal plains, but here pictured on a New England seamount, where they often have brittle stars on top.

The dynamic sediment

The abyssal seafloor is covered with tracks, burrows, imprints, and traces left by animals moving across and within the soft sediment (see page 86). The sediment is constantly reworked and turned over by the mobile animals living there. Some of the most spectacular marks are spoked-wheel patterns created by large spoon worms (subclass Echiura), which reach out in different directions, creating traces radiating from the central burrow opening. Animals landing or browsing on the seafloor can throw up clouds of fine sediment, burying old traces and creating new ones.

The dynamic nature of the soft sediment environment creates a problem for sessile animals—such as sponges (phylum Porifera); sea pens, sea fans, corals, and anemones (phylum Cnidaria); and tunicates, or sea squirts (subphylum Tunicata). These creatures cannot move, might easily be buried, and must be attached to a solid surface. Hard surfaces are rare in the abyss.

At high latitudes, rocks falling from melting icebergs can provide abyssal hard substrate. Coal-fired steamships of the past deposited vast quantities of solid coal waste known as "clinker," forming deep-sea corridors beneath the old shipping lanes where sessile animals can thrive. Natural patches of hard rock are found at fracture zones between tectonic plates, and improved seabed surveys are revealing that there may be as many as 25 million abyssal hills punctuating the plains and providing hard surfaces to which animals may attach. On the abyssal plains of the Pacific Ocean, mineral-rich, potato-shaped rocks called manganese nodules are scattered across the seafloor, and some are richly colonized by attached animals. Any hard surface in the abyss is rapidly exploited, testifying to the dispersal and search capabilities of the larvae of sessile creatures. Some animals search for hard surfaces to which they can attach their eggs. Cameras retrieved from the deep sea sometimes have egg masses laid by snailfishes (family Liparidae) attached to them. Despite the dominance of soft sediments, the comparatively rare hard substrates contribute greatly to diversity of life in the abyss.

Sea pen

Rooted

Some organisms, such as sea pens (order Pennatulacea), left, can develop a rootlike anchor in the sediment to support an erect stalk above the surface.

Abyssal glass sponge
Hyalonema sp.

Bottom: tulip-shaped glass sponge standing on a stalk attached to the abyssal floor of the northeast Pacific Ocean at 13,125 ft. (4,000 m) depth. The stalk is colonized by tiny anemones (*Epizoanthus stellaris*) and near the base there is a large mobile anemone.

Trachymedusa
Crossota sp.

Deep-sea jellyfish photographed at 15,750 ft. (4,800 m) on the Porcupine Abyssal Plain of the northeast Atlantic Ocean. They spend their entire lives in the water column but can pick up food from the seafloor.

Because of the reduced food supply, abundances of individuals in the abyss are much lower than on the continental slopes or mid-ocean ridges. Related to this, body size tends to decrease with increasing depth of occurrence, especially among invertebrates living on and within the sediments.

Owing to its remoteness, the abyssal ocean is largely free of direct human disturbance, except for the occasional cable laid across the deep-sea floor, scattered shipwrecks, and the rain of litter including plastics that pervades the entire planet. Human influence may increase substantially if tentative plans for deep-sea mining are fulfilled.

TRENCHES

Life in the hadal zone beyond the abyss

The hadal zone, defined as depths greater than 19,700 ft. (6,000 m), encompasses all the deepest parts of the ocean, down to the maximum depth of nearly 36,000 ft. (11,000 m) found at the Challenger Deep in the Mariana Trench in the western Pacific Ocean. The trenches represent a small part of total ocean area but harbor unique species and pose particular challenges for protection and conservation. This zone accounts for 45% of the depth range of the world's oceans but occupies less than 0.5% of total ocean area. Most of it is confined to a great chain of trenches surrounding the Pacific Ocean: Kermadec, Tonga, Mariana, Japan, and Kuril–Kamchatka Trenches in the west; Aleutian Trench in the north; and Peru–Chile Trench in the east. The Java Trench is in the Indian Ocean, and the Puerto Rico and South Sandwich Trenches are in the Atlantic Ocean.

Thus, in contrast to the global continuity of the abyssal zone (see pages 146–153), the hadal zone is made up of isolated pockets of deep water that are long, narrow, and V-shaped in cross section. The deepest parts are sediment-filled basins occupying less than 10% of the trench area. The trenches are located around the margins of the oceans, so despite the remoteness of their great depth, they are close to land and are regularly enriched by earthquake-triggered avalanches of organic matter as well as human-generated pollutants and litter. The water is well oxygenated and, despite extremely high pressures, some types of animal life thrive even at the greatest depths.

All the main animal groups familiar from the abyssal depths occur in the hadal trenches, but the number of species declines with increasing depth. Cosmopolitan abyssal species generally extend into the margins of the trenches down to depths of about 22,970 ft. (7,000 m). At greater depths, specialist hadal species occur, many of which are unique to each trench system. Among the tiny meiofauna (see pages 148–149), foraminifera and nematode worms are dominant, including four species of forams discovered from the Challenger Deep. Hadal forams tend to have soft shells because calcium carbonate dissolves in seawater at the high pressures found in trenches.

Over 100 species of polychaetes, or bristle worms, have been recorded from hadal trenches, including endemic species found at depths greater than 32,800 ft. (10,000 m), some of which have been observed swimming in the water above the seafloor.

Hadal trenches

Locations and depths of the ocean's deepest trenches.

1 Mariana Trench 35,840 ft. (10,925 m)
2 Tonga Trench 35,500 ft. (10,820 m)
3 Kuril–Kamchatka Trench 34,587 ft. (10,542 m)
4 Philippine 34,580 ft. (10,540 m)
5 Kermadec Trench 32,963 ft. (10,047 m)
6 Izu Ogasawara Trench (Izu-Bonin Trench) 32,190 ft. (9,810 m)
7 Tropical Southwest Pacific trenches — deepest: New Britain Trench 29,990 ft. (9,103 m)
8 Japan Trench 27,598 ft. (8,412 m)
9 Puerto Rico Trench 27,490 ft. (8,526 m)
10 South Sandwich Trench 27,116 ft. (8,265 m)
11 Peru–Chile (Atacama) Trench 26,427 ft. (8,055 m)
12 Aleutian Trench 25,663 ft. (7,822 m) (biggest trench by area)
13 Ryukyu Trench 24,476 ft. (7,460 m)
14 Java (Sunda) Trench 23,920 ft. (7,290 m)

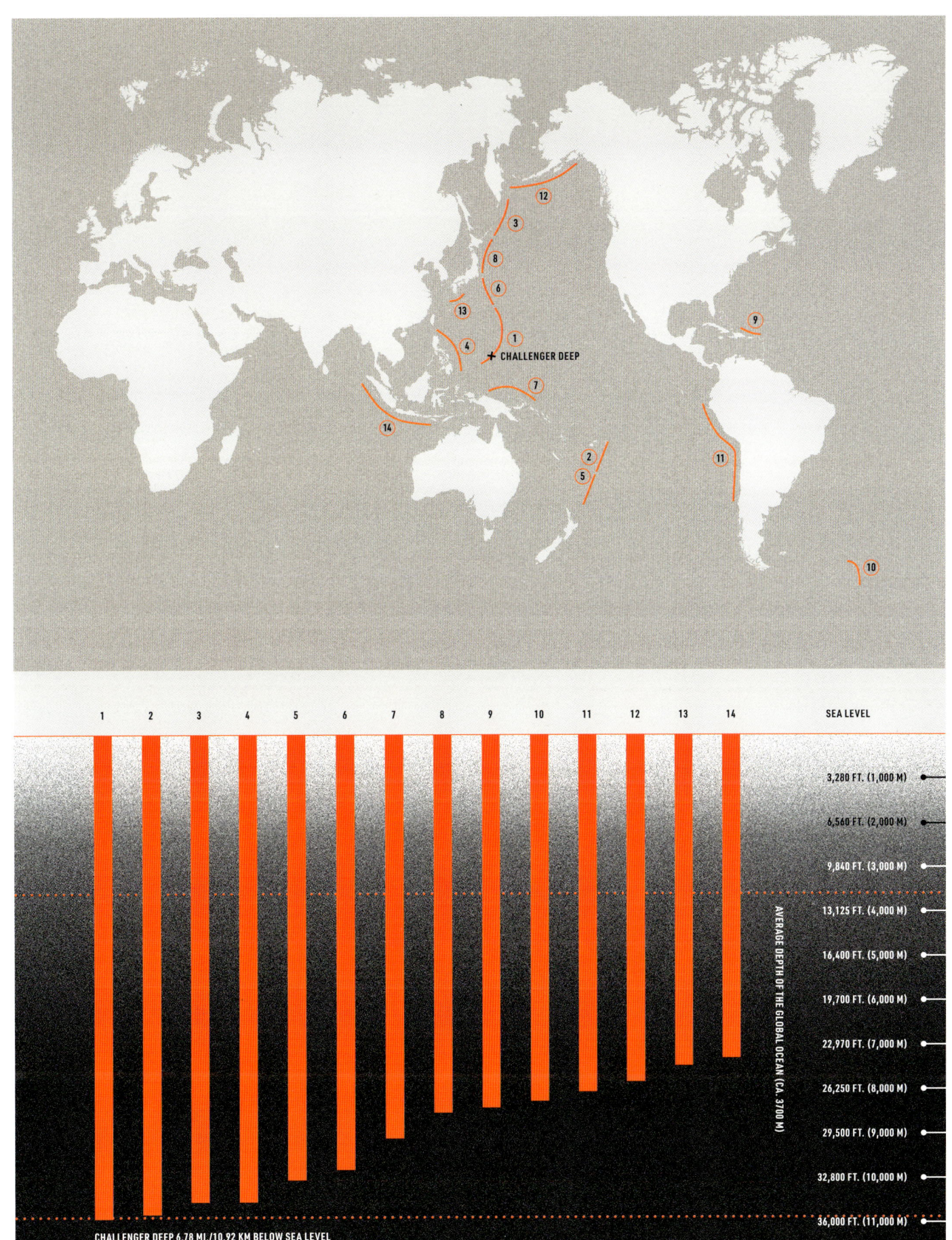

SEA LEVEL

3,280 FT. (1,000 M)

6,560 FT. (2,000 M)

9,840 FT. (3,000 M)

13,125 FT. (4,000 M)

16,400 FT. (5,000 M)

19,700 FT. (6,000 M)

22,970 FT. (7,000 M)

26,250 FT. (8,000 M)

29,500 FT. (9,000 M)

32,800 FT. (10,000 M)

36,000 FT. (11,000 M)

AVERAGE DEPTH OF THE GLOBAL OCEAN (CA. 3700 M)

CHALLENGER DEEP 6.78 MI./10.92 KM BELOW SEA LEVEL

Mollusks, echinoderms, and crustaceans

The hadal fauna includes members of the phylum Mollusca: gastropods (snails), bivalves (clams), and cephalopods (octopods and squids). The gastropods are usually quite small, less than 0.5 in. (1 cm) in diameter, but larger predatory whelk species have been observed to move across the seafloor at speeds of a few centimeters per minute, attracted to bait left on the bottom. They have been found down to full ocean depths, where the shells are invariably soft and fragile in the absence of calcium carbonate. Bivalves tend to live together in patches of very high numbers, possibly utilizing seepage of fluids from the seafloor or sunken debris such as trees. Finned or dumbo octopods (*Grimpoteuthis* sp., suborder Cirrata) and bigfin squids (family Magnapinnidae) have recently been documented in the upper areas of some trenches.

The echinoderms, including sea lilies (class Crinoidea), sea stars (class Asteroidea), brittle stars (class Ophiuroidea), urchins (class Echinoidea), and sea cucumbers (class Holothuroidea), are a very conspicuous component of the hadal fauna. Urchins are found only around the trench margins down to 24,600 ft. (7,500 m) depth. Brittle stars occur down to just over 26,250 ft. (8,000 m) depth, whereas sea stars and sea lilies are found to almost 32,800 ft. (10,000 m). Sea cucumbers are found down to maximum ocean depth, where they can account for over 90% of the animal biomass beyond 32,800 ft. (10,000 m). Like their abyssal cousins, they continuously graze over the sediment surface, rapidly removing deposited organic matter from the seafloor.

Over 350 species of crustaceans have been recorded from hadal depths. An important characteristic is that deeper-living scavenging crustaceans tend to be bigger than their shallow-water relatives, a phenomenon known as gigantism. This is most extreme in the supergiant amphipod *Alicella gigantea*, which is the world's largest-known amphipod, growing to a total length of 13.5 in. (34 cm), compared with the typical 0.5 in. (1 cm) length of related beach-dwelling sandhoppers. *Alicella gigantea* has been found down to a depth of 22,970 ft. (7,000 m), where it aggressively dominates the scavenging community. The trend of increased size depth in the ocean is found in at least four orders of Crustacea.

1CM

Hadal starfish
Family Freyellidae

Top left: with long, thin arms these deep-living sea stars are similar in appearance to brittle stars (order Ophiurida) and are found throughout abyssal depths of the oceans, usually on soft sediment. This one was discovered at 21,325 ft. (6,500 m) depth in the Java Trench of the east Indian Ocean.

Hadal amphipod
Hirondellea gigas

Bottom left: from 35,850 ft. (10,927 m) depth in the Challenger Deep, 2 in. (5 cm) long. Swarms of this species are attracted to food falls reaching the deepest point in the world's ocean.

Octopod in the trench
Grimpoteuthis sp.

Below: these cirrate (or "dumbo") octopods have been seen foraging on the seafloor down to 22,825 ft. (6,957 m) depth in the Java Trench in the Indian Ocean. This is the deepest known occurrence of a cephalopod.

Mariana Snailfish
Pseudoliparis swirei

Bottom: the world's deepest fishes feeding at mackerel bait at 24,600 ft. (7,500 m) depth in the Mariana Trench. This species is found down to 26,745 ft. (8,152 m).

Feeding in the trenches

When a submersible descends to the bottom of a trench, life can appear very sparse, but if a piece of fish is placed on the seafloor as bait, the scene quickly changes. The odor from the bait is carried on the bottom current, and typically within 10 minutes, the first amphipod arrives, swimming upstream following the odor plume. The numbers increase rapidly; there are 100 amphipods present after one hour and 1,000 after eight hours. The trench floor is teeming with life that responds rapidly to the artificial food fall. The number of amphipods attracted at maximum depth is 100 times greater than at 19,700 ft. (6,000 m) on the trench margins. In the Kermadec and Tonga Trenches of the southwest Pacific, the dominant amphipod species at maximum depth is *Hirondellea dubia*, whereas in the northwest Pacific, in the Mariana, Japan, and Kuril–Kamchatka Trenches, it is *H. gigas*. In the eastern Pacific, three different species of *Hirondellea* have been found in the Peru–Chile Trench.

Fishes are also attracted to bait; however, none are observed at depths greater than 26,900 ft. (8,200 m), which seems to be physiologically the maximum depth for fishes. Around trench margins (19,700–22,970 ft./6,000–7,000 m deep), abyssal species such as grenadiers and cusk eels are most common, but at greater depths within the trench proper, snailfishes predominate, often arriving in groups to feed. Snailfishes are small, less than 10 in. (25 cm) long, with a rudimentary sucker on the belly and pale, transparent skin. Species are endemic to particular trench systems: *Pseudoliparis swirei* in the Mariana Trench, *Notoliparis kermadecensis* in the southwest Pacific, *Pseudoliparis amblystomopsis* and *Pseudoliparis belyaevi* in the northwest Pacific, and at least one as yet unnamed snailfish in the Peru–Chile Trench and another in the South Sandwich Trench.

CHEMOSYNTHETIC ECOSYSTEMS

Creating the energy for life without sunlight

Life requires energy. In most ecosystems, sunlight is the source of this energy. Plants grow by using the energy from sunlight to convert carbon dioxide to sugars through a process known as photosynthesis. Plants then form the basis of the food web on which most other organisms depend. In some ecosystems, where light is not available, the energy needed to build sugars can be derived from the oxidation of certain simple "reduced" (the chemical opposite of oxidized) compounds, including hydrogen sulfide (H_2S) and methane (CH_4). The production of organic compounds using the energy from chemical oxidation is known as chemosynthesis.

Chemosynthesis is not confined to the deep sea, but it is certainly important there, not least because of the abundance of hydrogen-rich chemicals that leak from the seafloor. Where continental plates separate at mid-ocean ridges or are forced under an adjoining plate at subduction zones (see pages 46–47), or where the crust is weakened through a volcanic hot spot, hot fluids rich in hydrogen sulfide emanate from hydrothermal vents. These vents are fissures in the seafloor. Seawater percolates through the seafloor, becoming superheated by the hot magma in the earth's mantle and enriched with inorganic compounds such as H_2S, and then, because of the increased pressure caused by heating, the hot H_2S-enriched seawater is forced upward through the vents. Cold seeps also often occur at plate boundaries but result where hydrocarbons such as methane, formed from the decay of microorganisms over millions of years, leak upward through the seafloor at ambient temperature. Thus both hydrothermal vent and cold seep environments are rich in reduced chemicals conducive to chemosynthesis.

Chemosynthesis occurs within specialized microbes, which can be either bacteria or archaea (members of the domains of single-celled microorganisms Bacteria and Archaea, respectively), whose biochemistry has evolved to oxidize specific compounds. Thiotrophic (which translates from Greek as "sulfur-eating") microbes oxidize hydrogen sulfide; methanotrophic (methane-eating) microbes oxidize methane. Most microbes are free living and simply draw everything they need to build sugars (carbon dioxide, a source of oxygen, and the reduced compounds) from the surrounding seawater, but some form a symbiotic relationship with a host organism such as a worm, a shrimp, or a snail. In this scenario, the microbes produce sugars for the host, but the host provides the microbes with ideal living conditions—a rich oxygen supply, carbon dioxide, and protection. The emblematic animal of hydrothermal vents is the Giant Tube Worm (*Riftia pachyptila*), which grows to over 6.5 ft. (2 m) tall and hosts thiotrophic microbes in a modified organ, the trophosome. The deep blood-red color of the worm's flesh is due to its very specialized hemoglobin, which allows it to carry large amounts of oxygen to the microbes despite the presence of sulfide, which inhibits the function of ordinary hemoglobin. The free-living chemosynthetic microbes often form dense mats that are grazed by animals without symbionts.

A wide variety of other organisms visit chemosynthetic habitats to exploit the rich food source or the increased temperatures. Octopods have even recently been found brooding their eggs in warm vent waters, and the Pacific White Skate (*Bathyraja spinosissima*) appears to do the same.

Black smoker and Giant Tube Worms
Hydrothermal vent in the northeast Pacific on the Juan de Fuca plate, with colonies of the Giant Tube Worm *Riftia pachyptila*.

Hydrothermal mussels
Bathymodiolus mussels can grow to 16 in. (40 cm) long, and there may be millions of individuals at any one site. Here, near Champagne Vent in the Mariana Arc region of the western Pacific, they occur in a dense assemblage together with shrimps and limpets. The mussels host bacterial symbionts in their gills that chemosynthesize sugars, but they can also feed by extracting suspended food particles (mostly bacteria) from the surrounding water. Different species of *Bathymodiolus* mussels occur at cold seeps.

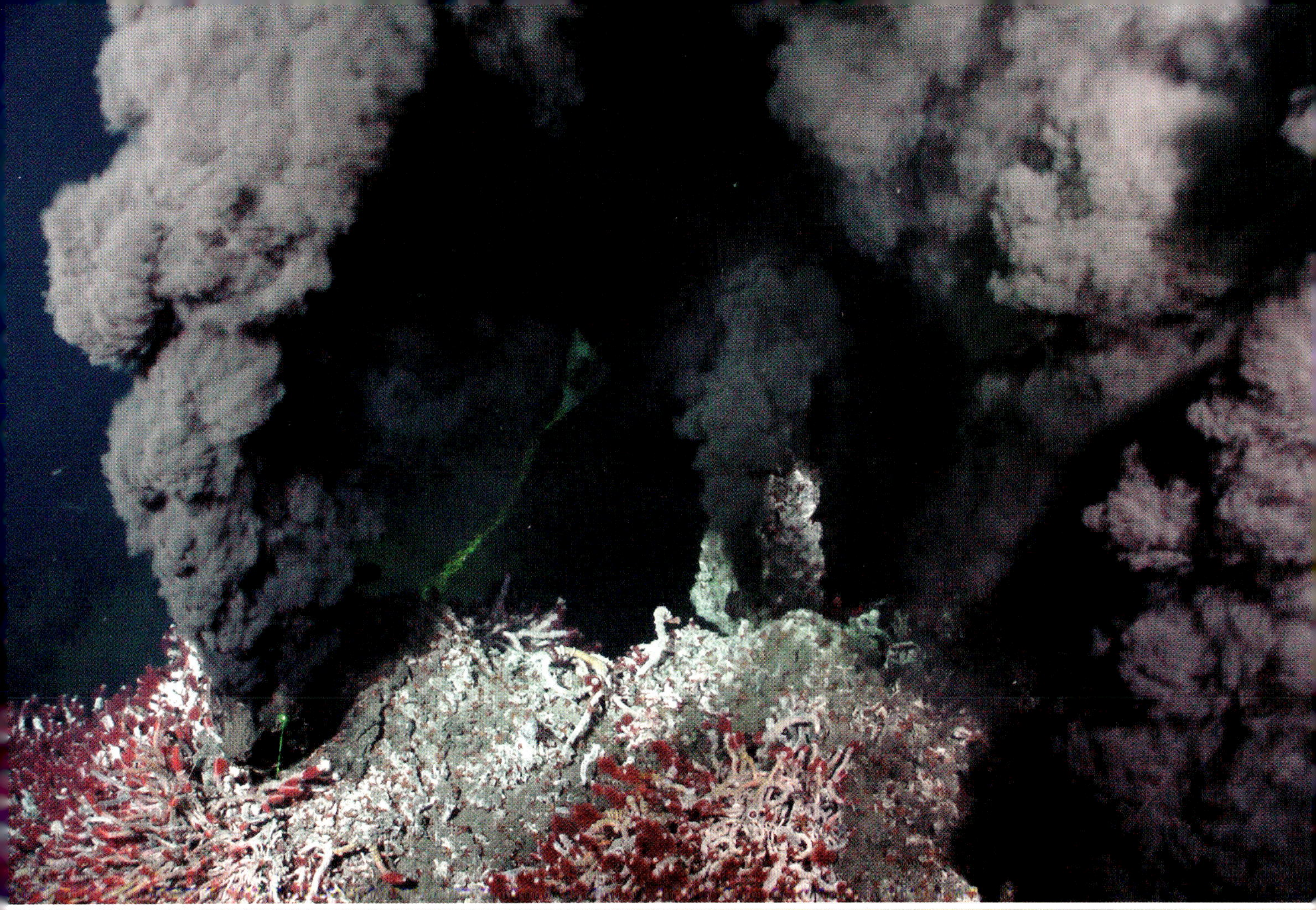

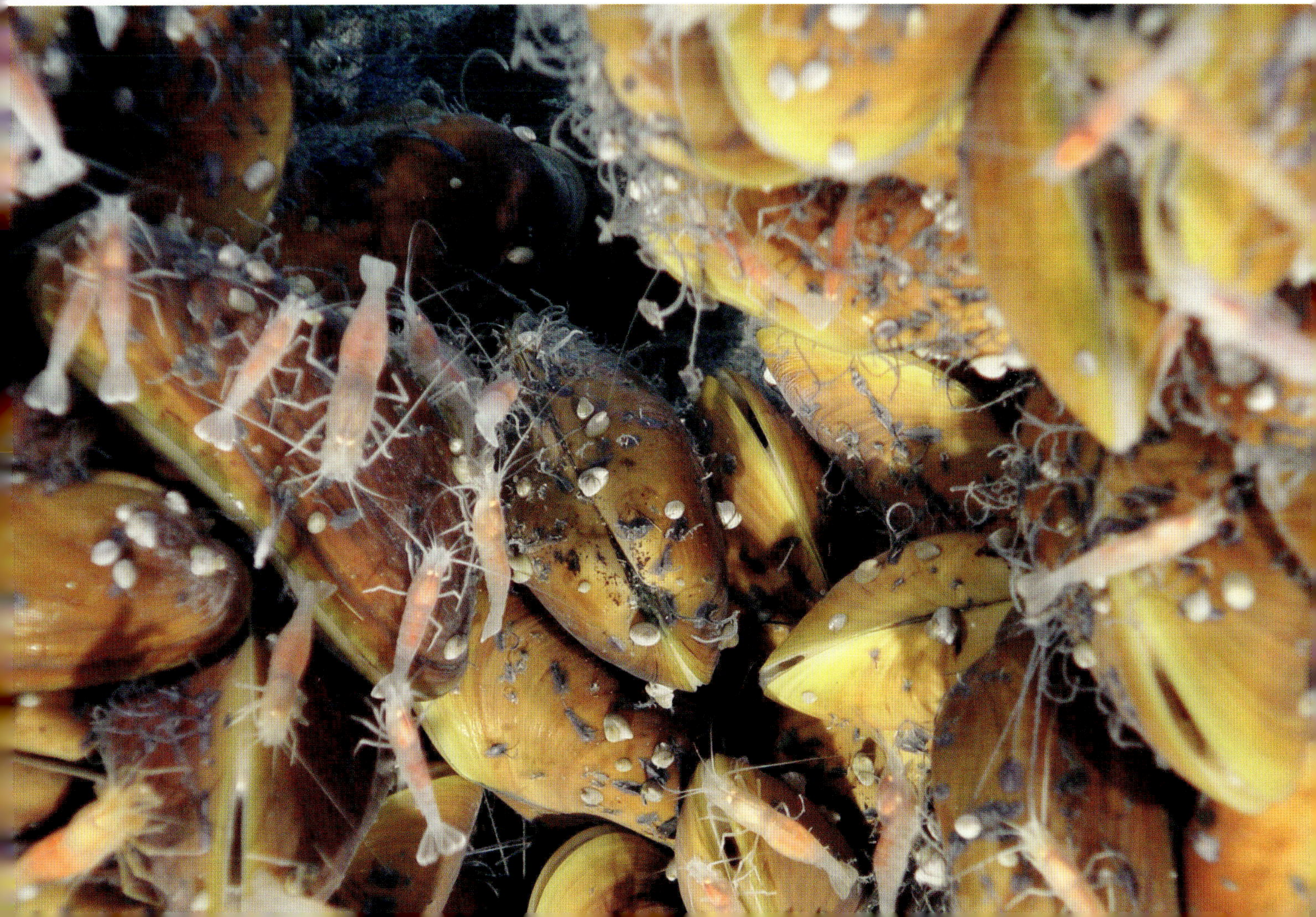

Giant Tube Worms
Riftia pachyptila

An assemblage of Giant Tube Worms at the low temperature Tempus Fugit Vent Field. Here, on the eastern part of the Galápagos Rift in the eastern Pacific Ocean at around 8,200 ft. (2,500 m) depth, anemones are also abundant.

Hydrothermal vents

Hydrothermal vent habitats are often dominated by tall chimneys. These are formed from the precipitation of mineral sulfides from the venting seawater enriched with inorganic compounds. In the Atlantic Ocean, these chimneys are often covered in swarms of hydrothermal vent shrimps. At deeper vents of the Mid-Atlantic Ridge, the eyeless vent shrimp *Rimicaris exoculata* swims along chimney walls in the gradient between hydrothermal fluids and cold oxygenated ambient seawater. Although described as "eyeless," the vent shrimp has a wing-shaped photoreceptor, which contains a pigment similar to the rhodopsin in our own eyes and can detect extremely faint light sources. It is not detecting light as we normally think of it but rather thermal radiation and possibly also sonoluminescence, produced when microbubbles of gas collapse. The photoreceptor probably helps the shrimp to avoid moving too close to venting water that is too hot for it or to find the active venting if it drifts too far.

Like Giant Tube Worms, vent shrimps host symbiotic bacteria. Their carapace wraps around their gills and forms a branchial (gill) chamber, open to the environment and stuffed with bacteria. Although vent shrimps can, and do, graze on the chimney walls, much of their nutrition comes from sugars produced by their bacterial symbionts.

Life at vents occurs in a gradient—only the hardiest organisms can live closest to the venting fluid and tolerate both the temperature and toxicity of the sulfide-laden water. One animal that has evolved unique adaptations to these extreme conditions in Pacific vents is the Pompeii Worm (*Alvinella pompejana*). Pompeii Worms live in large colonies on the walls of active vents, each worm in a paper-thin tube with its tail in temperatures over 212°F (100°C). Pompeii Worms keep their heads in cooler waters than their tails, and they peep out of their tubes to feed on free-living bacteria. The worms' hairy backs are covered in different bacteria, in a layer that may be as thick as 0.5 in. (1 cm). In this symbiotic relationship, it is the bacteria that are fed by the worm. Pompeii Worms produce a sugary mucus for the bacteria, and the thick layer of bacteria in return provides insulation from the extreme heat. The bacteria also metabolize sulfur, lead, zinc, calcium, and copper from the vents, making the Pompeii Worm's home tube less toxic.

Dealing with toxicity is crucial for those animals adapted to live nearest to the venting fluid. The Scaly-foot Snail (*Chrysomallon squamiferum*), sometimes called the Sea Pangolin, has made the strangest use known of the metal-rich waters: it grows scales fortified with iron sulfide nanoparticles. These scales are presumed to protect the snail from predators, but it has also been suggested that their evolution may have been linked to processes that moderated the snail's internal sulfide levels. Sea Pangolins need sulfur because they derive their nutrition from thiotrophic bacteria housed in a gland off their esophagus, but too much is likely toxic. Sea Pangolins at the Solitaire hydrothermal vent field, where iron levels are very low, have white scales, which are not iron fortified. Sea Pangolins are known from very few locations, some of which are threatened by deep-sea mining. Consequently, the species is listed as Critically Endangered on the International Union for Conservation of Nature (IUCN) Red List of Threatened Species.

Pompeii Worm
Alvinella pompejana
Top left: a Pompeii Worm removed from its tube to show its hairy back that attracts insulating and-detoxifying bacteria. Top right: a Pompeii Worm observed by ROVs in the Gulf of California.

Scaly-foot Snail
Chrysomallon squamiferum
Three Scaly-foot Snails with foot scale colors that reflect the different water properties of the vent fields they inhabit.

Vent shrimps
Rimicaris exoculata
Swarms of *Rimcaris* shrimps on the walls of hydrothermal vent chimneys (bottom left). *Rimicaris exoculata* clearly showing the broad carapace over the gills under which the symbiotic bacteria are hosted, and the dorsal wing-shaped photoreceptor (right).

Tall chimneys
Seawater percolates through the seafloor, becomes superheated by magma from below and infused with inorganic compounds, and rushes upward. Tall chimneys form as mineral sulfides precipitate out of these hot vent fluids.

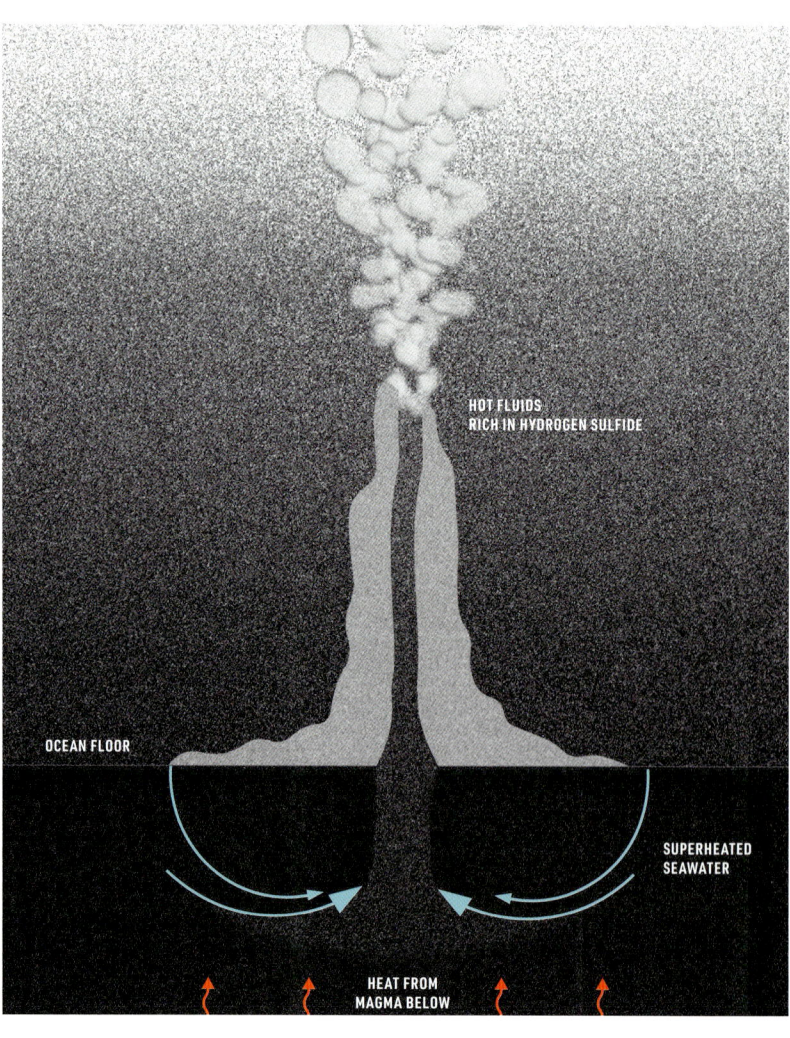

HOT FLUIDS RICH IN HYDROGEN SULFIDE

OCEAN FLOOR

SUPERHEATED SEAWATER

HEAT FROM MAGMA BELOW

Cold seeps

While cold seeps are home to animals that are very similar to (and in some cases the same as) those found at hydrothermal vents, they often look very different. Cold seeps lack the chimneys that are produced by precipitation of minerals from hot vent fluids, and they often feature extensive calcium carbonate (chalky) outcrops. This is "authigenic" rock—rock made on-site. One of the key processes that occurs at cold seeps is the anaerobic oxidation of methane by bacteria and archaea in the sediments. The microbes oxidize methane (CH_4) using the oxygen from sulfate ions (SO_4^{2-}) dissolved in seawater. This process provides the energy the microbes need to build sugars, but also produces carbonate ions (HCO_3^-) as a byproduct. These carbonate ions react with calcium ions also dissolved in seawater and calcium carbonate precipitates. Finding slabs of calcium carbonate on the continental slope is a good indication that there may be nearby methane seepage.

Like hydrothermal vents, cold seeps often have very fragile communities. In Gulf of Mexico cold seeps, tube worms in the genus *Lamellibrachia* form huge colonies. Each worm can live hundreds of years and grow several meters tall. Like the hydrothermal vent tube worm *Riftia*, they house symbiotic bacteria in their specialized trophosome and derive their energy from sugars produced by their bacterial symbionts. The worms are preyed upon by mobile predators such as crabs, but as well as being a source of food for other animals, they also provide a home—the hidden spaces between their tubes provide habitat for a myriad of other creatures, and hence their presence significantly increases the biodiversity of the area.

Specialized shrimps with bacterial symbionts in their branchial chambers abound, as do large bivalves that house chemosynthetic symbionts in their gills. The dark brown *Bathymodiolus* mussels and white *Calyptogena* clams can grow over 1 ft. (30 cm) long. Both are also found at hydrothermal vents and sometimes at whale falls, which can also provide an environment rich in reduced chemicals (see pages 166–167).

Methane seeps

Methane gas percolates upward from reserves below the seabed providing an energy source for bacteria, which form a mat on the surface of the seafloor. Animals such as clams and polychaetes with chemosynthetic symbionts colonize the soft sediment. Other animals feed on the bacterial mat. Slabs of calcium carbonate form from the carbonate ions produced in the oxidation of methane by bacteria, and these provide hard substrate for a variety of other organisms.

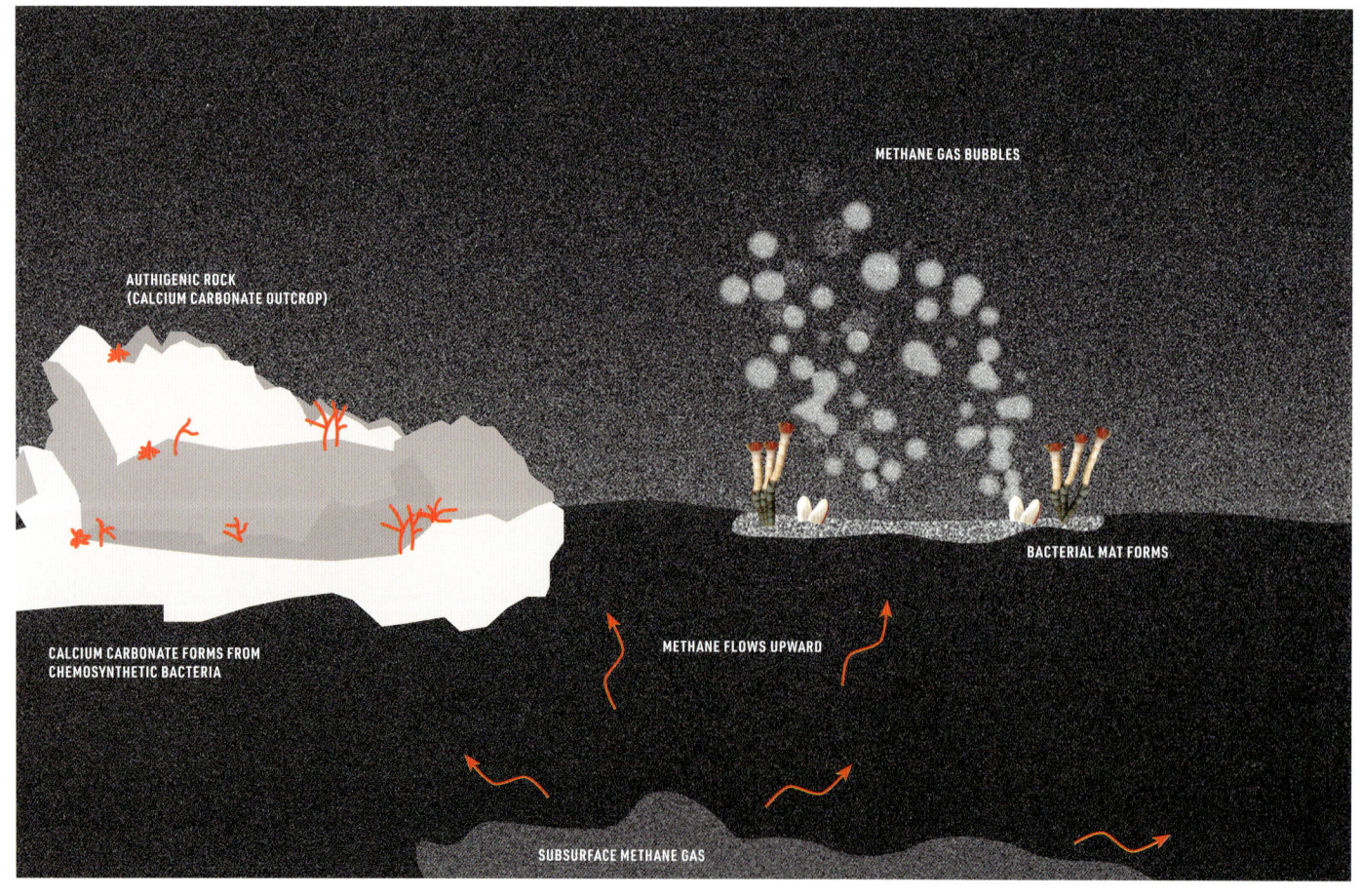

AUTHIGENIC ROCK
(CALCIUM CARBONATE OUTCROP)

METHANE GAS BUBBLES

BACTERIAL MAT FORMS

CALCIUM CARBONATE FORMS FROM
CHEMOSYNTHETIC BACTERIA

METHANE FLOWS UPWARD

SUBSURFACE METHANE GAS

Gulf of Mexico cold seep

Top left and right: Gulf of Mexico cold seep tube worms (*Lamellibrachia* sp.) form a habitat for a myriad of other small creatures.

Methane clathrates

Occasionally at methane seeps, methane mixed with seawater crystallizes, forming methane clathrates. In this incredibly inhospitable methane ice, a very special worm is found. *Hesiocaeca methanicola* appears to feed on specialized bacteria and archaea that use the clathrate for chemosynthesis.

Whale fall

When even a small whale dies, it provides a potential feast for thousands of organisms. The carcass may become engorged with gases and float, scavenged by seabirds and animals that dwell near the ocean surface. But most whales that die over deeper waters eventually sink to the seafloor and become food for large deep-sea scavengers such as sharks and hagfishes (see pages 88–89), and then, as the carcass is consumed and spread, for a succession of smaller deep-sea animals that clean the bones and even digest them.

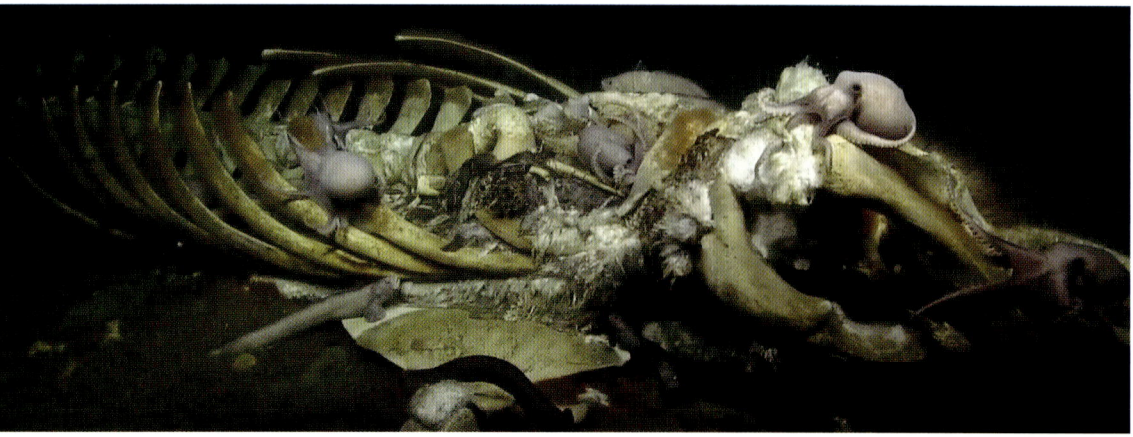

A whale fall off California
The top image shows the filter-feeding baleen plates decomposing. The bottom image captures octopuses scavenging on a whale-fall community, here the dark "fuzz" on the ribs are *Osedax* worms.

Hagfishes arrive rapidly. These primitive fishes, similar to the earliest vertebrates, lack a lower jaw and backbone. This makes them flexible enough to worm their way easily to the interior of a carcass, and their two rows of unique keratin teeth—the same strong material in our own fingernails—are ideal for rasping at flesh. Most remarkable is their ability to slime—a defense against predators that also eases their passage through the carcass. They tie themselves in knots, easing the knot down their body, to clean away excess mucus.

Hagfishes, sharks, and then smaller scavengers such as amphipod crustaceans, the insects of the sea, with their remarkably complex jaws, are all messy eaters. Enthusiastic scavenging, together with defecation, spreads flesh, blubber, and digested organic matter over the seafloor, opening up the carcass to worms and octopods and also enriching the surrounding sediment, providing food for deposit feeders such as sea cucumbers. Slowly the scavengers clean the bones, allowing settlement of *Osedax*, the unique whale-fall specialist bone-eating worm. Called the zombie worm, *Osedax* has no eyes, mouth, or gut. Instead, it secretes acid, sinking roots into the whale bones. These roots house symbiotic microbes, which decompose the whale-bone collagen and provide *Osedax* with all the nutrients it needs. Its red, hemoglobin-packed, feathery plumes absorb oxygen from the water and transport it to the oxygen-hungry microbes to power their work.

Meanwhile, all around, anaerobic microbes in the sediment and on the bones use sulfate (SO_4^{2-}) dissolved in seawater as a source of oxygen to break down fats and lipids in the bone marrow, producing hydrogen sulfide as a byproduct. In the sulfide-enriched sediments, clams and tube worms, like those seen at hydrothermal vents, hosting chemosynthetic symbionts, build a secondary ecosystem.

In time, the rich nutrients are all exhausted, production of sulfide stops, and all that remains are nutrient-depleted bones. Finally, the whale fall acts as a reef—a site for settlement of deep-sea corals and other organisms that seek out hard substrate for their home.

Zombie worm
Osedax sp.

Female *Osedax* using symbiotic bacteria to extract nutrients from whale bones. Male *Osedax* are dwarf, little more than sperm factories, each settling on a female, and relying on her for its nutrients.

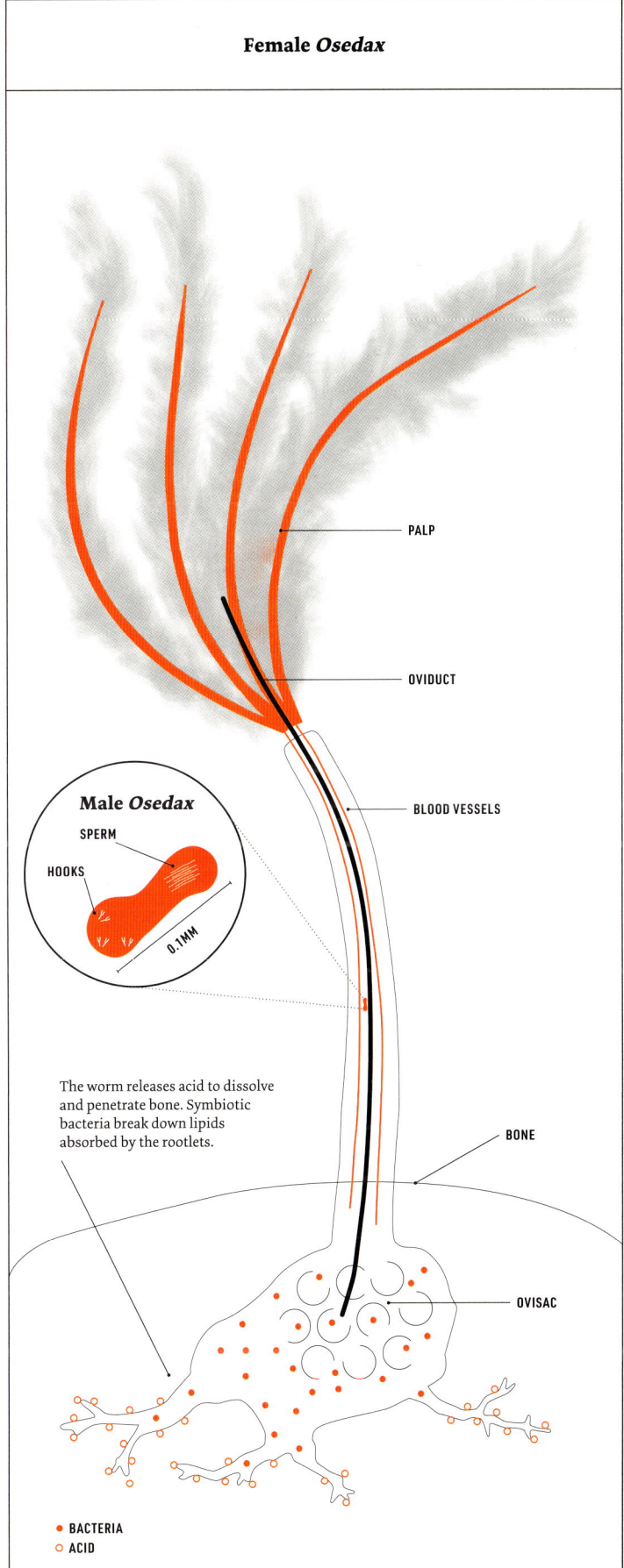

Female *Osedax*

PALP

OVIDUCT

BLOOD VESSELS

Male *Osedax*

SPERM

HOOKS

0.1MM

The worm releases acid to dissolve and penetrate bone. Symbiotic bacteria break down lipids absorbed by the rootlets.

BONE

OVISAC

● BACTERIA
○ ACID

THE WATER COLUMN

Layers of open water and how they differ

A lot of deep water, over a billion cubic kilometers, or roughly three-quarters of the total volume of the ocean, covers most of the hard parts of this planet. Other than the ice of the polar regions, no structure is visible in all that water. However, structure is there, in the physics and chemistry of the water. As is presented elsewhere in this book, some of this structure is apparent in the geographic differences in temperature and chemistry among the major basins of the global ocean (see pages 66–75). Smaller-scale horizontal differences are often related to the major ocean currents, the frontal structures at their boundaries, and the large eddies that spin hundreds of meters down through the water (see pages 54–55). However, the most dramatic and consistent differences in this world of water are vertical, resulting in a layered structure to the water. Because the pattern of these vertical layers is very consistent around the world, it is conceptually useful to think of the miles of water above a particular location on the seafloor as the "water column," with epipelagic, mesopelagic, bathypelagic, abyssopelagic, and hadopelagic layers piled on top of each other. This is also called the "midwater" environment, although that term is considered by some to mean just the mesopelagic zone.

Black Swallower
Chiasmodon niger

Top: a Black Swallower has an extendable gut so that if it encounters a large animal to eat, the prey can be swallowed whole. Some Swallowers have been caught with prey larger than themselves curled in a stomach extended much more than in this ROV video framegrab.

Comb jelly
Bathocyroe sp.

Above: this lobate comb jelly (ctenophore) is often encountered during deep midwater exploration. Unlike the classical locomotion for ctenophores, resulting from the synchronized beating of the cilia in their comb rows, these lobates swim by coordinated pulsing of their large lateral lobes. This looks somewhat like the "frog kick" used in the breast stroke of human swimming competitions.

Water layers and permanent thermocline

Temperature is extremely important in establishing and maintaining the layered structure of the deep sea. This graphic represents a stylized profile of bottom depths (see also page 16). The thin orange line shows a typical mid-latitude profile of the change in temperature with depth, illustrating the zone of decreasing temperature known as the "permanent thermocline." Transitions between the layers are gradients rather than the abrupt change indicated by the graphic design here.

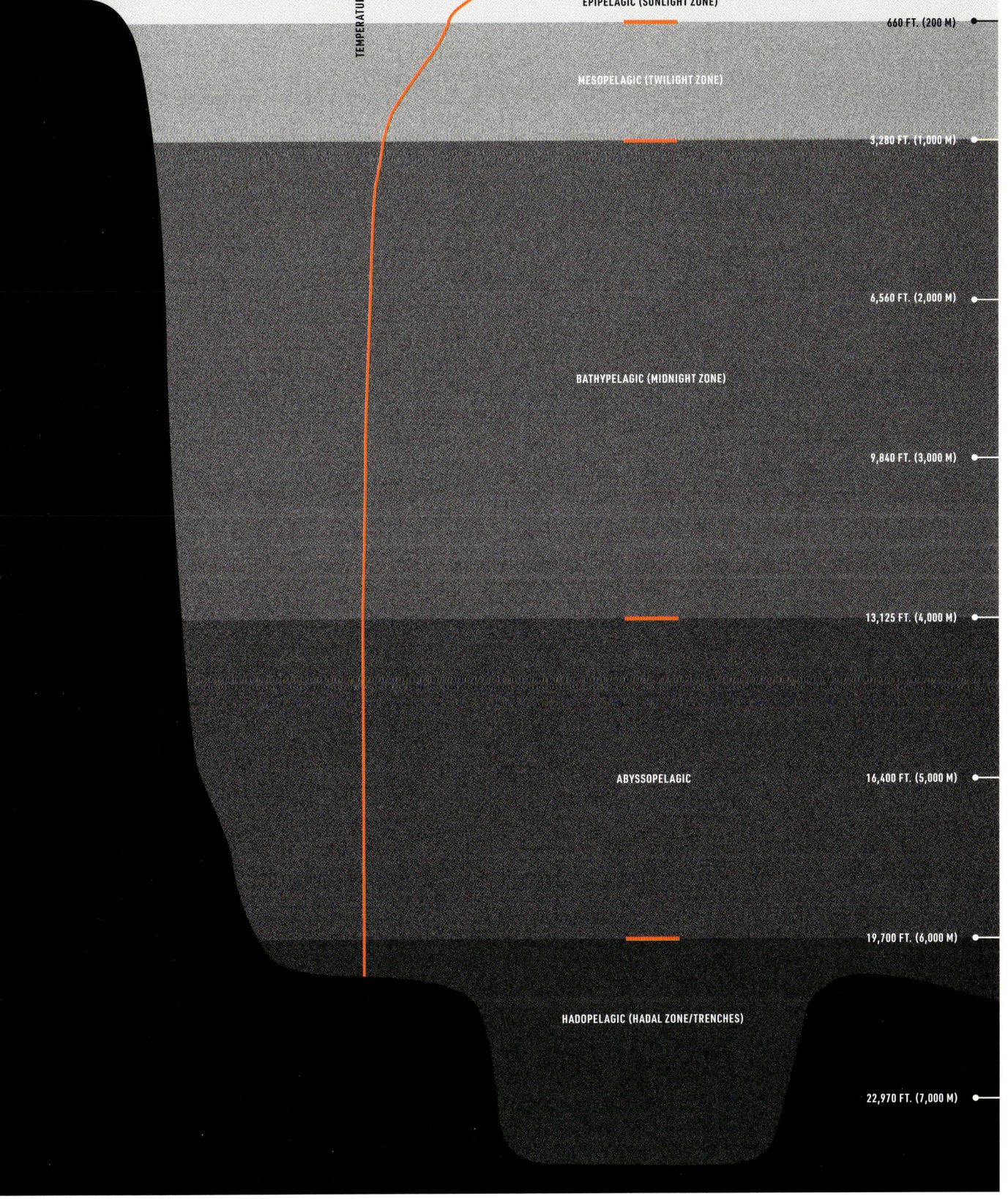

°C

0 5 10 15 20 25 30

TEMPERATURE

EPIPELAGIC (SUNLIGHT ZONE)

660 FT. (200 M)

MESOPELAGIC (TWILIGHT ZONE)

3,280 FT. (1,000 M)

6,560 FT. (2,000 M)

BATHYPELAGIC (MIDNIGHT ZONE)

9,840 FT. (3,000 M)

13,125 FT. (4,000 M)

ABYSSOPELAGIC

16,400 FT. (5,000 M)

19,700 FT. (6,000 M)

HADOPELAGIC (HADAL ZONE/TRENCHES)

22,970 FT. (7,000 M)

The mesopelagic layer

The mesopelagic zone, extending from about 660 to 3,280 ft. (200 to 1,000 m) is defined by the penetration of sunlight. The regional depth limits vary with water clarity. At midday, without heavy clouds, substantial sunlight reaches down from the sea surface, but generally not enough at 660 ft. (200 m) to support photosynthesis. At about 3,280 ft. (1,000 m), the last of the sunlight dies out. Because of the dim light (and memories of the old mystery TV series), this is often called the "twilight zone." Coincident with this vertical gradient of light, temperatures decrease with increasing depth, forming the permanent thermocline, the transition from warm productive surface waters (the epipelagic zone) to the universal permanent cold of the really deep waters. The mesopelagic zone functions as an interface between the warm epipelagic and the dark and cold of the vast bathypelagic and deeper ecosystems.

Aside from curiosity about occasional weird animals caught in nets towed fairly deep in the open ocean, interest in the twilight zone really began with the invention of sonar for detection of submarines. Naval researchers were surprised to find thick layers that scattered the sound pulses of sonar, layers under which submarines could hide. Sonar showed that these "deep scattering layers" (DSL) move up and down with time of day.

Of course, sunlight penetration goes away at night. Most mesopelagic animals follow their preferred light levels upward as the sun sets and back downward the next morning, creating the daily phenomenon of the largest migration on earth. These migrators feed in the productive epipelagic waters during the darkness of night, avoiding visual predators, and return to the cooler deeper water to digest their meals. This huge migration of DSL animals is now known to include the planet's largest biomass (total weight) of fishes. People are beginning to take notice of this biomass by targeting it for harvest. This, in turn, has increased scientific interest in the mesopelagic ecosystem.

Another characteristic of mesopelagic animals is that almost all of them produce light (bioluminescence; see pages 118–121) for various reasons. These reasons include: (1) luring prey close enough to catch; (2) hiding in the clear water by matching the light coming down from the surface; (3) startling a predator with a bright flash to escape from it; and (4) attracting a mate in the vast open waters.

In addition to the animals that we think of as typical of the mesopelagic zone, recently developed technology is making it increasingly clear that animals generally considered to reside near the surface regularly use the twilight zone. Tags attached to large animals that can record information such as temperature, pressure, and ambient light have tracked Leatherback Sea Turtles (*Dermochelys coriacea*), Great White Sharks (*Carcharodon carcharias*), and bluefin tunas (*Thunnus* spp.) through the mesopelagic and even deeper waters.

Ctenophore
Ctenophora sp.

Top left: a comb jelly is one of the most primitive types of animal, but not a true jellyfish. The red color is often seen in mesopelagic animals because almost all of the light at those depths is blue, which does not reflect off of a red animal.

Pram bug amphipod
Phronima sedentaria

Top right: the crustacean amphipod—seen here gripping a salp from below—takes over the body of a salp, hollowing it out, propelling it around, and living inside where it lays its eggs.

Siphonophore
Forskalia asymmetrica

Bottom left: a very common colony of jellyfish relatives. Individuals in the colony are specialized for swimming, feeding, or reproduction. Many species possess very powerful stingers.

Deep-sea squid
Histioteuthis sp.

Bottom right: the mesopelagic squids in the family Histioteuthidae are commonly known as jewel squids or strawberry squids. This view shows the ventral (bottom) surface covered with prominent photophores (bioluminescent light organs) used to break up its silhouette to make it less obvious to predators looking up from below.

Oxygen-minimum zones

All oxygen in the ocean comes from the surface, either produced by photosynthesis in the epipelagic layer or absorbed from the atmosphere across the ocean surface. Therefore, oxygen dynamics in the deep ocean consist of consumption. In some parts of the ocean, the biological production in the epipelagic layer is so great that much of it either sinks passively or is actively transported by animals down into the mesopelagic. There, respiration of mesopelagic animals and bacterial breakdown of the organic matter uses up so much oxygen that the water becomes sufficiently depleted (hypoxic) to limit what can live there. These oxygen-minimum zones (OMZs) are very extensive regionally and becoming more so with the changing global climate.

Very few species of animals can live in water with low oxygen concentrations. In general, less than 2 ml of oxygen per liter of seawater can be physiologically limiting to many animals. Some vertical migratory pelagic animals can spend the daytime in the OMZ layer by greatly reducing their metabolism, and then become active at night by migrating into the well-oxygenated surface layer. The low oxygen presumably protects these specialists from most predators. However, deep-diving air breathers, such as toothed whales or Leatherback Sea Turtles, would not be constrained by hypoxic conditions in the mesopelagic zone.

Along the edges of continents and around oceanic islands, where OMZs are in contact with the bottom, benthic ecosystems are controlled by the hypoxia. This phenomenon is especially important throughout the eastern Pacific Ocean, in the northern Indian Ocean, and in isolated deep basins such as the Black Sea. As with pelagic animals, few benthic animals can survive in OMZs. This results in low diversity, although abundance can occasionally be high because of the lack of predators and competitors. Hypoxic microbes can form dense and extensive mats, which support grazing animals that can survive in those conditions.

The ability of water to hold dissolved oxygen decreases with increasing temperature. This, together with changes in oceanic circulation and surface productivity, is resulting in increasing size and intensity of OMZs as the global climate warms. This increase is not just in geographic coverage—the vertical extent of OMZs is increasing as well. A result is compression of the oxygenated surface zone, forcing the animals there to live in more brightly illuminated conditions during the day. Downward extension of OMZs in contact with the bottom pushes animals of continental and island slopes downslope, the overall effects of which are yet to be determined.

Humboldt Squid

Dosidicus gigas

Humboldt Squid are large (mantle lengths reaching 3 ft./0.9 m) voracious predators. They can spend daytime in waters with very low oxygen content, as deep as 3,940 ft. (1,200 m), and migrate upward at night to the epipelagic layer, which has plenty of oxygen, to reoxygenate and feed.

Average oxygen levels globally

Global distribution of strong OMZs, where O_2 is less than, or equal to, 1.4 ml per l of seawater. Color indicates the thickness of the OMZ; white lines show the extent of areas with severely hypoxic conditions (0.5 and 0.2 ml O_2 per l). (Source: Moffitt et al., 2015.)

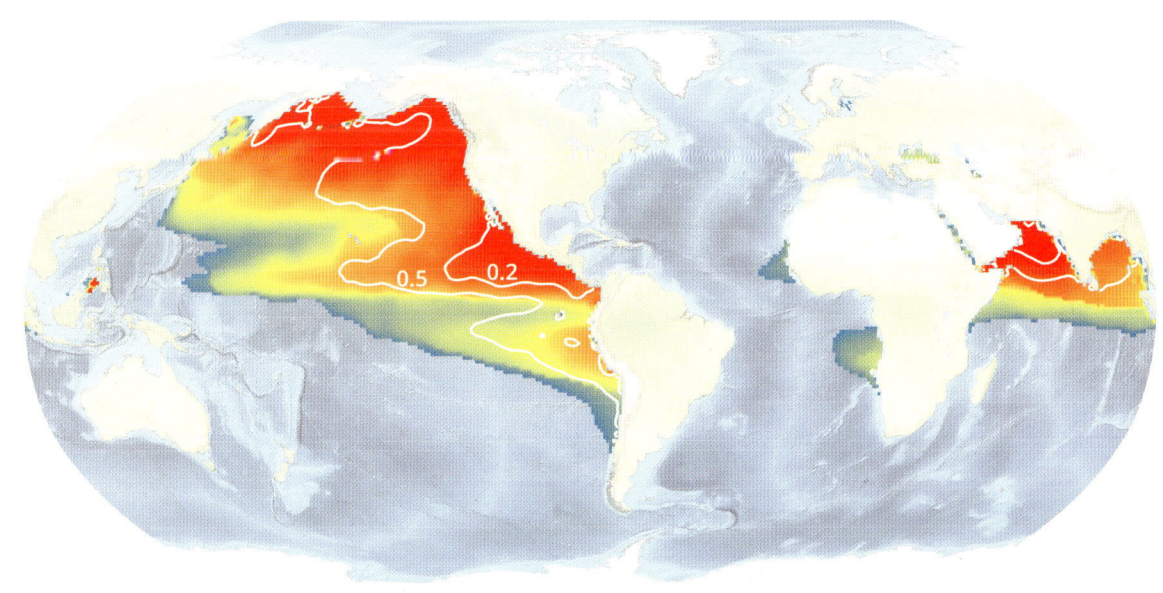

Thickness (M) of OMZ, O_2 <= 1.4 ml L

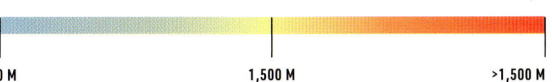

0 M   1,500 M   >1,500 M

Bathypelagic and abyssopelagic layers

Sometimes called the "midnight zone" as a comparison with the shallower twilight zone, the bathypelagic environment has conditions somewhat more analogous to polar winter. Whereas midnight implies that daytime is just hours away, the ocean below 3,280 ft. (1,000 m) is permanently cold and lacking sunlight, although myriad points of biological light are present. The mesopelagic zone, at depths of 660–3,280 ft. (200–1,000 m), is five times the thickness of the productive surface layer, the epipelagic, at 0–660 ft. (0–200 m). In comparison, the bathypelagic zone, extending from the bottom of the mesopelagic to beyond the average depth of the global ocean (about 12,140 ft./3,700 m—an average that includes continental margins, ridges, seamounts, and trenches), is 27 times the thickness of the epipelagic. Generally, the bathypelagic is defined as open-water depths of 3,280–13,125 ft. (1,000–4,000 m). In some animal groups that have been examined (e.g., fishes), a shift in species composition is found at about 13,125 ft. (4,000 m) depth. Since this coincides in general with the depths of the abyssal plains, the pelagic environment from 13,125 ft. (4,000 m) down to 19,700 ft. (6,000 m) (upper limit of the hadal or hadopelagic zone) is termed "abyssopelagic." Major characteristics of the mesopelagic, such as directional downward penetration of dim sunlight and a strong permanent thermocline, disappear in the bathy- and abyssopelagic zones, leaving the consistent increase of hydrostatic pressure with depth (about one atmosphere per 33 ft./10 m) as the major structuring characteristic. Remember, though, that a change in pressure/depth has a much greater proportional effect closer to the surface than at greater depth. For example, a change of 33 ft. (10 m) depth from 295 to 330 ft. (90 to 100 m) (10 to 11 atmospheres) is a 10% increase in pressure, whereas the same distance from 3,250 to 3,280 ft. (990 to 1,000 m) (100 to 101 atmospheres) is just a 1% increase. The percentage of change is what is important to an animal moving vertically.

Elongated Bristlemouth
Sigmops elongatus

Found throughout most of the world's oceans in midwater at 330–4,920 ft. (100–1,500 m) depth, the Elongated Bristlemouth grows to 10 in. (25 cm) length, feeding on smaller fishes and crustacea.

Pelagic decapod shrimp
Systellaspis sp.

A worldwide predator in the mesopelagic zone, generally deeper, down to 2,950 ft. (900 m) during the day, and shallower at night.

The hadopelagic zone

The hadal ecosystems (depths greater than about 19,700 ft./6,000 m or 21,325 ft./6,500 m, depending on whom you ask) of the ocean trenches, those extreme geomorphic gashes in the seafloor, are very difficult to study (see pages 154–157). Of course, those trenches are all filled with water. If sampling the bottom of a trench is difficult, sampling the pelagic animals in the water of the trenches is even more so. As a result, we know very little about hadopelagic ecosystems. Each trench is very limited in area, and the combined area of the 37 hadal trenches is only about 1% of the seafloor, but the depth range, and therefore change in pressure, of trenches (19,700–36,000 ft./6,000–11,000 m) is almost half of the total for the global ocean. The pressure is so great that temperature increases slightly with depth, reversing the general trend throughout the deep sea.

The benthic boundary layer

It is often the case in ecology that the interface between two habitats is biologically enriched by both. Several aspects of this enrichment are seen in the deep ocean. As was mentioned in the introduction to pelagic animals, some benthic animals swim up into the water column (see page 94). Conversely, when mesopelagic organisms, migrating vertically at depths of 660–3,280 ft. (200–1,000 m), are transported by currents over the continental slopes, they encounter at the lower end of their vertical migration something beyond their normal experience: the bottom, inhabited by many predators for which a pelagic food subsidy is welcome. This is sometimes referred to as a "benthic kill zone" for the pelagic animals.

In areas where currents move across the bottom, friction between the moving water and the benthic substrate creates turbulence that resuspends sediment, most of which has been deposited out of the water column. This "benthic boundary layer," also called a "nepheloid layer," extends tens of meters above the bottom, depending on current speed and sediment composition. It provides a relatively food-rich environment for near-bottom plankton as well as both benthic and pelagic predators that feed on them. Many animals and colonies have evolved to feed on this enriched layer of plankton and detritus.

Two-Arm Ctenophore
Duobrachium sparksae
A genus and species of ctenophore (comb jelly) recently discovered and described from archived video recordings of deep dives of an ROV very near bottom at 12,800 ft. (3,900 m) depth in Guajataca Canyon, north-northwest of Puerto Rico.

Cusk Eel
Eretmichthys pinnatus
Currently, only a single species is recognized as valid in this genus, which has an unusual arrangement of the fins on its sides. This one was recorded on video from an ROV at Fryer Guyot (flat-topped seamount) near the Mariana Trench.

GLOBAL
PATTERNS

CONNECTED ECOSYSTEMS

How oceanic communities interact

Previous pages: **Soft coral** *Anthomastus ritteri*

Ecosystems are geographically defined areas of biological communities interacting with their environment. The spatial scale is not fixed; scientists may talk about the deep-ocean ecosystem as a whole or consider the component parts of the deep ocean as individual ecosystems, such as the abyssal plain ecosystem, the mesopelagic ecosystem, or chemosynthetic ecosystems. In particularly isolated or well-defined areas, the spatial scale may be reduced further and pertain to a single location, such as the Lucky Strike vent ecosystem. As this loose definition implies, these ecosystem classifications are all artificial. The natural world does not have sharp boundaries, and ecosystems interact with one another regardless of how we define them.

SUNLIGHT
PHYTOPLANKTON
ZOOPLANKTON
NEKTON
MARINE SNOW
SCAVENGERS AND DEPOSIT FEEDERS
FOOD FALLS

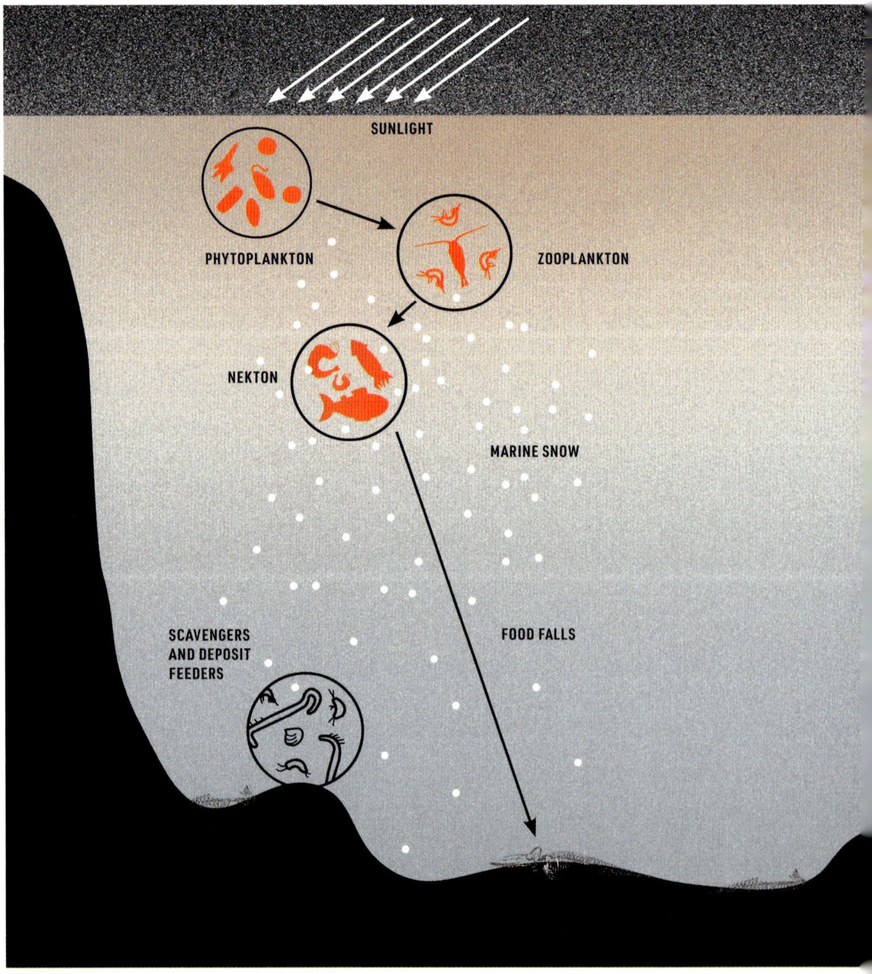

Food supply to the deep sea

Photosynthesis in surface waters drives the deep-sea food supply. Currents and water masses connect ecosystems and local topography can affect food availability, particularly in submarine canyons and around seamounts.

Food supply

The food reaching all deep-sea habitats except chemosynthetic ecosystems (see pages 158–167) originates in surface waters, where photosynthetic phytoplankton in sunlit surface layers of the ocean form the base of the marine food chain. Phytoplankton are consumed by zooplankton, and the diel vertical migration of the animals of the deep scattering layer (DSL) moves these food resources throughout the mesopelagic zone (see pages 106 and 170–171). Some animals migrate to deeper habitats as they grow and mature, and organic carbon that originated as newly formed sugars in phytoplankton through photosynthesis slowly reaches the deep ocean. Life also produces waste: phytoplankton and zooplankton die, crustaceans and other organisms cast off their exoskeletons, and all animals defecate. This waste material sinks downward, forming clumps of marine snow as it descends. It is a slow passage: marine snowfalls reach their peak on the deep-ocean floor about six weeks after a plankton bloom in temperate waters. Marine snow, despite being waste, contains enough carbon and nitrogen to satisfy the needs of many scavengers and deposit feeders, while its seasonal influx means that even the deepest ecosystems can experience seasonality. Nutrients may also reach the deep-ocean floor through large food falls—carcasses of large fishes and whales may bring to a localized area of the seafloor a feast that can last for years (see pages 166–167).

The vertical connections between ecosystems vary. In canyons, where material cascades from shelf waters, the influx of nutrients is greater. Around seamounts, various processes lead to an influx of food from the surrounding pelagic waters. On the abyssal plains, in the middle of oceans, the influx can be reduced, because a lack of iron, which reaches the open ocean by being blown off land, limits the strength of the phytoplankton bloom in the surface waters.

Nutrients that originated in shallow waters also reach chemosynthetic ecosystems, but their importance is diminished. However, horizontal connections may be greater in these ecosystems as mobile predators amass to take advantage of the rich oasis of life.

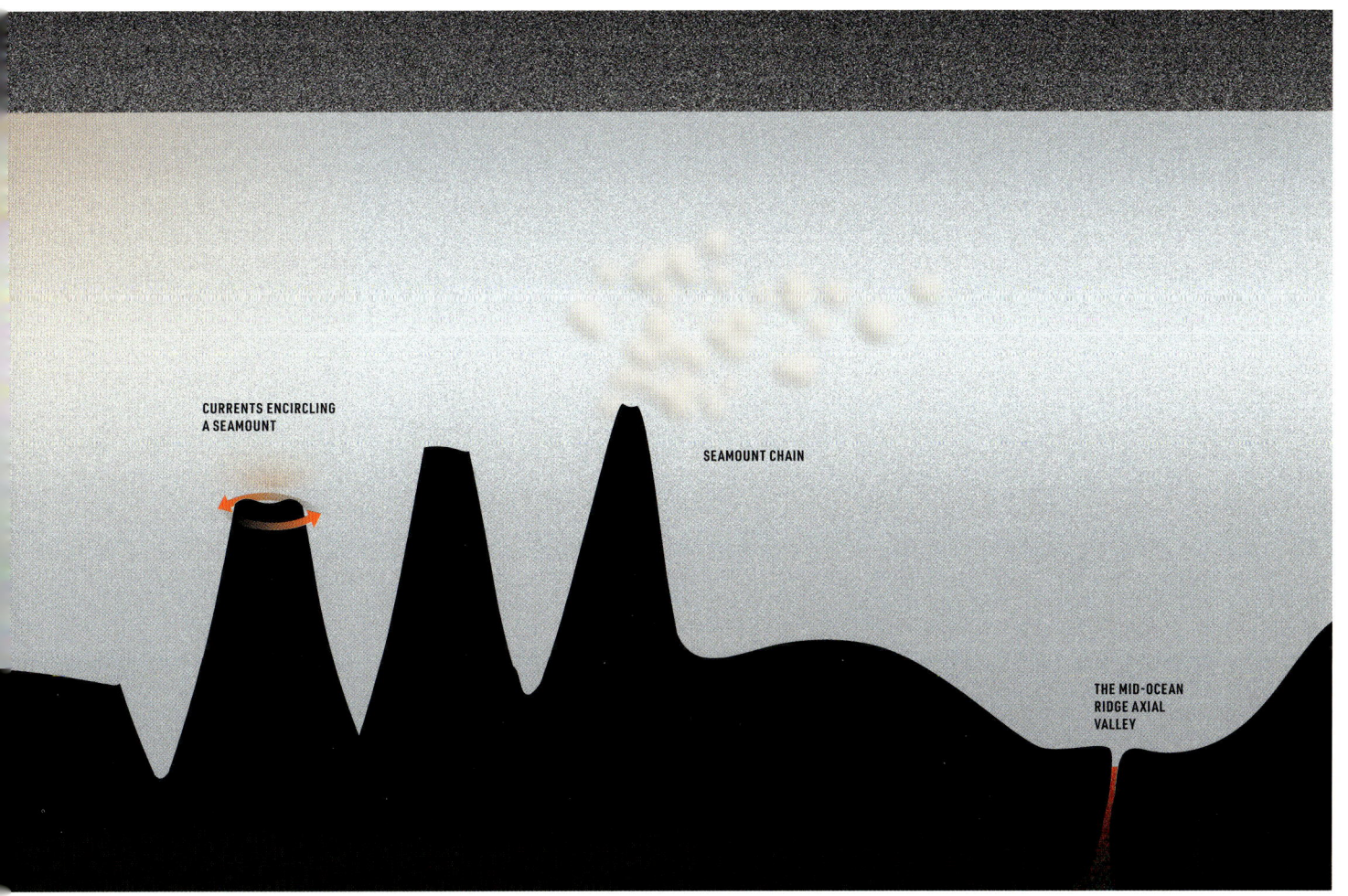

CURRENTS ENCIRCLING
A SEAMOUNT

SEAMOUNT CHAIN

THE MID-OCEAN
RIDGE AXIAL
VALLEY

Ocean currents and fronts

Currents operate at different spatial scales across our oceans. On the largest scale is the thermohaline circulation, also sometimes known as the global ocean conveyer belt, which is driven by the sinking of cold, dense oxygen-rich surface water at the poles (see also page 56). While a full global cycle of this circulation takes between 1,000 and 2,000 years to complete, the deep oxygen-rich water masses formed by this process flow over the deep ocean floor and bring essential oxygen to deep-sea habitats.

Currents also disperse eggs and larvae. Many oceanic species are found in more than one ecosystem, and some species are cosmopolitan, with distributions that extend around the globe. In larger, mobile species, adult migrants may ensure genetic mixing within the species. For small and/or sessile animals, however, the dispersal of early life stages as planktonic larvae is crucial to maintaining a wide, genetically mixed distribution. How far such a species disperses depends on its planktonic larval duration (PLD). PLDs vary widely, from a few days for *Osedax* worms (see pages 166–167), for example, to up to a year for *Bathymodiolus* mussels. PLD is affected by larval type. Lecithotrophic (yolk-eating) larvae have no feeding apparatus but instead are supplied with yolky resources to sustain them until they metamorphose. This puts an absolute time limit on lecithotrophic larval duration, although in cold waters, where metabolism tends to be slow, resource use can also be slow, and dispersal can still be considerable. Planktotrophic (plankton-eating) larvae feed in the water column, a strategy that theoretically can allow them to be very long lived but may also require them to move vertically through the water column to access sufficient food resources. The movement of larvae on horizontal currents depends on their buoyancy and their swimming ability.

As well as connecting ecosystems, currents can also separate them, by entraining larvae locally in gyres and restricting connectivity. For example, modeling of the dispersal of larvae from tube worms in Gulf of Mexico cold seeps suggested that the larvae never leave the gulf. Oceanic fronts can also act as barriers to larval dispersal. The lack of mixing of water bodies can prevent larvae from crossing the front, or those larvae that do penetrate the front may not have the physiological tolerance for the potentially different temperature, salinity, and oxygen saturation in the adjacent water mass.

Cosmopolitan species

The Jewel Enope Squid (*Pyroteuthis margaritifera*) has an almost worldwide distribution (red) occurring in all but the most southerly and northerly seas and oceans.

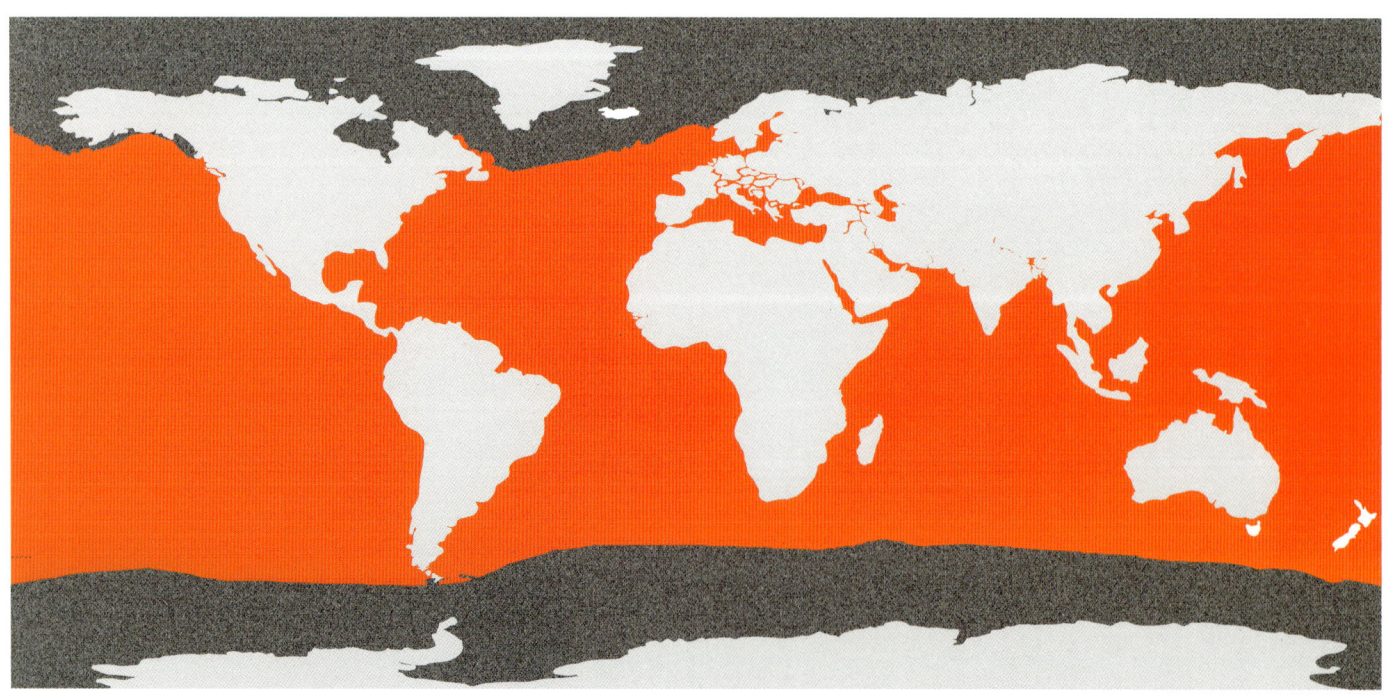

Planktonic larval duration (PLD)

PLD varies. The vent barnacle *Neoverruca* sp. has a PLD of about 95 days. The larva begins as a nauplius and develops into a cyprid after about six naupliar stages and prior to settling and metamorphosing into a juvenile. The bathyal urchin *Aspidodiadema jacobyi* has a PLD of 150 days. The seep tube worm *Lamellibrachia satsuma* has a PLD of around 45 days.

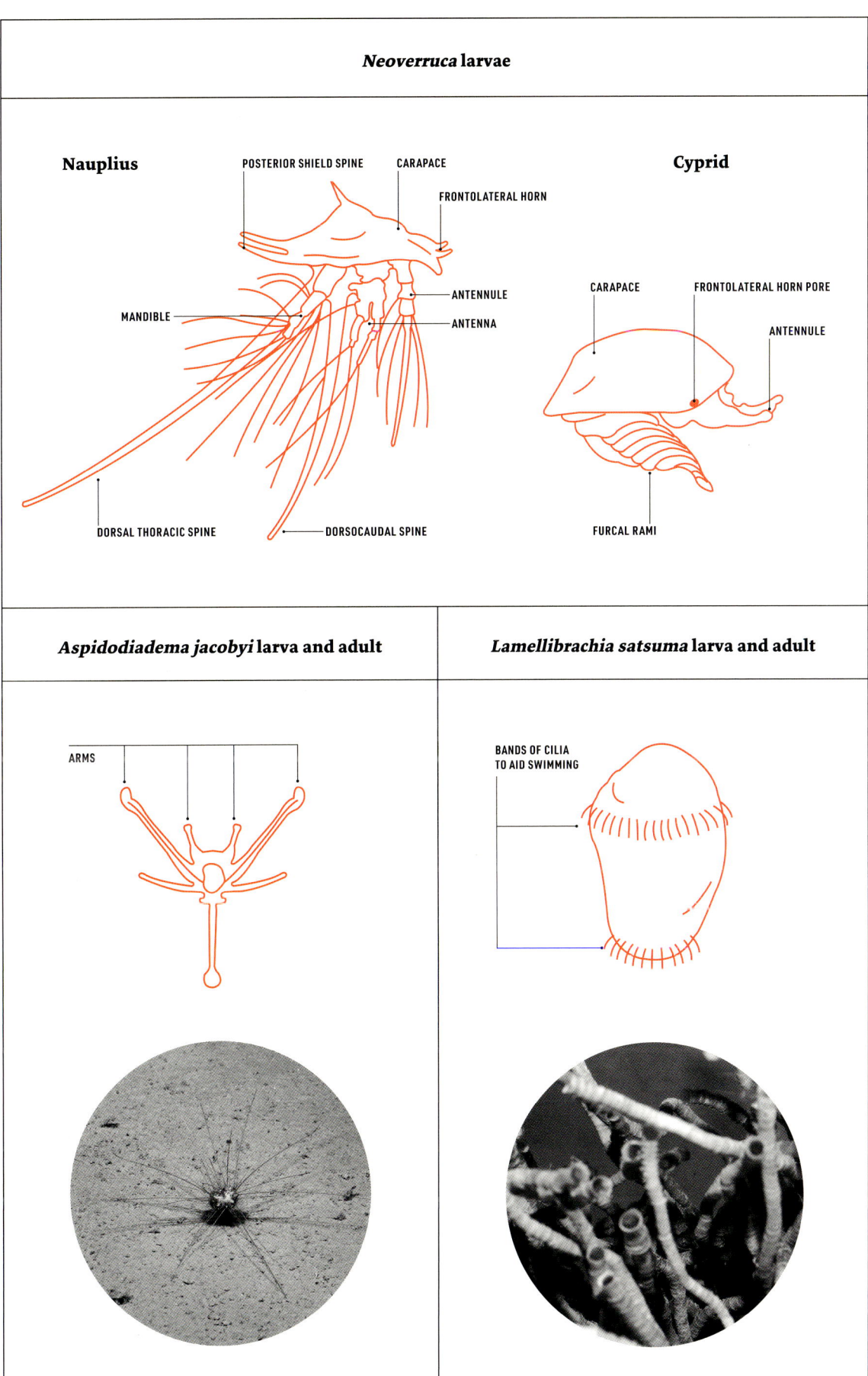

Neoverruca larvae

Nauplius

POSTERIOR SHIELD SPINE — CARAPACE — FRONTOLATERAL HORN

ANTENNULE

MANDIBLE — ANTENNA

DORSAL THORACIC SPINE — DORSOCAUDAL SPINE

Cyprid

CARAPACE — FRONTOLATERAL HORN PORE — ANTENNULE

FURCAL RAMI

Aspidodiadema jacobyi larva and adult

ARMS

Lamellibrachia satsuma larva and adult

BANDS OF CILIA TO AID SWIMMING

Deep-sea stepping-stones

Some deep-sea habitats are widely spaced; examples include submarine canyons and cold seeps, both found mostly only on continental margins, and hydrothermal vents, which exist predominantly at centers of geological spreading. Modeling of the dispersal of larvae in currents often suggests a lack of connectivity between distant areas known to harbor the same species. This has led to "stepping-stone" hypotheses.

For deep-water corals (phylum Cnidaria), which occur at both submarine canyons and seamounts, seamounts provide ideal stepping-stones. Although individual larvae may never travel more than one step, the net effect of larvae released from all the stepping-stones is to maintain a single species across entire oceans. Genetically distinct populations may still exist on individual seamounts, as is the case with the reef-building coral *Solenosmilia variabilis*, which seems to rely more on asexual reproduction through fragmentation than on the release of gametes and larvae through sexual reproduction to maintain its population at any particular seamount. Alternatively, seamount stepping-stones may allow genetically mixed populations to exist over much wider areas, as is the case with the solitary coral *Desmophyllum dianthus*.

It has been suggested that whale falls may serve as stepping-stones for cold seep and hydrothermal vent fauna, although there is some controversy about this. There is noted similarity in the fauna at these three chemosynthetic ecosystems (see pages 158–167), but so few whale falls have been studied in detail that insufficient data exist to really test the stepping-stone hypothesis. It has also been suggested that whale falls might have played a more substantial role as stepping-stones prior to the decimation of whale populations through hunting. A greater abundance of whales in earlier times would have meant a greater

Distinct populations

Top left: the hermatypic (reef-building) coral *Solenosmilia variabilis* on a vertical wall at around 2,625 ft. (800 m) depth in the Caribbean Sea just southeast of Puerto Rico.

Connected populations

The solitary coral *Desmophyllum dianthus* is a common sight on vertical walls. The large polyps are often more than an inch across. Here it is living in association with (top right) *Acesta excavata* bivalves, a brisingid sea star, and an octopod of the genus *Graneledone*, and (right) with the distinctive anemone *Actinernus michaelsarsi*.

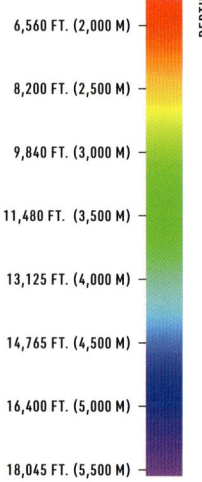

Stepping-stones

The New England Seamount Chain extends east-southeast from the edge of the continental shelf off Massachusetts providing stepping-stones of habitat for over 620 mi. (1,000 km).

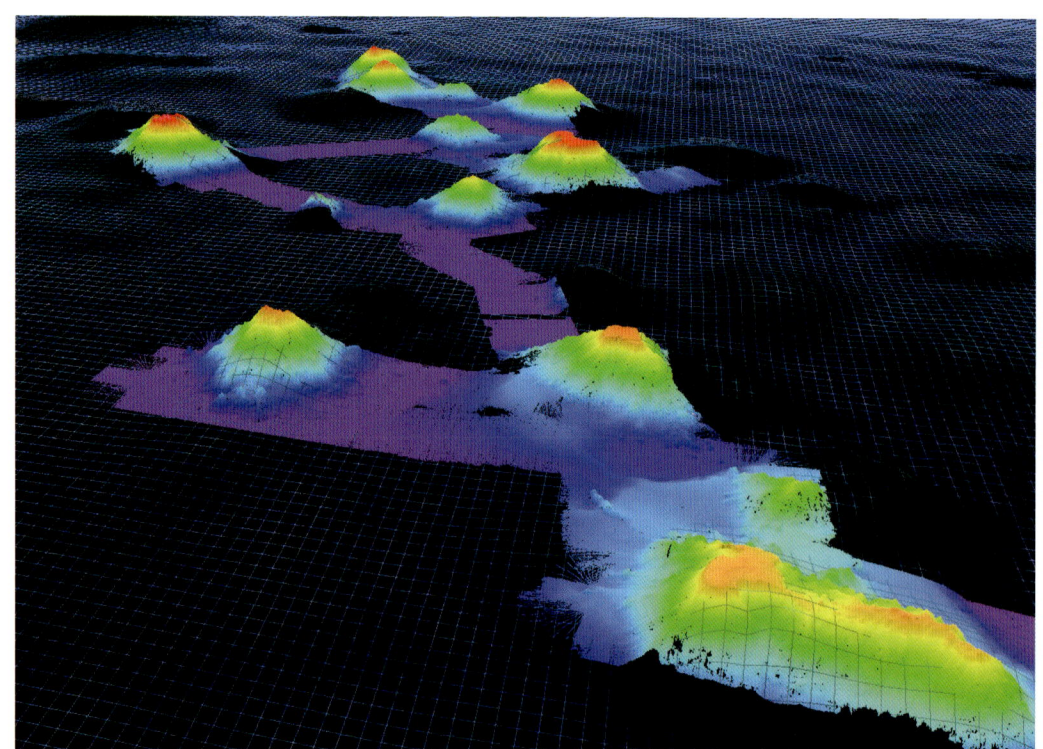

DEPTH

6,560 FT. (2,000 M)
8,200 FT. (2,500 M)
9,840 FT. (3,000 M)
11,480 FT. (3,500 M)
13,125 FT. (4,000 M)
14,765 FT. (4,500 M)
16,400 FT. (5,000 M)
18,045 FT. (5,500 M)

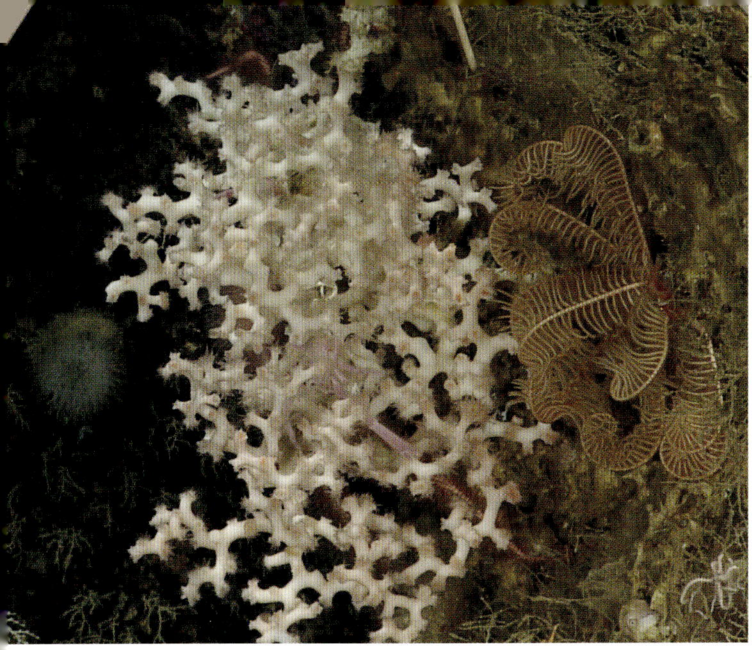
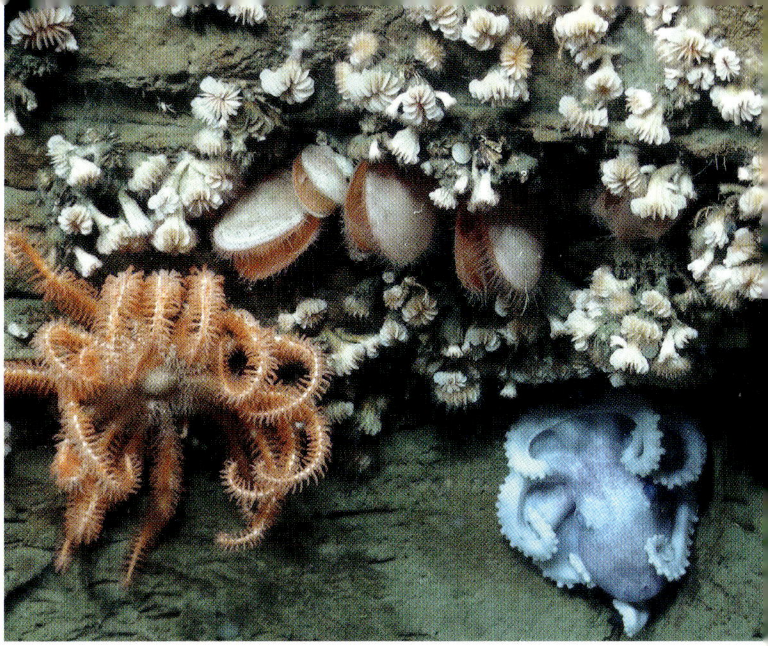
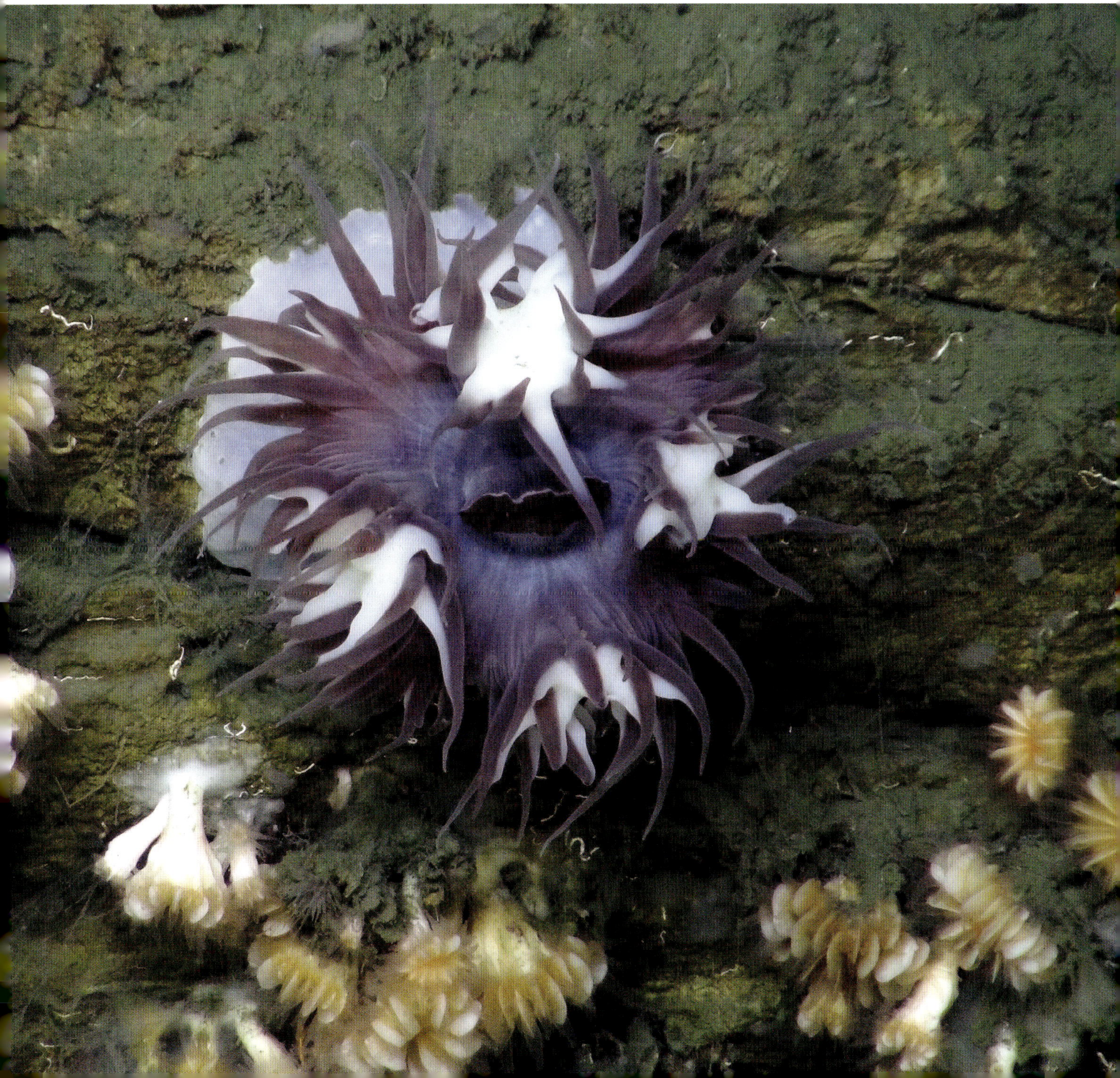

Classifying ecoregions with environmental data

Although some oceanic species have an almost global distribution, many do not, and therefore even similar ecosystems vary in different parts of the world. Because environmental conditions play a key role in determining which species thrive where, dividing the ocean into units that have similar physical and chemical conditions can help us understand the distinctiveness of various parts of our oceans. Scientists use parameters such as temperature, salinity, oxygen saturation, and nutrient availability because they are known to play a role in determining the diversity and abundance of animals in any given area. Other parameters could be useful as well, but there are few for which we have good global knowledge.

Attempts to classify mesopelagic waters (660–3,280 ft./ 200–1,000 m depth) based on environmental parameters tend to show bands of water masses arranged by latitude. Of course, the water column is not homogeneous, and each of the bands shown begins and ends at different depths, and those starting and ending depths vary by location—so that while a two-dimensional picture of the ecological units looks simple, the three-dimensional reality is complex.

The separation of water masses is driven by oceanographic processes, which are outlined in more detail in the Oceanography chapter (pages 42–75). In the Atlantic, Pacific, and Indian Oceans, gyres driven by complex seafloor topology and major wind systems play important roles. Farther north, similar processes separate Arctic waters from those of the Atlantic and Pacific Oceans.

To the south, the subtropical front and a series of fronts associated with the Antarctic Circumpolar Current, the planet's largest current, distinguish the water masses. The region of the subtropical front, where warmer subtropical waters meet cooler, denser subantarctic waters, is well oxygenated and highly productive. There are marked differences between the species of zooplankton and micronekton (such as krill and small cephalopods and fishes; see pages 98–100) found here and the species found farther south.

Southward lie the subantarctic front and the polar front, south of which are the coldest polar waters. The most southerly mesopelagic water masses appear to be less distinct and less isolated than the epipelagic waters in this region. For example, of all the many species of mesopelagic lanternfishes (family Myctophidae) that occur here, only one, *Electrona antarctica*, has its distribution concentrated south of the polar front.

CTD measuring instrument

A CTD rosette being recovered. It is lowered through the water column to measure conductivity (salinity proxy) and temperature at different depths. The bottles collect water samples at discrete depths for detailed analysis.

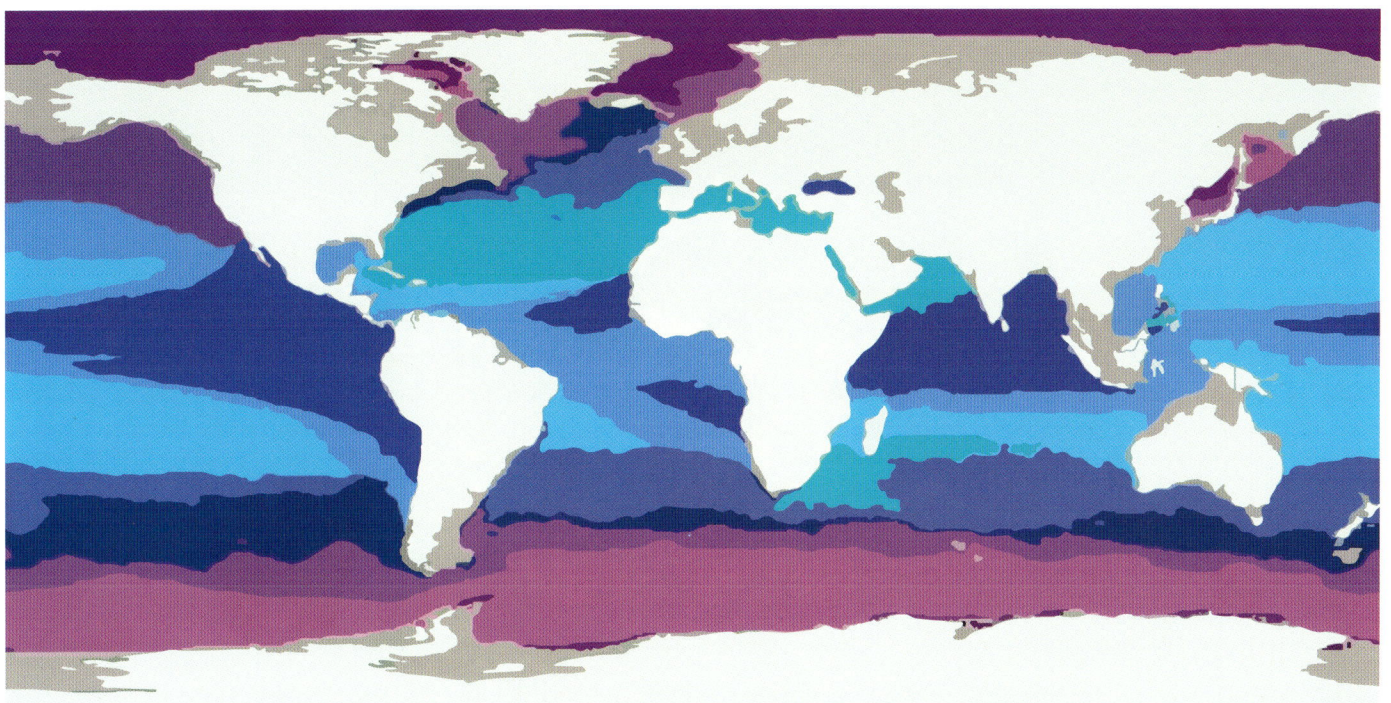

Water masses

Attempts to classify mesopelagic waters based on environmental parameters tend to show bands of water masses arranged by latitude. (Source: Sutton et al., 2017.)

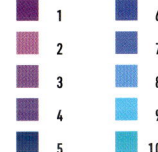

1	6
2	7
3	8
4	9
5	10

Polar fronts

The subtropical front marks the northern boundary of the cooler, less saline, subantarctic waters. The Antarctic Circumpolar Current, isolating the Southern Ocean from the Atlantic, Pacific, and Indian Oceans, follows the path of the subantarctic and polar fronts.

Subtropical front (yellow)
Subantarctic front (purple)
Polar front (red)

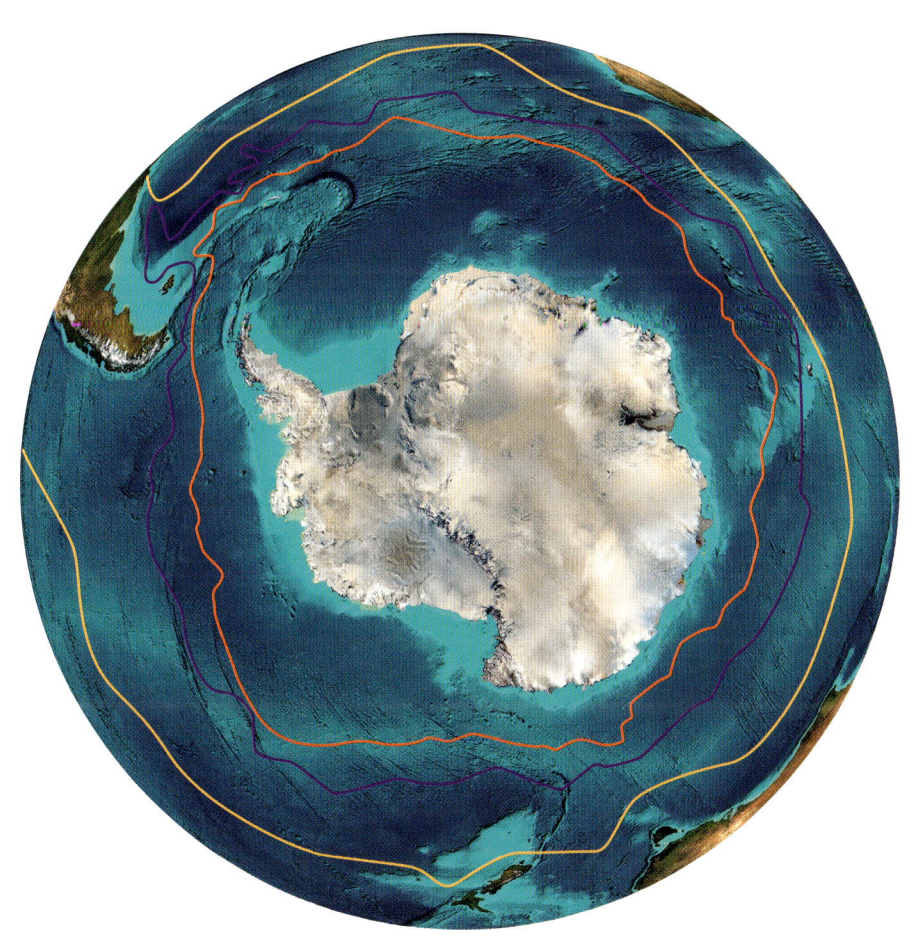

Classifying ecoregions with faunal data

Mesopelagic ecosystem classifications that also take into account the fauna tend to have more subdivisions than those based on environmental data alone (pages 180–181). Expert knowledge of the fauna of particular regions, particularly breaks in faunal distributions, can thus provide more detailed ecosystem classifications with greater numbers of subdivisions than those based on physical and chemical data alone.

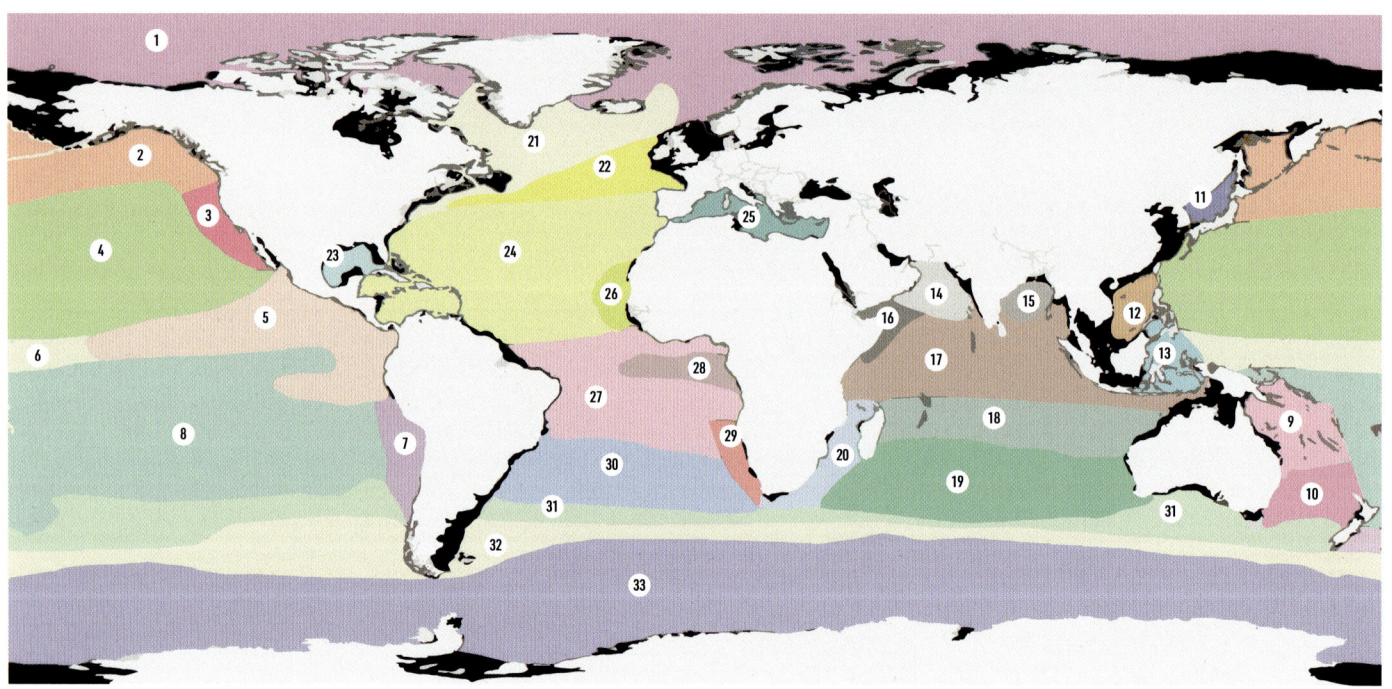

Marine ecoregions

An ecoregion classification for mesopelagic depths that takes into account the faunal composition and the environmental characteristics of the water masses yields 33 ecoregions compared with just 10 regions when the classification is based on water masses only. (Source: Sutton et al., 2017.)

1   Arctic
2   Subarctic Pacific
3   California Current
4   Northern Central Pacific
5   Eastern Tropical Pacific
6   Equatorial Pacific
7   Peru Upwelling/Humboldt Current
8   Southern Central Pacific
9   Coral Sea
10 Tasman Sea
11 Sea of Japan
12 South China Sea
13 Southeast Asian Pocket basins
14 Arabian Sea
15 Bay of Bengal
16 Somali Current
17 Northern Indian Ocean
18 Mid-Indian Ocean
19 Southern Indian Ocean
20 Agulhas Current
21 Northwest Atlantic Subarctic
22 North Atlantic Drift
23 Gulf of Mexico
24 Central North Atlantic
25 Mediterranean
26 Mauritania/Cape Verde
27 Tropical and West Equatorial Atlantic
28 Guinea Basin and East Equatorial Atlantic
29 Benguela Upwelling
30 South Atlantic
31 Circumglobal Subtropical Front
32 Subantarctic waters
33 Antarctic/Southern Ocean

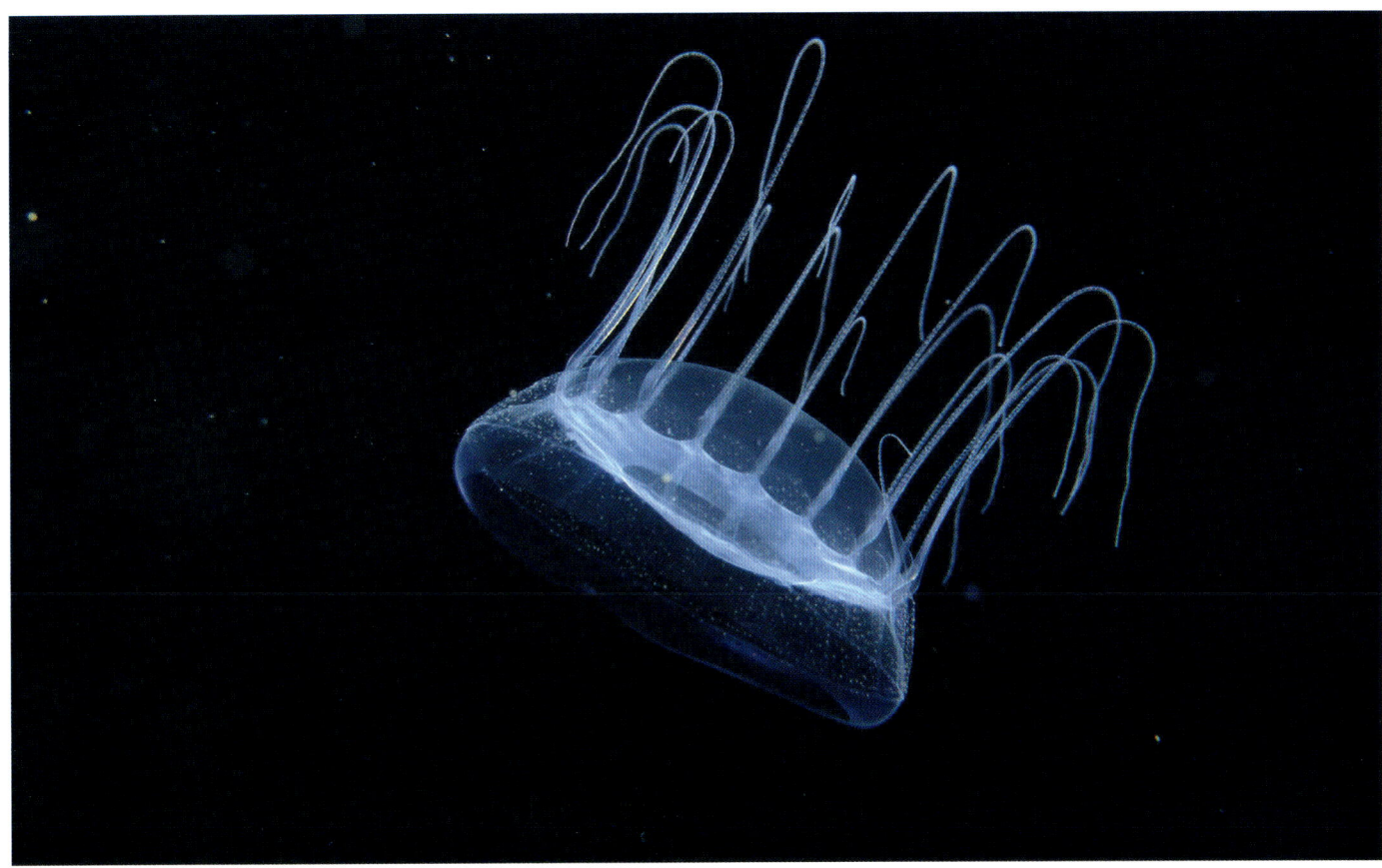

Endemic species

The Mediterranean endemic jellyfish *Solmissus albescens* is found from about 165 ft. to 3,280 ft. (50–1,000 m) depth. It is an active swimmer that migrates vertically at night, with swimming speeds estimated at up to 165 ft. (50 m) per hour.

Some of the additional subdivisions derived by adding in faunal data are intuitive. For example, even though water masses in the tropical and subtropical Indian, Atlantic, and Pacific Oceans may have similar properties, there are barriers to animal dispersal that prevent the same biological communities from being present in all three oceans. In geological history, there were connections between these oceans, and there are still dispersal routes around the south of South America and Africa for species sufficiently mobile and tolerant, but many species' distributions are restricted to just one ocean.

Enclosed and semi-enclosed seas, such as the Mediterranean and the Gulf of Mexico, are also physically separated from adjoining water bodies. The Mediterranean is uniquely species poor. It is a young sea, little more than 5 million years old, and is connected to the Atlantic Ocean by a narrow strait just 8 mi. (13 km) wide. Its mesopelagic fauna is mostly a subset of the Atlantic fauna, but it also has endemic species: examples include the jellyfish *Solmissus albescens* and the fish *Cyclothone pygmaea*. In contrast, the Gulf of Mexico is recognized as one of the most diverse and productive ecoregions. Fed by the Loop Current, which enters through the Caribbean Sea and exits through the Straits of Florida, the abundance and richness of mesopelagic fishes in the Gulf of Mexico is greater than in neighboring seas.

There are other reasons that the fauna in a particular area might be unique. The mesopelagic Eastern Tropical Pacific ecoregion is an oxygen-minimum zone (see pages 172–173), experiencing very low oxygen levels, and has a unique fauna adapted to this challenging habitat. The Humboldt Squid (*Dosidicus gigas*), an apex predator endemic to the eastern Pacific, has metabolic adaptations that allow it to live in the low-oxygen waters. Some species present in the Eastern Tropical Pacific are also distributed throughout the Equatorial Pacific ecoregion. Ecoregions do not represent hard divisions where the entire biological community changes; rather, the distribution of each species depends on its physiological adaptations, which govern how tolerant it is to various conditions, and its dispersal ability.

Benthic ecoregions

Scientists studying the seafloor found that most seafloor shallower than 2,625 ft. (800 m) was closely aligned to continental shelf, but the bathyal depths from 2,625 to 11,480 ft. (800–3,500 m) had a more interesting distribution, associated with mid-ocean ridges and seamount systems. Abyssal seafloor (11,480–21,325 ft./ 3,500–6,500 m) accounts for 65% of the total seafloor and for most of the habitat between seamounts and ridges, while hadal depths (>21,325 ft./6,500 m) account for a tiny proportion of the seafloor closely associated with continental plate boundaries. Applying expert judgment on faunal boundaries overlaid with environmental parameters, scientists have identified 14 bathyal, 14 abyssal, and 10 hadal ecoregions or "provinces."

There are limited data with which to test the accuracy of these benthic ecoregion delineations, particularly at the greatest depths, where species records are scarce. Scientists have recently emphasized the connection between even the deepest depths and surface waters— finding coconut shells and other terrestrial plant debris in hadal trenches—but many of the species observed at hadal depths are new and await taxonomic description before tests of biogeography can be conducted. At abyssal depths, data are lacking for a global analysis, but various studies on bivalves, polychaete worms, and crustaceans suggest a change in fauna at some of the proposed biogeographic boundaries.

Studies on demersal (bottom-dwelling) fishes around seamounts have found some support for the proposed bathyal regions, although limited sampling means most boundaries are not tested per se. Nonetheless, changes in demersal fish species support boundaries in the North Atlantic and between Australia and New Zealand, where the greatest amount of research has been conducted.

Benthic animals

Top: a deep-sea lizardfish on the continental slope (bathyal) in the western Atlantic Ocean; center: a sea cucumber *Psychropotes longicauda* on the abyssal plain in the Pacific Ocean; bottom: an acorn worm in a hadal trench in the east Indian Ocean.

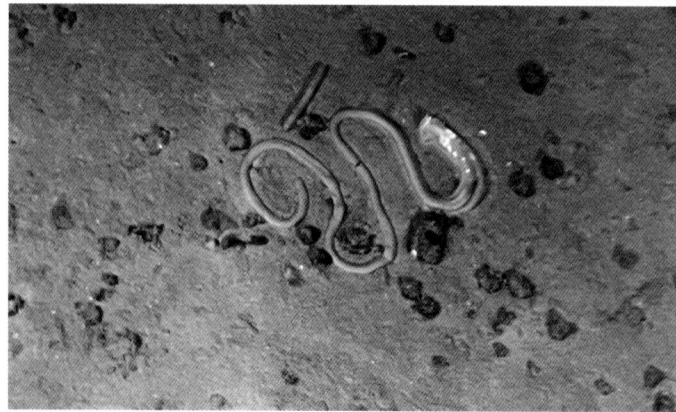

Bathyal provinces

Areas with bathyal depths are often widely separated, but even where physically connected, the fauna can be distinct due to changes in oceanographic conditions.

1 Arctic
2 Northern Atlantic Boreal
3 Northern Pacific Boreal
4 North Atlantic
5 Southeast Pacific Ridges
6 New Zealand–Kermadec
7 Cocos Plate
8 Nazca Plate
9 Antarctic
10 Subantarctic
11 Indian
12 West Pacific
13 South Atlantic
14 North Pacific

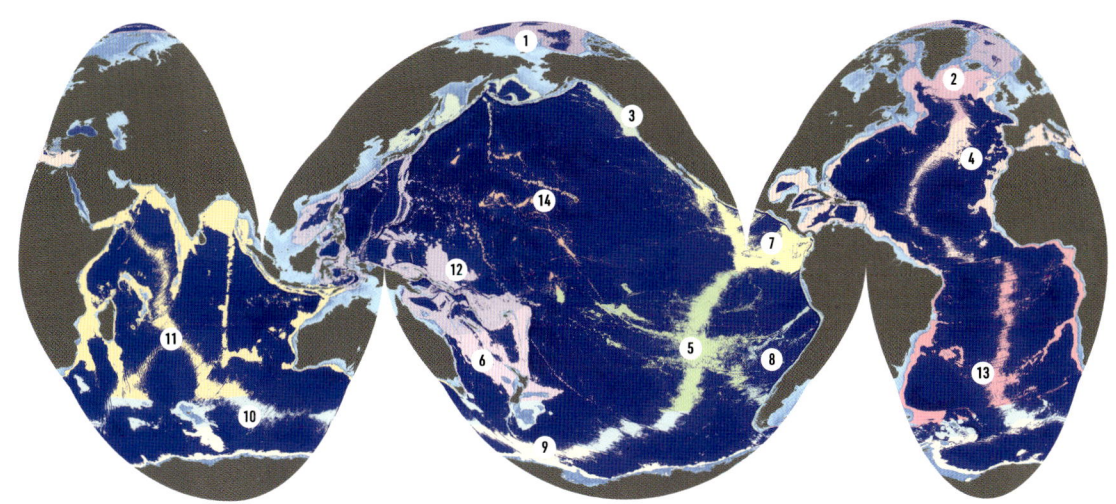

Abyssal provinces

While some abyssal ecoregions are separated by ocean ridges, many have no obvious physical boundaries, but experience different environmental conditions that determine the fauna that flourishes there.

1 Arctic Basin
2 North Atlantic
3 Brazil Basin
4 Angola, Guinea, Sierra
   Leone Basins
5 Argentine Basin
6 Antarctica East
7 Antarctica West
8 Indian
9 Chile, Peru, Guatemala Basins
10 South Pacific
11 Equatorial Pacific
12 North Central Pacific
13 North Pacific
14 West Pacific Basins

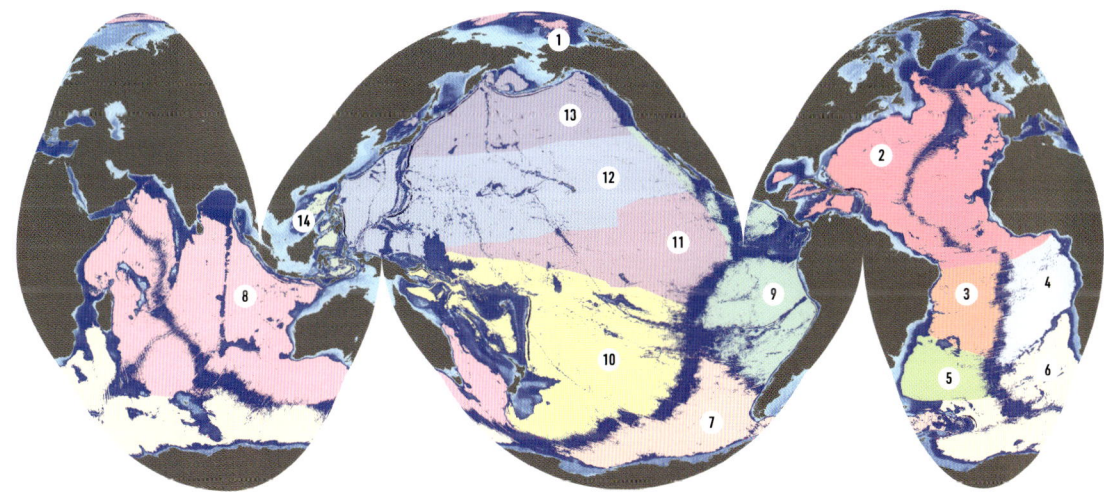

Hadal provinces

The hadal trenches, the very deepest parts of the ocean, are widely separated and little explored.

1 Aleutian–Japan
2 Philippine
3 Mariana
4 Bougainville-New Hebrides
5 Tonga–Kermadec
6 Peru–Chile
7 Java
8 Puerto Rico
9 Romanche
10 Southern Antilles

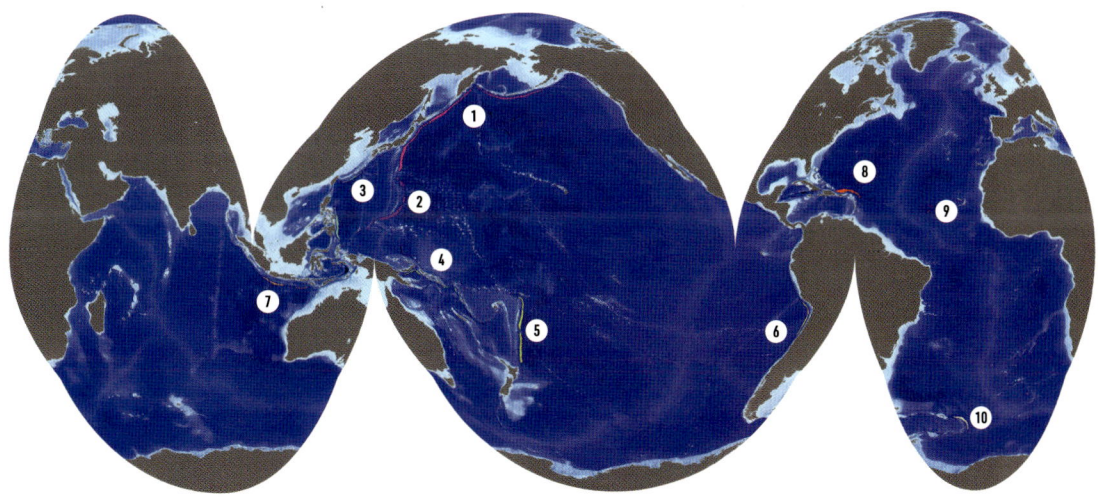

Hydrothermal vent ecoregions

Scientists have also proposed hydrothermal vent ecoregions, known as provinces, reflecting knowledge of the charicteristic fauna found at these oceanic communities across the globe. Because most species occur at more than one vent province, and only a few are unique to a particular location, putting an exact figure on the number of vent provinces is tricky: estimates range from 6 to 11, but haven't included data from Arctic Mid-Ocean Ridge vent fields such as Loki's Castle and Aurora.

There are clear visual differences among communities from different regions. Communities in the northeast Pacific, for example, are dominated by the smaller tube worm *Ridgeia piscesae*, whereas those farther south on the East Pacific Rise are dominated by the Giant Tube Worm (*Riftia pachyptila*). In contrast, the most abundant groups of animals at western Pacific vents are stalked barnacles (subphylum Crustacea), and limpets and sea snails (class Gastropoda). Atlantic vent fauna varies by depth, with shallow vents dominated by *Bathymodiolus* mussels, and deeper vents characterized by the eyeless vent shrimp *Rimicaris exoculata*. Recently discovered vents north of the Azores (e.g., Moytirra vent field) also have different communities, with *Mirocaris* shrimps and limpets most common. Communities on the Indian Ocean Ridge have a mix of Atlantic and western Pacific fauna, with shrimps, mussels, and gastropods present.

In the Southern Ocean, vents on the East Scotia Ridge are dominated by yeti crabs (*Kiwa tyleri*), stalked barnacles, limpets, and sea snails. Other species of yeti crabs are also known from cold seeps, but the genus *Kiwa* has a southerly distribution—known from the Galápagos Rift southward. In the Arctic, Loki's Castle vent field seems to comprise locally adapted fauna derived from Atlantic cold seeps and Pacific vents, and potentially represents a separate vent ecoregion.

Genetic studies of particular species show that there are additional barriers to connectivity in some locations. For example, populations of the Pompeii Worm (*Alvinella pompejana*), as well as those of Giant Tube Worms, are genetically different between the northern East Pacific Rise and the southern East Pacific Rise, supporting the idea of two separate biogeographic regions in this area. Understanding how this fauna evolved is challenging, but lost ancient connections between oceans likely played a role. The Panamanian Seaway—closed by the Isthmus of Panama around 3 million years ago—linked the Atlantic and Pacific Oceans. The Tethys Sea likely linked the Atlantic and Pacific in much earlier times (until c. 100 million years ago). The Kula oceanic ridge, now subducted below the Alaskan landmass, linked the western and eastern Pacific. The ephemeral nature of hydrothermal vents—each vent system having a brief life span (decades) before eruptions cease—has likely also played a role in determining the present-day distribution of species.

Vent fauna
Characteristic fauna of vents globally. Top left, skinny tube worms *Ridgeia* from the northeast Pacific; top right, a swarm of *Rimicaris exoculata* and *Mirocaris fortunata* shrimp from Rainbow hydrothermal vent on the Mid-Atlantic Ridge; center, mussels from the Mid-Atlantic Ridge off the coast of the Azores; bottom, vent barnacles from the western Pacific.

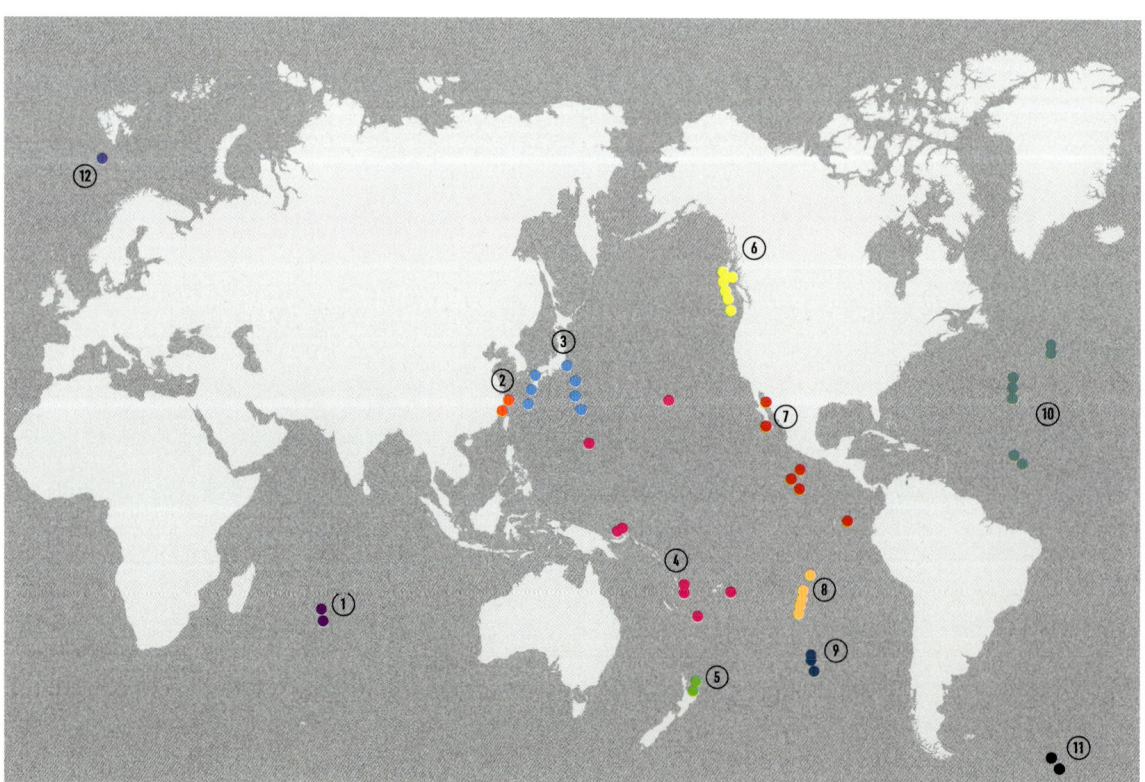

Hydrothermal vent ecoregions
Up to 11 vent ecoregions have been proposed based on analyses of the composition of faunal communities. The Arctic region has never been included in such analyses, but may represent a separate ecoregion.

1 Indian Ocean
2 Western Pacific
3 Northwest Pacific
4 Central Southwest Pacific
5 Kermadec Arc
6 Northeast Pacific
7 Northeast Pacific Rise
8 Southeast Pacific Rise
9 South of the Easter Microplate
10 Mid-Atlantic Ridge
11 East Scotia Ridge
12 Arctic Mid-Ocean Ridge

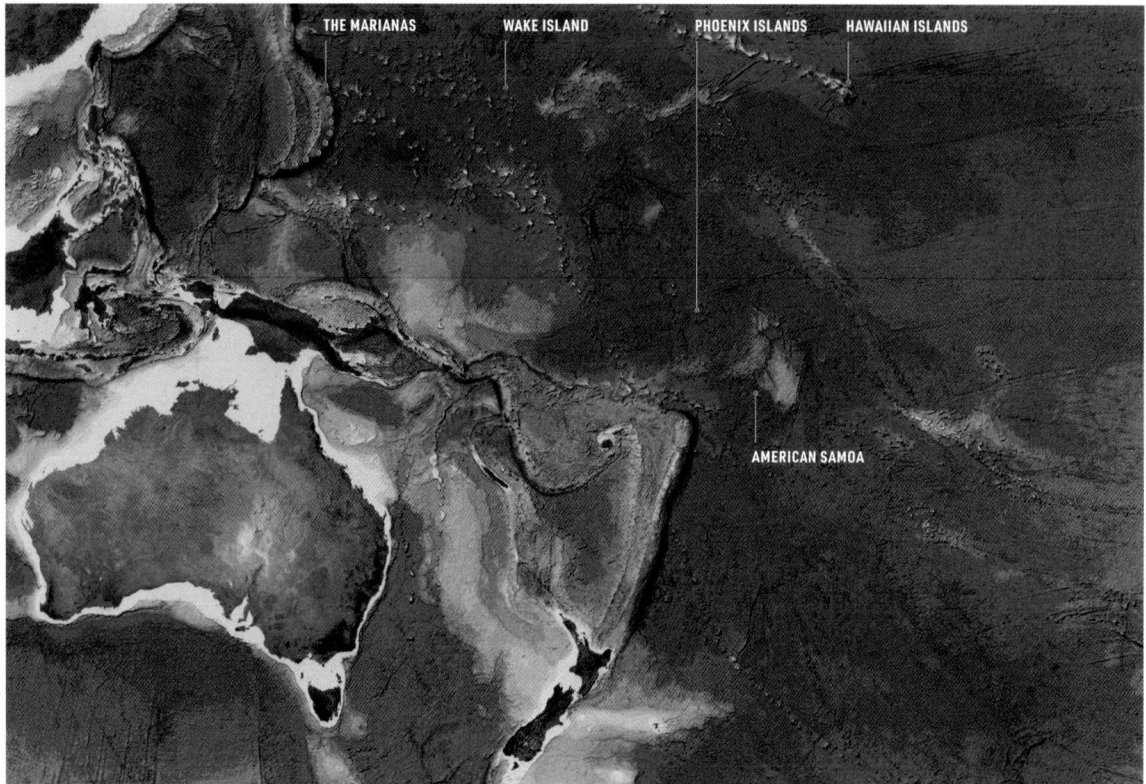

THE MARIANAS   WAKE ISLAND   PHOENIX ISLANDS   HAWAIIAN ISLANDS

AMERICAN SAMOA

Widespread coral genera
Left: many genera of coral
found during ROV dives at
these remote locations in
the western Pacific are also
known from the Atlantic
Ocean even though the
actual species may
be different.

How different is different?

Biogeographic analyses, such as those described in the
preceding pages, highlight connections and barriers
within and between our oceans, and they can also help
us understand just how much these regions vary. Some
regions, for example, the unique polar seas (see pages
196–211), can differ substantially from others, while
some regional differences are subtler.

In many cases the species compositions of communities
in biogeographic regions may differ, although the
communities may look remarkably similar. For example,
many deep-water coral genera are cosmopolitan and
have a nearly global distribution. Distinct species within a
genus may occupy discrete locations, but these species
play similar roles wherever they are found. Over a three-
year period, the National Oceanic and Atmospheric
Administration (NOAA) conducted 187 remote-operated
vehicle (ROV) dives around American Samoa, the Phoenix
Islands, the Hawaiian Islands, Wake Island, and the
Marianas. About 85% of the coral genera that they saw
most frequently (those spotted more than 1,500 times
on the ROV video) are also known from the Atlantic.

Among fishes, many of the same families are present
across the globe. There are endemic families, but
certain families, such as Myctophidae (lanternfishes),
dominate the mesopelagic zone. The demersal fish
fauna of seamounts is dominated by (among others)
roughies and redfishes (family Trachichthyidae), oreos
(Oreosomatidae), grenadiers (Macrouridae), and morids
(Moridae). While these families occur in all oceans, their
relative abundance varies, with roughies and oreos more
abundant in the Pacific, and grenadiers and morids more
so in the Atlantic.

A study of oceanic and deep-sea cephalopods
(class Cephalopoda) in the Atlantic, which explored
their diversity from north to south, found that the same
six squid families accounted for half the cephalopod
diversity everywhere except at polar latitudes. Many squid
species have extended distributions. Examples from the
glass squid family (Cranchiidae) include *Taonius pavo*,
whose distribution extends from about 50°N to 40°S in
the Atlantic, and *Sandalops melancholicus*, which has a
similar latitudinal extent but in the Atlantic, Indian, and
Pacific Oceans. In contrast, the glass squid *Teuthowenia
megalops* is confined to subarctic and northern
temperate Atlantic waters, and *Teuthowenia maculata*
to the tropical and subtropical Atlantic. So an overall
change in species composition occurs as biogeographic
boundaries are crossed, but it may be subtle.

Broad distributions

Top: the Orange Roughy
(*Hoplostethus atlanticus*) is a
long-lived deep-sea species
that has been heavily
overfished. These fish live
to over 140 years and don't
mature as adults for 30
years, meaning they are
slow to reproduce. Fishing
them has been compared
to mining because the
stocks are so slow to
replenish, you can only
fish them once. Bottom:
False Boarfish (*Neocyttus
helgae*)—a species of oreo,
here observed by ROV
around the summits
of the New England
seamount chain in the
northwest Atlantic.
It lives at 3,250–3,940 ft.
(990–1,200 m) depth.

THE POLAR DEEP SEA: ANTARCTICA

Unique conditions and highly specialized fauna

The connectivity of the Southern Ocean with the surrounding oceans, from which it is separated by the Antarctic Circumpolar Current, varies with depth (see page 187). Antarctic shelf waters are much colder than those of other continental shelves, whereas meso-, bathy-, and abyssopelagic waters each vary less. Hence, there are bigger differences between Antarctic shelf fauna and other shelf faunas and smaller differences between deeper-water faunas. The Antarctic shelf is also much deeper than other shelves. It has gradually deepened over the past 34 million years due to the weight of ice sheets and through erosion by glaciers. Its average depth is over 1,475 ft. (450 m), and it reaches 3,280 ft. (1,000 m) in places where glaciers have excavated deep basins. Thus, for the first time in this book, we consider the shelf waters and fauna part of the deep ocean.

Antarctica
At the southern extreme of our planet, Antarctica—isolated from the Pacific, Atlantic, and Indian Oceans by fronts and currents—offers a unique environment in which highly specialized fauna has evolved.

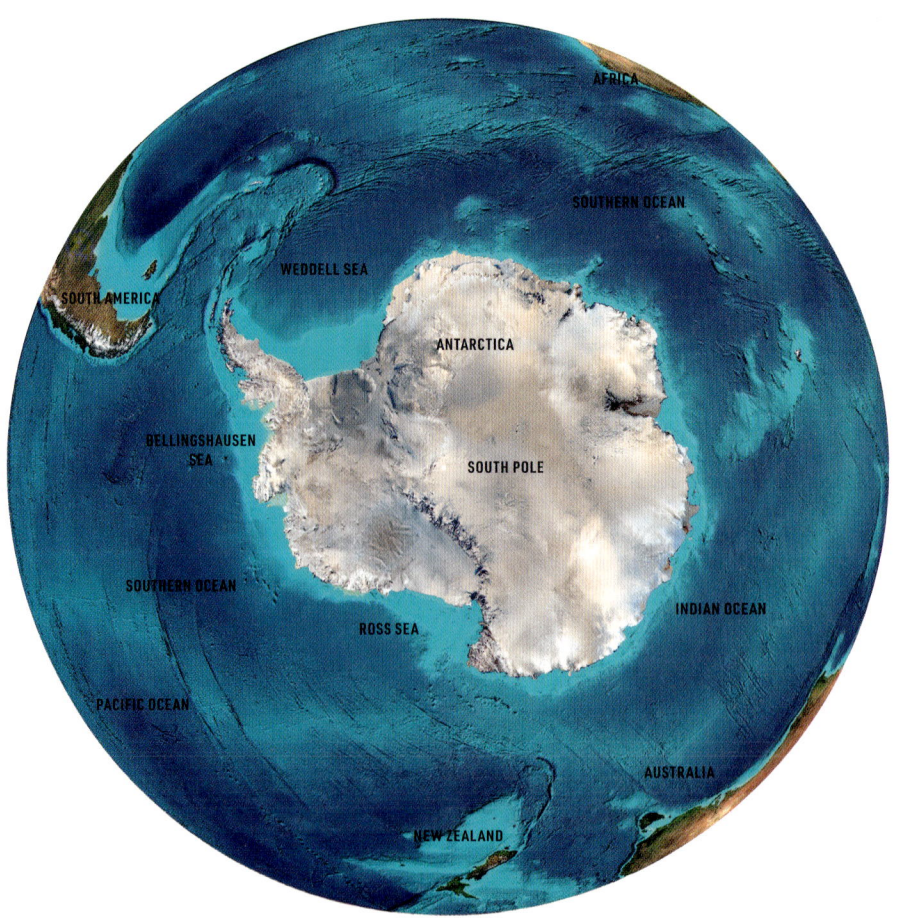

Antarctic invertebrates

The low temperatures have led to a highly specialized endemic fauna, which has evolved in situ since the Oligocene cooling began. Among invertebrates, 55% of sea spiders in the Southern Ocean are endemic, as are 75% of sea snails and 44% of sponges (for invertebrate classifications, see page 79). High endemism has also been noted in squids and octopuses. Some taxa are more species rich here than they are in other oceans, for example sponges and amphipods, while others, such as bivalves and isopods, are more species poor. Certain groups, for example decapod crustaceans, are almost completely absent, although fossil records indicate their presence in the Eocene prior to the onset of glaciations, and their decline reflects the rise of conditions unfavorable to their physiology. Gigantism (see also page 156) is seen in some groups—possibly due to the high oxygen saturation of the cold Antarctic waters.

The deepest parts of the shelf are typically dominated by mobile scavengers and deposit feeders. Elsewhere the fauna is dominated by filter and suspension feeders, and long-lived sponges form structurally complex habitats that provide substrate, refuges, and nursery grounds for other organisms. As in other oceans, there remain strong connections with the waters above. For example, epipelagic krill (*Euphausia superba*) produce thousands of eggs that sink to the seafloor, some hatching at depths of several thousand meters, with larvae slowly ascending the water column as they develop through a series of instars (stages) and molts.

Pycnogonids

Top: some 20% of all known sea spider species are found around Antarctica.

Amphipods

Upper center: hyperiid amphipods are pelagic predators, and use their large eyes to locate their prey.

Gigantism

Lower center: the giant Antarctic isopod *Glyptonotus antarcticus* grows to 3–4 in. (7–10 cm) in length, and occasionally larger. Most isopods are less than 0.5 in. (1 cm) long.

Suspension feeders

Bottom: sessile corals and sponges (suspension and filter feeding respectively) form a thicket on the Antarctic seafloor. Other suspension feeders, such as brittle stars and feather stars, sit atop these, taking advantage of an elevated position in the currents.

Following pages:

Southern Ocean amphipods

Endemic radiation has led to an abundance of highly diverse amphipod species in the Southern Ocean.

Antarctic fishes

The Southern Ocean has a unique fish fauna. Perhaps most famous are the crocodile icefishes of the family Channichthyidae. The only vertebrates that lack hemoglobin, these fishes rely on the high oxygen saturation of the cold Antarctic waters and their low metabolic rates to survive, moving little and ambushing their prey. Crocodile icefishes are part of a larger group of fishes, notothenioids (suborder Notothenioidei), that dominate Southern Ocean shelf and slope waters. Nearly all demersal, they lack a swim bladder, which would give them buoyancy in the water column. Their evolutionary radiation in the Southern Ocean was facilitated by the evolution of antifreeze glycoproteins that have given them an adaptive advantage in the frigid Antarctic waters.

Although notothenioids are not wholly deep-water species, many have a depth distribution that extends to 2,625–3,280 ft. (800–1,000 m). The toothfishes (genus *Dissostichus*), also known as cod icefishes, extend even deeper, to beyond 6,560 ft. (2,000 m). They are heavily fished and often sold as "Chilean sea bass."

There are few species of pelagic notothenioids. The distinctive Antarctic Herring (*Pleuragramma antarcticum*), which achieves neutral buoyancy through a reduced skeleton, is the most abundant. It has a wide depth distribution from near-surface waters to around 2,300 ft. (700 m). Other fish families dominate the meso- and bathypelagic zones. Lanternfishes (family Myctophidae) are common, but few species are endemic to Antarctic polar waters. In contrast, eelpouts (family Zoarcidae), which reached the Southern Ocean from the North Pacific during the Miocene, have undergone a recent radiation, leading to many endemic species. Snailfishes (family Liparidae) are mostly benthic, commonly found on the seafloor at depths of 985–6,560 ft. (300–2,000 m), but abyssal species in the genera *Paraliparis* and *Notoliparis* extend to depths below 16,400 ft. (5,000 m).

Several families common in other oceans, for example the grenadiers (family Macrouridae) and morids (family Moridae), are present in the Antarctic but are low in diversity. Although it's often said that sharks are absent from the Antarctic, there are a few small species of lantern sharks (family Etmopteridae) in deep water, and a handful of ray species (order Rajiformes) occur on the Southern Ocean continental slope. However, fossils indicate a greater presence of sharks up to the Eocene, prior to the Oligocene cooling, as with decapod crustaceans (see page 197).

Crocodile icefish
Chaenocephalus aceratus

Top left: this species of crocodile icefish inhabits benthic waters around the Antarctic Peninsula and Scotia Sea, where, as an adult, it mostly feeds on other small species of icefish. They grow to about 1 ft. (0.3 m) long, with females growing slightly larger.

Rock Cod
Notothenia sp.

Top right: there are several species of the genus *Notothenia* in the Southern Ocean, which vary in size and length. Also benthic, they feed mostly on marine invertebrates and other fishes.

Snailfish
Paraliparis sp.

Bottom right: snailfishes are soft and without scales and are abundant in very deep waters worldwide, including in Antarctica.

Antarctic megafauna

As emphasized previously (see pages 196–197), there are strong vertical connections between shallow- and deep-water ecosystems. Nowhere is this better illustrated than in the Southern Ocean by the region's deep-diving animals.

Emperor Penguins (*Aptenodytes forsteri*) can hold their breath for more than 30 minutes and dive to 1,640 ft. (500 m) to forage on Antarctic Herring, krill, and mesopelagic squids such as *Psychroteuthis glacialis* and *Moroteuthopsis longimana*. They dive mostly in the epipelagic zone (0–660 ft./0–200 m), but anatomical and physiological adaptations such as solid bones (in contrast to the hollow bones that facilitate flight in most birds) and specialized hemoglobin facilitate deeper forays. The penguins tend to make these deeper dives during the day, relying on sight to find their prey. Southern Ocean water is much clearer than that of other oceans: a standardized visual Secchi disk (used to measure water transparency) can be seen at 260 ft. (80 m) depth in the Weddell Sea—as compared to 16.4–33 ft. (5–10 m) in the North Sea. Adélie Penguins (*Pygoscelis adeliae*) and King Penguins (*Aptenodytes patagonicus*) also occasionally dive below 660 ft. (200 m) depth but only to the very upper regions of the mesopelagic zone.

Several Southern Ocean seals—for example, the Crabeater Seal (*Lobodon carcinophagus*) and the Antarctic Fur Seal (*Arctocephalus gazella*)—also foray into the upper mesopelagic zone. More impressive, Weddell Seals (*Leptonychotes weddellii*) can dive for longer than an hour and reach depths below 1,970 ft. (600 m) in their hunts for squids and fishes, while the deepest dives of the Southern Elephant Seal (*Mirounga leonina*), which regularly reaches 1,640 ft. (500 m), can extend below 3,280 ft. (1,000 m).

The Southern Ocean provides rich foraging grounds for many whale species. These include known deep divers such as Sperm Whales (*Physeter macrocephalus*), which can spend as long as two hours underwater in their search for squids and toothfishes, reaching depths of 6,560 ft. (2,000 m). Arnoux's Beaked Whale (*Berardius arnuxii*), found from temperate Southern Hemisphere waters to the high Antarctic, makes hour-long dives, although the depth reached by this little-studied species is unknown.

Crabeater Seal
Lobodon carcinophagus

Crabeater Seals are one of the most abundant seal species. They feed predominantly on krill and have specially shaped (multicuspid) teeth that strain krill from the seawater. Although they mostly feed in the top 165 ft. (50 m) of the water column, dives can last longer than 20 minutes and reach depths below 1,640 ft. (500 m).

Emperor Penguin
Aptenodytes forsteri

There are around quarter of a million breeding pairs of Emperor Penguins in very high latitudes of the Antarctic. Their lifestyle is closely associated with the sea and pack ice and they are considered vulnerable to climate change.

Secchi disk
A researcher lowers a Secchi disk with contrasting black and white sectors into the ocean to measure light penetration.

Weddell Seal
Leptonychotes weddellii

More than half a million Weddell Seals inhabit the waters of Antarctic and associated subantarctic islands. They grow to around 9 ft. (2.75 m), and live for up to 25 years.

THE POLAR DEEP SEA: ARCTIC OCEAN

An isolated basin, sea ice, and marine mammals

The Arctic Ocean is characterized by a deep central basin with many shallow marginal areas, some very broad. Even these shallow areas have conditions similar in some ways to those in the deep sea. The entire ocean is permanently cold, with little temperature stratification of the water. It is also dark for very long periods because of both the polar winter north of the Arctic Circle and the shading under the extensive snow-covered sea ice.

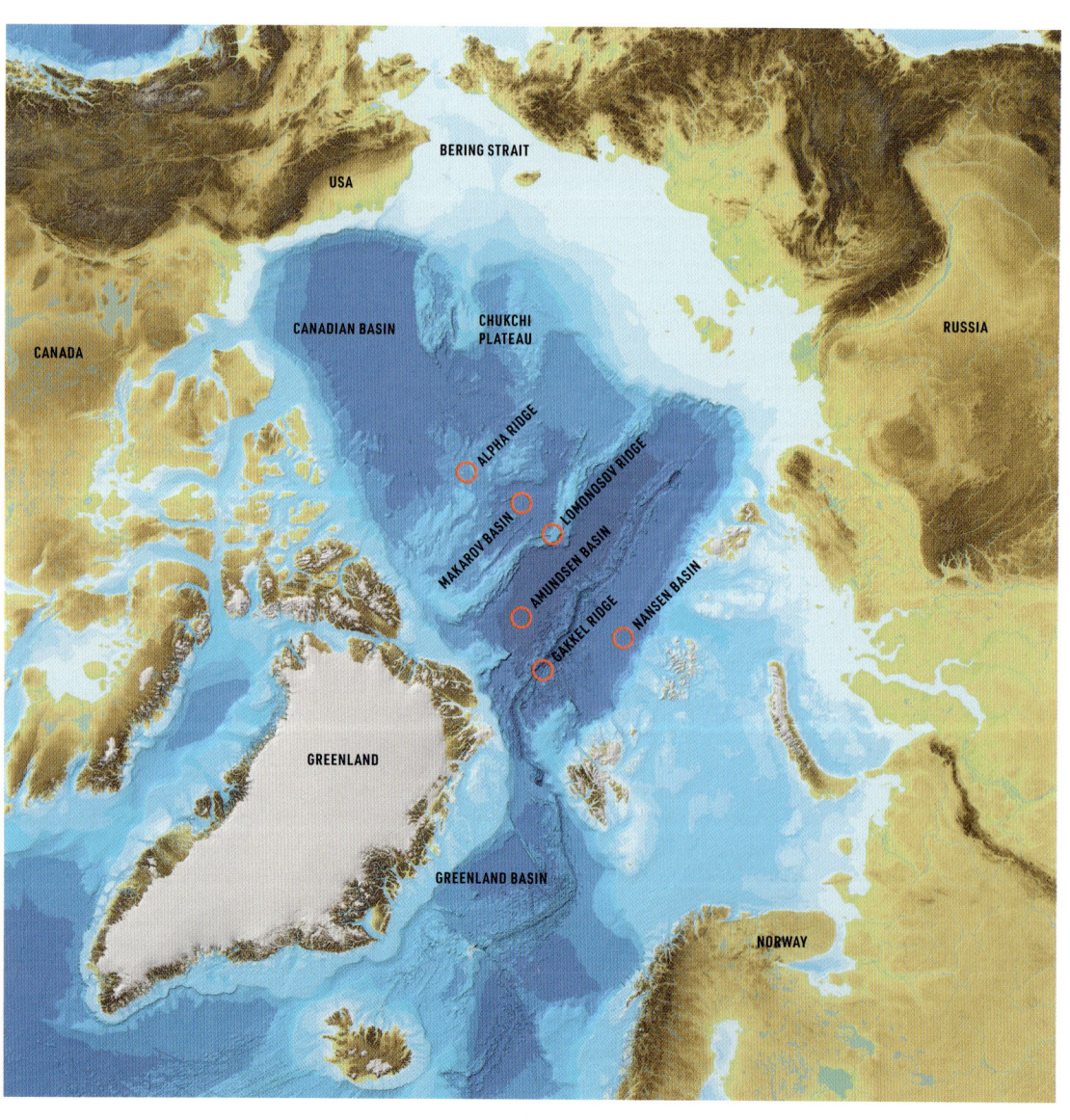

Arctic exploration

Top: a research vessel conducting ROV operations in Arctic sea ice at midnight. Because of 24-hour sunlight in the summer and darkness in the winter, the dynamics of light penetration in polar waters is very different than in most of the world's oceans.

Arctic Ocean basins

Left: unlike the Southern (Antarctic) Ocean, the Arctic Ocean is surrounded by land, with very limited connections to the other oceans. The continents drain fresh water into it. Greenland is covered by thick glaciers, somewhat like a mini-Antarctica, which shed icebergs, including the one that sank the *Titanic*. The deep basins of the Arctic are separated by parallel ridges.

Such conditions are also reminiscent of the Southern Ocean around Antarctica (see pages 196–197). Indeed, many people think the polar regions are very similar, but important differences distinguish them. Whereas the Antarctic is land surrounded by ocean, the Arctic is ocean surrounded by land. Connections between the Arctic Ocean and other oceans are very limited. The Bering Strait, connecting it to the Pacific between North America and Asia, is both narrow and shallow. The only extensive connection, which is also the major deep-water connection, is with the Atlantic between Greenland and Scandinavia. The Antarctic Circumpolar Current separates the Southern Ocean from the Atlantic, Pacific, and Indian Oceans; the Arctic, in contrast, doesn't have a circumpolar current. Surface waters flow through the Arctic from the Bering Strait toward the Atlantic, but deep flow in the other direction is very limited by ridges that divide the deep Arctic into several distinct subbasins. The Canadian Basin is among the most isolated areas of the deep sea. Unlike the broad, shallow shelves, the deep, isolated, ice-covered basins of the Arctic are typically very low in productivity and biodiversity, both pelagic and benthic. While humans have occupied and exploited the Antarctic for only two centuries, they have been permanent residents of the Arctic, harvesting marine resources, for millennia.

Ice dynamics

In the Arctic, ice dynamics are dominated by sea ice, which is formed from seawater (in contrast to glacial icebergs, formed from precipitation). In contrast, the Southern Ocean has, in addition to sea ice, extensive ice shelves that calve, forming flat-topped icebergs. Icebergs can drag across the bottom in the Antarctic, whereas in the Arctic, the primary disturbers of benthic habitat are marine mammals.

Historically, much of the Arctic sea ice has survived through the summers and built up over multiple years. This is notably important for the Polar Bears (*Ursus maritimus*) that hunt on the ice and seals that breed on it, but also for the ice-associated community, ranging from ice algae to grazing amphipods to Polar Cod (*Boreogadus saida*) living under, and around, the ice. The ice algae from one-year-old ice are a major component of the photosynthetic primary production in the Arctic. As is seen elsewhere, the unique communities of deep-living animals, both pelagic and benthic, are dependent on patterns of primary production at the surface. These patterns are changing rapidly in the Arctic, because it is one of the most rapidly warming areas of the planet as it undergoes climate change. Although still essentially ice-covered in winter, the Arctic Ocean is becoming progressively more ice free by the end of summer, with a resulting shift from ice algae to phytoplankton production and follow-on changes to pelagic food webs and to delivery of food to benthic habitats.

When sea ice forms, it is as frozen fresh water because, as water molecules form the ice, salt is excluded, producing a salty brine of the water that has not yet frozen. The brine becomes progressively saltier, until it is so salty it will not freeze and is much denser than normal seawater. The dense brine sinks into deep water, leaving freshwater ice behind. When the sea ice melts, it results in a layer of relatively fresh water confined to the surface. Changes in sea-ice dynamics are therefore causing changes in salinity stratification of the Arctic water column, affecting the vertical and horizontal movement of phytoplankton, larvae, ice-associated animals, and vertically migrating pelagic animals as well as the water movements.

Sea ice

Arctic sea ice forms when seawater freezes, but the salt is removed by the process, resulting in freshwater ice. The light blue areas in the foreground are pools of fresh water formed when the ice melts in the summer sun.

Animals of the deep Arctic Ocean

Knowledge of Arctic biodiversity, especially for the deep sea, increased substantially during the first decade of the 21st century because of the Arctic Ocean Diversity Project (ArcOD) of the global Census of Marine Life. Deep-sea diversity is generally quite low, especially in the water column. An inventory of benthic species found slightly fewer than 1,000 multicellular species, mostly crustaceans, as well as polychaete and nematode worms. In contrast, a review of zooplankton found 174 oceanic species, most of which were crustaceans. Some other taxonomic groups, such as gelatinous megaplankton and swimming snails, can be very abundant. Some species, like the planktonic pteropod snail *Limacina helicina*, were once considered to be bipolar in distribution, but modern molecular sequencing has shown that distinct, although morphologically quite similar, species are found in the Arctic and Antarctic. In contrast, the fish fauna of the two regions is very different. Whereas Antarctic fishes are dominated by the unique and characteristic notothenioids (icefishes; see pages 200–201), the Arctic marine fishes are members of groups found in the North Atlantic, such as cods (order Gadiformes), snailfishes (family Liparidae), eelpouts (family Zoarcidae), flatfishes (order Pleuronectiformes), and sculpins (family Cottidae). The few Arctic endemic deep-sea species, such as the Glacial Eelpout (*Lycodes frigidus*), are members of those more widely distributed taxonomic groups.

Despite extreme seasonal changes in sunlight, diel vertical migration has been documented in the Arctic (and the Southern Ocean), even in winter. However, seasonal vertical migration is also very important. Many planktonic species spend the winter in deep water in a resting phase of their life cycle known as diapause. These species resume activities and move up in the water concurrently with the stimulation of primary productivity by the seasonal return of sunlight.

Because of a long history of hunting, both indigenous and from distant bases, marine mammals are among the best-known populations of Arctic animals. Truly Arctic mammals include a few seal species, the Walrus (*Odobenus rosmarus*), and the extremely long-lived

Bowhead Whale (*Balaena mysticetus*). Walruses are a major force in bioturbation (stirring or disturbance of sedimentary deposits by animals) of the Arctic sea bottom. Two of the strangest marine mammals, the Narwhal (*Monodon monoceros*) and Beluga (*Delphinapterus leucas*), are endemic to the Arctic, where they forage in deep waters for fishes, squids, and shrimps. The top predator of the Arctic is also a marine mammal, the Polar Bear. Although a swimmer, it hunts on top of the ice, which may not seem like a deep-sea phenomenon. Among its prey, however, are the deep-feeding Narwhal and Beluga, in addition to seals.

Atlantification of the Arctic

A problem with inferring biological changes in the deep Arctic is that it was very poorly studied before the current period of rapid climate change—no baseline exists against which to compare current conditions. In addition to endemics, Arctic marine fauna has long included some species more characteristic of the North Pacific (in shallow areas near the Bering Strait) and North Atlantic (at various depths near the Norwegian and Greenland Seas). These natural invaders have generally been considered to be nonreproductive in polar conditions. However, as global oceans have warmed and ice dynamics have changed, conditions have become more favorable for these southerly species. Especially around the more extensive connection with the Atlantic, there has been an increase in such species as well as evidence of reproduction. This process, called "Atlantification," has major potential ramifications for the Arctic, as competitors and predators of endemic species move in. An extreme example of this is increasing observations of Orcas (*Orcinus orca*) in Arctic waters, a trend that could be very bad news for the Narwhals and Belugas.

Sea Butterfly
Limacina helicina
The Sea Butterfly is a swimming snail known as a pteropod. This species is a major component of the Arctic zooplankton.

Walrus
Odobenus rosmarus
Once thought to be strictly shallow divers, Walruses have been shown to dive as deep as 1,640 ft. (500 m). Their foraging stirs up the bottom and disturbs benthic communities in addition to their prey of bivalves and other large invertebrates.

Bowhead Whale
Balaena mysticetus
Bowhead Whales, named for their curving upper jaw, have long been hunted by indigenous peoples and distant-water whalers. Some evidence indicates that, if not killed by humans, they may live for hundreds of years.

Beluga
Delphinapterus leucas
The white Beluga whales are sometimes called "canaries of the sea" because of the diverse patterns of sounds they make, including clicks, whistles, chirps, and squeals.

VERTICAL ZONATION OF BENTHIC ANIMALS

Changing life-forms from the shallows to the deep

From the earliest studies of the deep sea, it was apparent that there is vertical zonation in the occurrence of animal life, with different species living at different depths. Strange animals, such as stalked crinoids or sea lilies (phylum Echinodermata, class Crinoidea), previously only known from fossils, were dredged up from the depths. During the early 1800s, the naturalist Antoine Risso cataloged the natural resources of the Duchy of Savoy from the summits of the Alps to the coast and into the depths of the Mediterranean Sea. He observed how life changed with elevation, from lichens growing on bare rock on mountaintops to increasing vegetation along the tree line, and in meadows in the valleys to the rich variety of life-forms on the coast. Continuing underwater, down the continental slope into the abyss off the city of Nice, he found unique species at each depth zone. Zonation of species by elevation above sea level and depth below sea level is one of the fundamental properties of life on planet Earth. Abundance and diversity decrease with distance above or below sea level, becoming sparsest on the tops of mountains and at the greatest depths in the seas. In the ocean, zonation is driven by basic factors that change with depth: light, pressure, temperature, oxygen, and food supply.

Light

Availability of solar light decreases rapidly with depth, so that beyond 660 ft. (200 m), there is insufficient light to sustain photosynthesis. This defines the upper limit of the deep sea, below which algae, seaweeds, and phytoplankton cannot survive. Shallow-water corals in tropical regions have symbiotic algae cells (zooxanthellae) that provide most of their food requirements by photosynthesis. In the deep sea this is not possible, and corals living there are pure carnivores, capturing prey and food particles on their stinging tentacles. Some solar light penetrates to a maximum depth of around 3,280 ft. (1,000 m), and although this is insufficient for photosynthesis, it can be seen by the sensitive eyes of deep-sea animals. Bottom-living fishes, cephalopods, and crustaceans use this low light to hunt and capture prey. Many animals exhibit bioluminescence—the ability to emit light. Soft corals, such as gorgoniid whip corals (family Gorgoniidae) and bamboo corals (family Keratoisididae), and sea pens (order Pennatulacea) do not normally glow in the dark, but do emit patterns of blue light when disturbed by physical contact. Why they do this is not known, but it has been suggested that the light may act as a "burglar alarm," either confusing predators or attracting larger predators that might attack the primary predator. Beyond 3,280 ft. (1,000 m) depth, light is unlikely to influence zonation of animals, as the only light available derives from bioluminescence.

Stalked crinoid
Anachalypsicrinus sp.

A sea lily found at
3,940 ft. (1,200 m) depth
in the northwest Atlantic
Ocean off Florida.

**Bioluminescence in
bamboo coral**

A treelike branching coral
found on continental
seamounts and island
slopes. This specimen
from 2,130 ft. (650 m) depth
is emitting blue light in
response to stimulation.

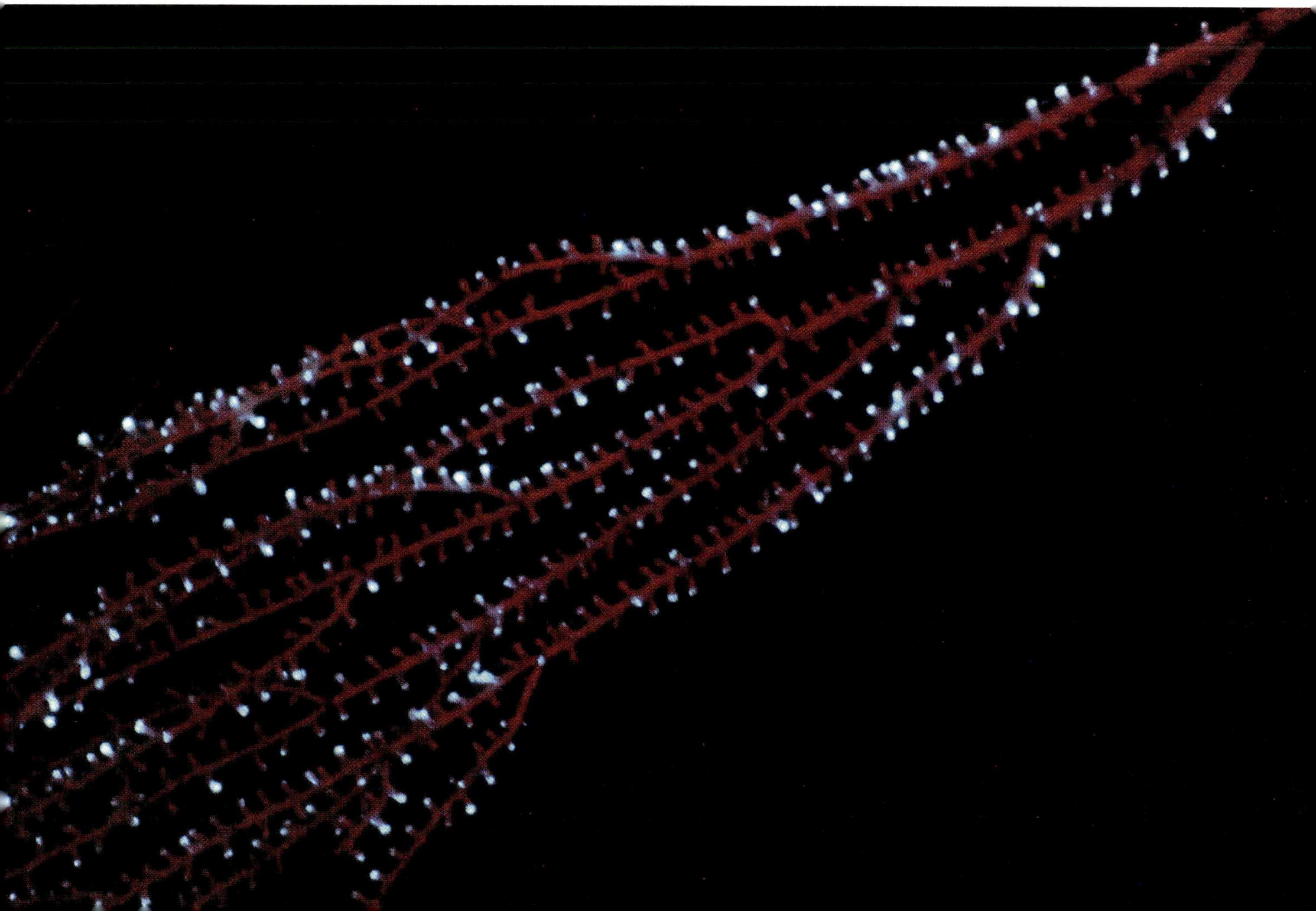

Pressure

The deep sea is subject to high pressures, increasing by approximately 1 atmosphere per 33 ft. (10 m) of depth, owing to the weight of water above. Despite the great pressure (400 atmospheres at average ocean depth), deep-sea animals are not crushed out of shape because their bodies are filled with fluids, which do not change in volume under pressure, unlike air or gases. Animals with gas- or air-filled buoyancy organs cannot withstand great pressure. The cephalopod mollusk *Nautilus* has a rigid air-filled shell that implodes at about 70 atmospheres, killing the animal. Consequently, *Nautilus* is restricted to depths above 2,300 ft. (700 m) and feeds mainly at 985–1,150 ft. (300–350 m), on slopes of oceanic islands and seamounts. In those bony fishes that have a swim bladder, the gas simply decreases in volume with no injury to the fish other than loss of buoyancy. However, pressure has invisible effects on animal function. A shallow-water species, if exposed to 30 atmospheres, experiences nerve function impairment, and pressures above 100 atmospheres may kill it. To survive in the deep sea, creatures must either have adaptations to the structure of key biomolecules to resist pressure, or they may accumulate protective chemicals known as piezolytes. Modification of biomolecules involves the stiffening of the molecular structure with additional cross bonding, which has the side effect of slowing down the rate of chemical processes. Thus, muscles of animals adapted to high pressure contract slowly, and metabolic rate is low. This is an excellent adaptation to the deep sea, but these creatures cannot compete with fast-moving species at shallower depths. Accumulation of piezolytes also has limits. The concentration required increases with depth and eventually reaches a maximum, beyond which further increase is not possible. Adaptations to pressure therefore restrict species to preferred depth zones where their functioning is optimal.

A distinct effect of pressure is its influence on solubility of calcium carbonate, an important constituent of hard parts of animals, including shells of mollusks, exoskeletons of crustaceans, plankton shells, and structures of corals, sponges, echinoderms, and fishes. At the sea surface, calcium carbonate forms a strong, solid insoluble material, but at about 11,480 ft. (3,500 m) depth it becomes soluble in seawater. At the pressures prevailing beyond this so-called carbonate-compensation depth, the rate of dissolving begins to exceed the rate of supply of calcium carbonate, and it becomes increasingly difficult for animals to maintain a hard shell. At hadal depths (exceeding 19,700 ft./6,000 m), the deepest-living sea snails indeed have soft shells, and reef-building corals cannot survive. The carbonate-compensation depth is variable, occurring at around 13,125 ft. (4,000 m) in the Pacific Ocean and 16,400 ft. (5,000 m) in the Atlantic.

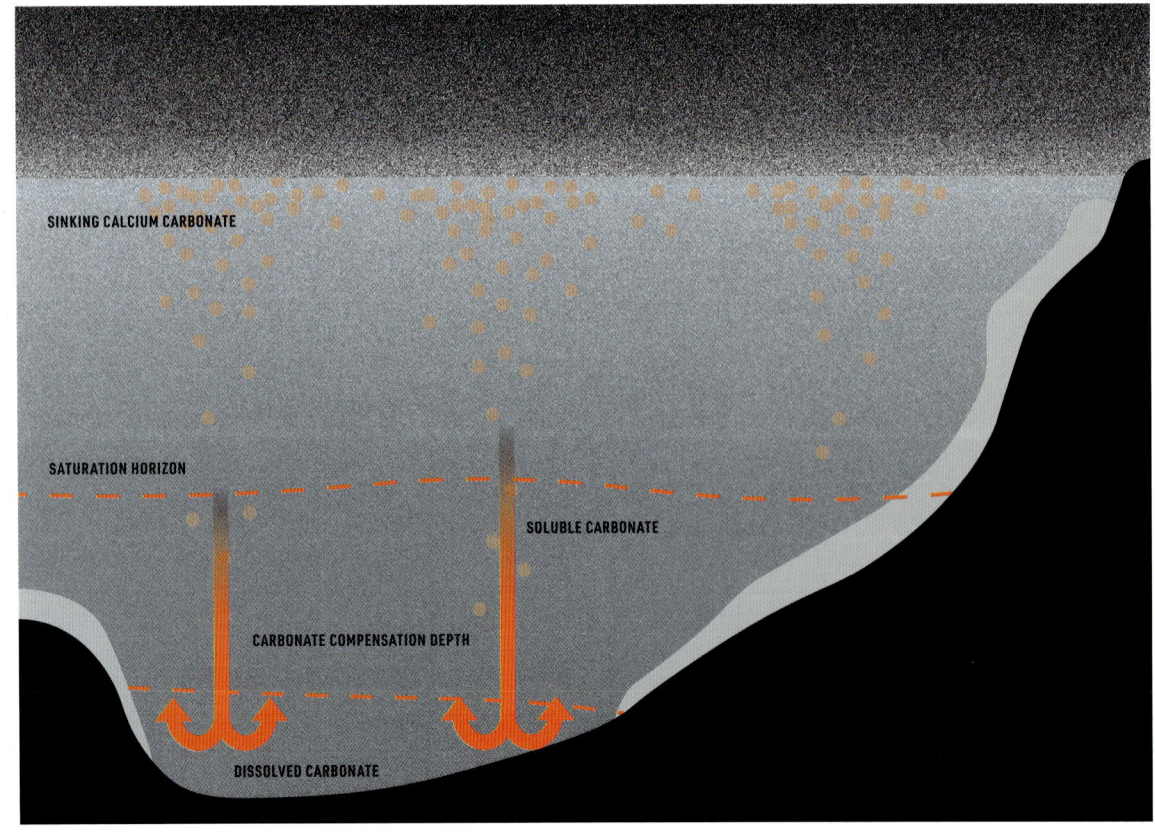

SINKING CALCIUM CARBONATE

SATURATION HORIZON

SOLUBLE CARBONATE

CARBONATE COMPENSATION DEPTH

DISSOLVED CARBONATE

Carbonate compensation

Calcium carbonate, which is essential for animals to maintain a hard shell, starts to dissolve beyond the carbonate-compensation depth.

Chambered Nautilus
Nautilus pompilius
The rigid chambered
shell is filled with air that
keeps the animal neutrally
buoyant down to 2,300 ft.
(700 m) depth.

Anoxic conditions
An oxygen-minimum
zone (OMZ) can form
at intermediate depths
between oxygenated
surface waters and
oxygenated deep waters.

Temperature and oxygen

In the deep sea, temperature is generally rather
constant, typically less than 39°F (4°C) at depths greater
than 4,920 ft. (1,500 m). In tropical regions, deep-living
species avoid the warm surface-water layers, but in
polar regions they can be found nearer the surface. Low
temperature is not an essential feature for deep-sea life.
Throughout the history of the oceans, the deep sea has
alternated between warm (59°F/15°C) and cold (36°F/2°C)
periods; the last warm period ended 70 million years
ago, which is recent in evolutionary terms, so most of
the modern deep-sea life originated in relatively warm
conditions. The Mediterranean Sea remains warm, close
to 57°F (14°C) all the way down to maximum depth, and
has the typical zonation of animal life as seen in the
major oceans that are all cold at depth.

The great depths of the ocean are well oxygenated, with
dense cold water that sinks from the surface layers in
polar regions, so the deepest animals live with plenty of
oxygen for respiration. Problems occur at intermediate
depths in productive regions where there is high fallout
of organic matter from the surface. This results in a high
rate of oxygen consumption by microbes and animals
utilizing this bonanza, lowering the dissolved oxygen
concentration to nearly zero, creating an oxygen-
minimum zone (OMZ; see pages 172–173). OMZs are
found predominantly in the eastern Pacific Ocean,
the tropical eastern Atlantic, and monsoon areas of
the northern Indian Ocean, at depths of 490–4,920 ft.
(150–1,500 m). In extreme cases, there may be a
dead zone totally devoid of animal life. More often,
the OMZ has reduced abundance of animal life and
is dominated by creatures specially adapted to anoxic
(low-oxygen) conditions.

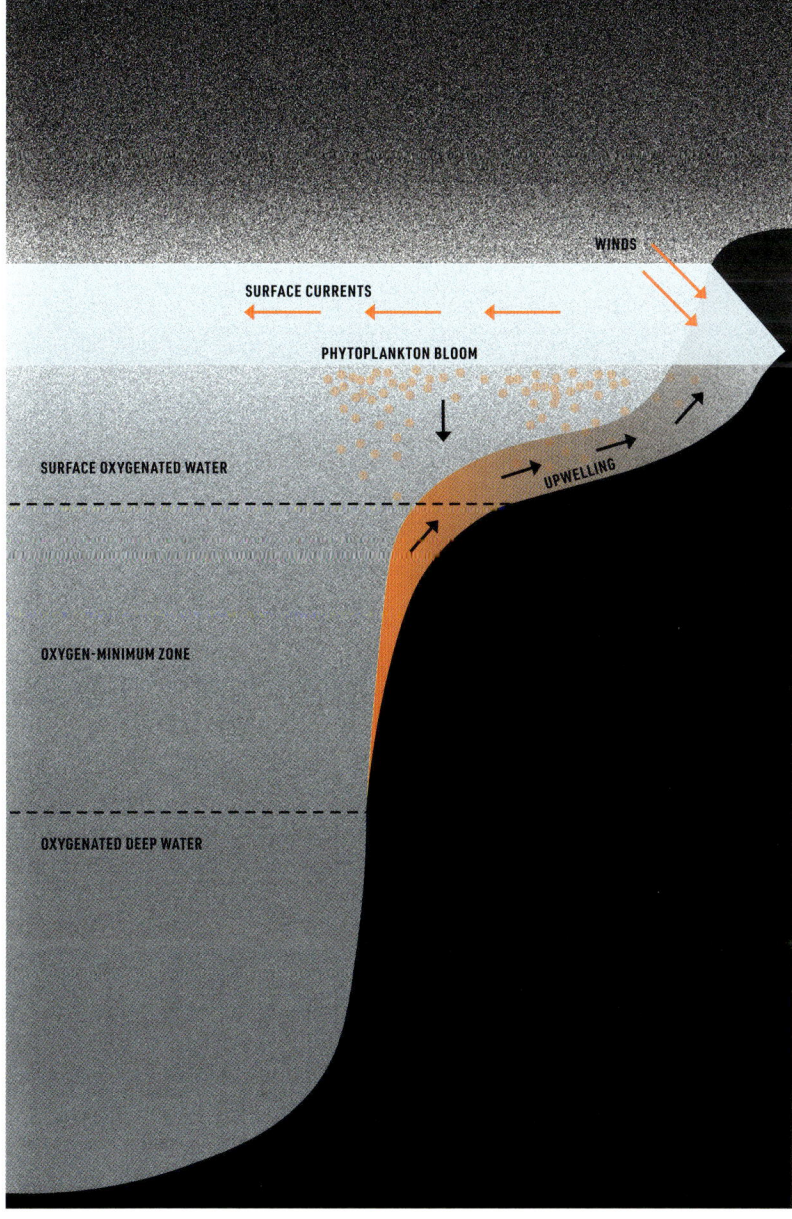

WINDS

SURFACE CURRENTS

PHYTOPLANKTON BLOOM

UPWELLING

SURFACE OXYGENATED WATER

OXYGEN-MINIMUM ZONE

OXYGENATED DEEP WATER

Food supply

The overwhelmingly dominant determinant of depth zonation is the gradual decrease in food supply, by approximately tenfold for every 6,560 ft. (2,000 m) of descent. Thus, at 6,560 ft. (2,000 m) depth, the food supply is 10% of the surface value, at 13,125 ft. (4,000 m) 1%, and at 19,700 ft. (6,000 m) 0.1%. There is no escaping the fact that at 19,700 ft. (6,000 m) depth, the amount of life present cannot be more than about one thousandth of that found near the surface. This is a huge contrast with the bounty of life on upper continental slopes, ridges, and seamount summits, at 660–3,280 ft. (200–1,000 m) depth, where there may be rich coral and sponge gardens with a wide diversity of mobile life, crustaceans, mollusks, echinoderms, and fishes. The abundance of life can be greatly influenced by changes of substrate and the current regime. Strong currents sweep away soft sediments, exposing rock on which filter feeders can thrive, sustained by food particles transported by those same currents. Below these depths there is generally a gradual decline in biomass and biodiversity. At around 8,200–11,480 ft. (2,500–3,500m) depth, a transition point is reached between the slope (bathyal) fauna and the abyssal fauna.

At abyssal depths, two kinds of species are found: (1) slope species that extend into the abyss, but at a population density too low to be sustainable and hence must be continuously replenished by downslope immigration from bathyal depths, and; (2) abyssal endemic species that have succeeded in adapting to the exceptionally food-sparse conditions. Among the latter are abyssal endemic holothurians, or sea cucumbers (class Holothuroidea), which browse on the organic debris deposited on the seafloor and respond rapidly to sudden episodic enrichment from the surface. Some crustaceans, such as isopods (order Isopoda), have been very successful in invading the abyss by adapting to low food supply, and in some regions they show maximum diversity at greatest depth. In the transition from slope to abyss there is also a decrease in size of most organisms; large creatures become rare, and the abyss is dominated by meiofauna, comprising tiny animals living within the sediments exploiting organic-matter deposits (see pages 148–149). Active scavenging animals that depend on food falls from the surface tend to become bigger with increasing depth because large body size enables them to survive long periods of starvation between rare feeding events. Examples include grenadiers of the genus *Coryphaenoides* and the supergiant amphipod *Alicella gigantea* (see pages 86 and 156, respectively).

Changes to the relative abundance of species over time also impact the competition for food sources. Many deep-sea species, particularly at greater depths, reproduce rarely, possibly once in a lifetime. In such cases, a reproductive outburst may make a particular species dominant in a local area for some decades. The presence of species down the depth gradient can therefore change from place to place and from time to time.

Supergiant amphipod
Alicella gigantea

Top left: the largest known species of amphipod, which can grow up to 13 in. (34 cm) in length, is a voracious scavenger at abyssal depths and in hadal trenches.

Nematode worm

Bottom left: deep-sea nematodes, typically less than 4 thousandths of an inch (100 micron) in length, occur in vast numbers in bottom sediments. There over 10,000 species.

Holothurian
Benthodytes sp.

Top right: a flattened purplish colored sea cucumber found at 2,530–23,790 ft. (770–7,250 m) depth. About 8 in. (20 cm) long, the gut is visible through the transparent body wall. It feeds on the seafloor but swims between food patches.

Harpacticoid copepods

Bottom right: these tiny crustaceans are less than 40 thousandths of an inch (1 mm) long and live in the sediment and feed on bacteria and other organic material.

DEEP-SEA BIODIVERSITY

Complex patterns and processes

Biological diversity is a fundamental characteristic varying among ecosystems. Protection of biodiversity has become a rallying cry for the environmental conservation movement. But what ecosystem has the most biodiversity? Answering this seemingly simple question raises immensely complex problems. First, what do we mean by "biodiversity"? Many people think of it as the number of species in the area (i.e., species richness). But even this is a difficult question to answer for most taxonomic groups in most areas of the deep sea. The best-known group is the fishes, and they are difficult to sample quantitatively. Most invertebrate and microbial groups, although some are easier to sample, remain poorly known due to lack of taxonomic experts. Whenever we get an opportunity, and funding, to study any deep-sea area in detail (e.g., the Gulf of Mexico after the *Deepwater Horizon* oil spill), species lists grow substantially, even for supposedly well-known groups like fishes and shrimps. The increase includes both newly discovered species and others, already described, that were not known to live in the studied area. This indicates gaps in our knowledge of other sparsely sampled geographic areas.

Components of biodiversity

Biodiversity includes other components beyond species inventories. A sample containing 10 species that are all from the same taxonomic family is arguably less diverse than a sample with 10 species that are members of several orders in multiple subclasses (demonstrating evolutionary/phylogenetic richness). If a sample has 100 animals divided evenly among 10 species, the diversity would be different from that of a sample with 91 animals in 1 species and 1 each of 9 other species (the "evenness" aspect of species diversity). Other components are the genetic diversity within each species, and how the species richness and evenness change from place to place ("ecosystem diversity" within an area). Function and connectedness within the food web should be considered as well. It is, in other words, a complex topic. Understanding patterns and processes is difficult even in terrestrial ecosystems with a long history of thorough sampling. Such studies are rudimentary at best for any deep-sea ecosystem because of the vastness of the ocean, the difficulty of studying it, and the relatively short history of such endeavors.

Calculating biodiversity

Detailed methods for calculating biodiversity as well as predictions of explanatory patterns have been developed, mostly from work on well-studied ecosystems on land or in shallow waters. Use of these methods is based on assumptions such as comprehensive coverage with samples that are directly comparable with each other (e.g., standard collecting gear, similar sample sizes, the same groups of organisms analyzed with the same methods). Until the 21st century, most studies in the deep sea have not fulfilled the basic assumptions of biodiversity comparisons (e.g., information theory, which considers both richness and evenness simultaneously) that are fundamental in other fields of ecology. As a result, most studies of diversity in the deep have relied on development of methods that are not as dependent on the assumptions mentioned above. (It is not within the scope of this book to detail the calculation of these methods.) Scientists are trying to determine patterns of biodiversity in the deep sea because it is important for understanding the ecosystems. Inferences of some general patterns are developing, but they are preliminary at best, because in most areas of the deep sea we are still in the exploratory, descriptive stage of science rather than the detailed testing of ecosystem-scale hypotheses.

Coral community

A colony of cold-water coral *Desmophyllum pertusum* with crinoids and soft corals in 2,460 ft. (750 m) deep water, Porcupine Seabight, off Ireland.

Studying and sampling

Scientists processing cold-water coral samples collected by the German operated submersible JAGO in Trondheim Fjord, Norway, North Atlantic Ocean. In order to protect important areas of the deep sea, we must know the diverse organisms that live there and the processes that sustain them.

Consistent patterns

The most standardized data for calculation of deep-sea biodiversity has been from accumulation of species in multiple samples of soft sediments at bathyal and abyssal depths (taken with box-core and epibenthic-sled collection equipment) and sorted into common taxonomic groups, such as copepod and isopod crustaceans, polychaete worms, or mollusks (for invertebrate groups, see pages 78–87). Broad-scale comparisons of such samples indicate a consistent depth-related pattern of maximum diversity at intermediate depths on the continental slopes. Such a unimodal pattern is consistent with a diversity prediction known as the "intermediate-disturbance" hypothesis. This states that diversity is kept low by frequent disturbances but also is limited by interspecific competition when disturbances are uncommon, resulting in the highest diversity at intermediate levels of disturbance. However, abyssal samples often contain a few relatively common species and many other species, represented by only one or two specimens, that are not seen anywhere else, even in nearby samples. This indicates that diversity of small animals in the mud of the abyss may be extremely high.

Determining diversity in the deep is not so easy. Of the few seamounts that have been sampled, out of the very many that are known to exist, sampling methods and animals examined have varied widely. Comparing diversity among these seamounts using different methods is like comparing apples with oranges. When you consider that different animal groups have been the focus of different seamount studies, it becomes more like comparing apples with beetles.

Another pattern that has been observed consistently in the deep is that when high abundance (or biomass) is encountered, the associated biodiversity is relatively low. This is the pattern encountered in chemosynthetic systems such as hydrothermal vents, cold seeps, and whale falls (see pages 159–167).

Determining large-scale patterns of pelagic diversity is even more difficult. Even when the gear is standardized, which limits the fauna sampled, it is very difficult to collect a standard size of sample. However, some patterns related to latitude and physical conditions such as temperature, phytoplankton productivity, and dissolved-oxygen content can be inferred.

The deep sea as a refuge

The deep sea may have provided refuge for some animals, enabling them to avoid some major extinction events in Earth's history. For example, when dinosaurs disappeared 66 million years ago at the end of the Cretaceous period, there was also major loss of some 90% of shark and ray species (class Chondrichthyes) in shallow seas—but deep-sea forms were little affected. A few coelacanths (*Latimeria*), above, which were common in shallow seas of the Cretaceous, also survived by contraction of their habitat into the deep sea. Relatives of ghost sharks (or chimaeras) were once widespread in freshwater and shallow seas; most disappeared by the end of the Permian extinction event 250 million years ago, but the lineage has persisted in the deep sea (see pages 88–89). Conversely, during periods when the deep sea became anoxic, the deep-sea fauna suffered the brunt of the related major extinction event (e.g., c. 182 million years ago during the warm Jurassic period).

Mid-slope peak diversity

Numbers of snail species collected at various depths on Atlantic continental slopes. The solid line is a nonlinear regression with gray showing the 95% confidence interval. This compilation illustrates the overall pattern of maximum diversity at intermediate depths on the slopes. $E(S_{50})$ is the expected number of species in a sample of 50 individuals. It is calculated using a biodiversity method known as "rarefaction," which is based on the accumulating number of species in progressively larger samples from an area. (Source: Rex et al., 1997.)

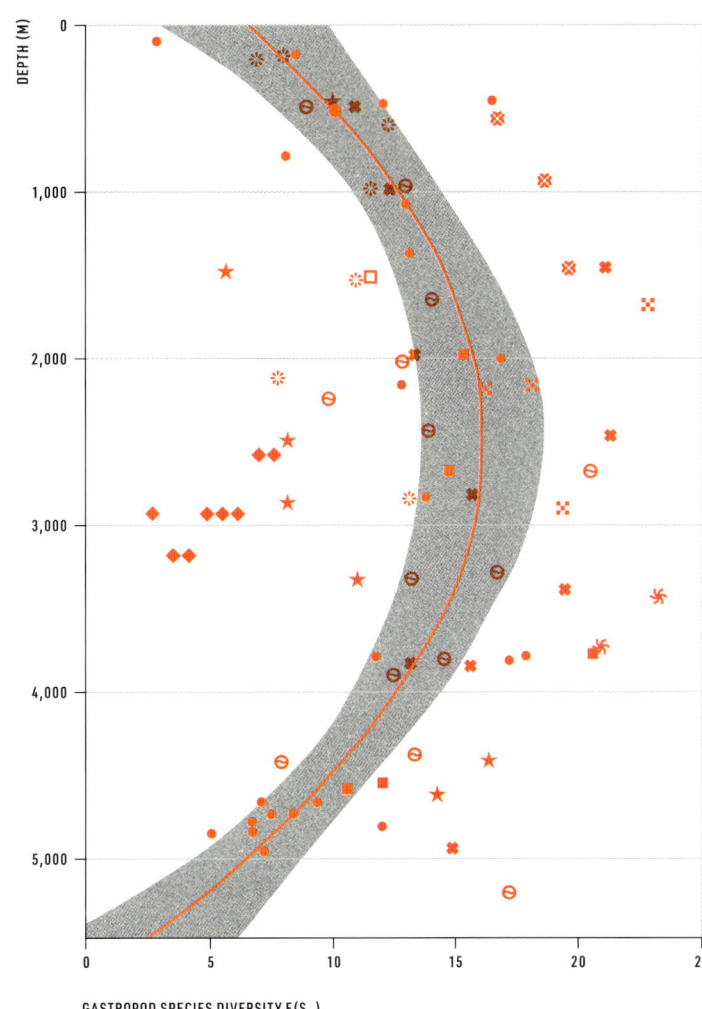

EQUATOR	
GAMBIA BASIN	
GUIANA BASIN	
NORWEGIAN SEA	
NORTH AMERICAN BASIN	
WEST EUROPEAN BASIN	
ARGENTINE BASIN	
CAPE BASIN	
ANGOLA BASIN	
BRAZIL BASIN	

GASTROPOD SPECIES DIVERSITY $E(S_{50})$

Midas Slit Shell

Bayerotrochus midas

This Midas Slit Shell is crawling along the underside of an overhanging rim of a rock wall at a depth of 1,400 ft. (425 m) off the Florida Keys.

The number of species in the deep sea

It is impossible to count all the species in the deep sea because most of the deep has not been explored, and many new species remain to be discovered. One way of estimating the total number of species is to count the increasing numbers discovered in an increasing number of samples. Initially, the number of species rises steeply because new species are easy to find; eventually, the rate of discovery begins to slow down, so it becomes possible to predict mathematically what the grand total is likely to be. In 1992, analysis of abyssal infauna (animals living in the sediment) from the northwest Atlantic yielded an estimate of possibly 10 million species in the global deep sea. Subsequent speculations have varied between 0.5 and 100 million. For fishes, the numbers are bit more certain; there are about 3,500 named deep-sea fish species, and the expected global total is about 5,500, indicating that over 64% of species have been discovered. Among pelagic species 76% have been discovered, among bottom-living species, 56%.

SPECIATION

How species evolve
in the deep sea

Scientists have long argued over the origins of deep-sea fauna. Not surprisingly, it did not all arise at the same time or in the same locations. The presence of animals such as sea lilies suggests the deep sea has an ancient fauna akin to that found in shallow Paleozoic seas, but DNA evidence suggests much of the fauna is younger than this. Although migrations from shallow to deep waters have probably occurred throughout geological time, much of the oldest fauna was extirpated by extinction events, particularly those associated with anoxia, the complete depletion of ocean oxygen other than very near the surface. (Anoxic waters often also have high sulfur content.) Notable global anoxic events occurred in the Late Permian, associated with the Permian–Triassic extinction, and in the Cretaceous, associated with other widespread extinctions. Not all deep-sea fauna was wiped out by these events. Some isopod groups appear to have originated before the Permian–Triassic boundary and have thus survived several extinction events. Many echinoderms found in the deep sea today, including various sea stars (class Asteroidea), brittle stars (class Ophiuroidea), and sea cucumbers (class Holothuroidea), have Mesozoic origins, indicating a colonization event shortly after the Permian–Triassic extinction and subsequent survival of the Cretaceous extinctions. Other groups are younger; for example, the ancestor of deep-sea scavenging amphipods can be dated to the Cretaceous–Paleogene boundary, with diversification of lineages dated at the Eocene–Oligocene boundary.

Anoxic events are mostly associated with global warming. In contrast, global cooling in the Paleogene led to recolonization opportunities that may have shaped much of the modern deep-sea fauna. Movement of major tectonic plates led to the separation of Antarctica from South America and Australia around the Eocene–Oligocene boundary, and this led to the development of a circumpolar current and subsequent cooling of Antarctica. The ensuing widespread Antarctic glaciation drove the development of the thermohaline circulation (see page 182) and its subsequent intensification in the mid-Miocene. This current system brings cold oxygen-rich water from the polar oceans to the deep seas of the world and appears to have provided a route for Southern Ocean fauna to colonize other deep seas. Some deep-sea octopods at lower latitudes have Antarctic relatives, as do some sea spiders, gastropods, and amphipods.

Once present in the deep sea, animal groups can speciate further through various mechanisms. Populations can be isolated by distance, by land masses, by ocean currents, or by seafloor topology, and each population can slowly evolve into a separate species. The evolutionary history of *Leiogalathea*, a genus of deep-sea squat lobsters with an ancestral distribution in the Atlantic and Indian Oceans, demonstrates how the movement of land masses can lead to speciation events. The Atlantic lineage of *Leiogalathea* split from its Indo-Pacific sister group after the closure of the Tethys Seaway, which linked the Indian Ocean to what is now the Mediterranean Sea.

Survivors

Top right: many echinoderms found in the deep sea are from lineages that have survived through multiple extinction events and are known from the fossil record.

Global events

Right: a geological time line showing extinction and anoxic events.

Atlantic lineage

Leiogalathea agassizii (left) is found in the Gulf of Mexico. The genus *Galatheia* has speciated into 18 species distributed around the globe.

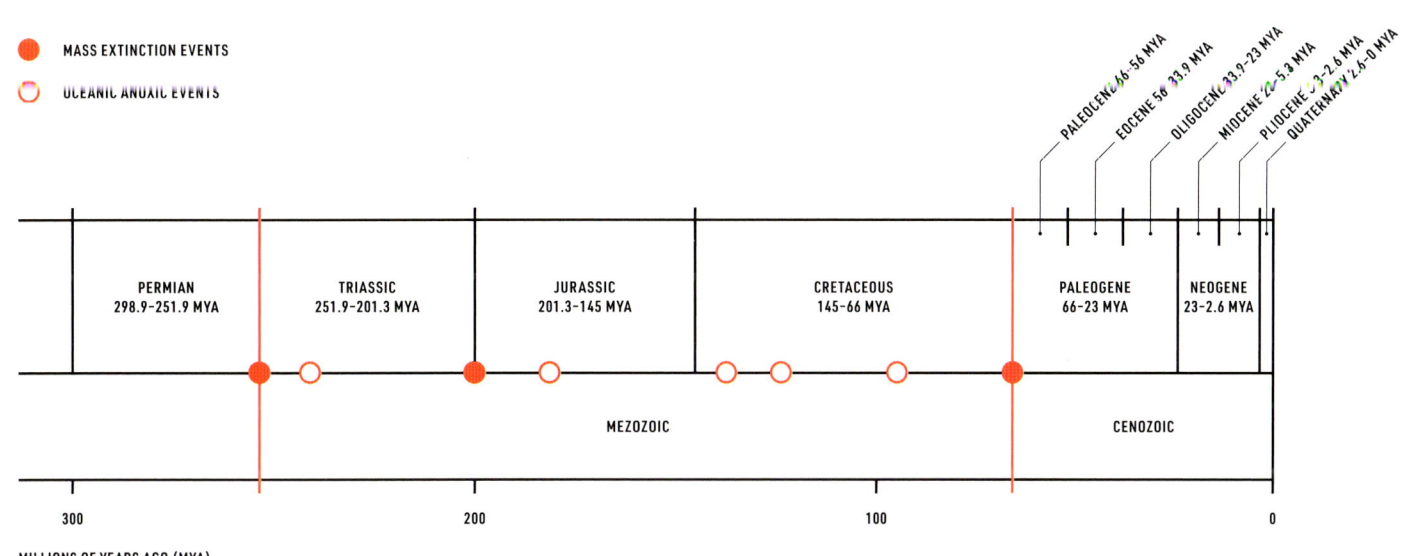

MASS EXTINCTION EVENTS

OCEANIC ANOXIC EVENTS

PALEOCENE 66-56 MYA

EOCENE 56-33.9 MYA

OLIGOCENE 33.9-23 MYA

MIOCENE 23-5.3 MYA

PLIOCENE 5.3-2.6 MYA

QUATERNARY 2.6-0 MYA

PERMIAN 298.9-251.9 MYA	TRIASSIC 251.9-201.3 MYA	JURASSIC 201.3-145 MYA	CRETACEOUS 145-66 MYA	PALEOGENE 66-23 MYA	NEOGENE 23-2.6 MYA

MEZOZOIC

CENOZOIC

300

200

100

0

MILLIONS OF YEARS AGO (MYA)

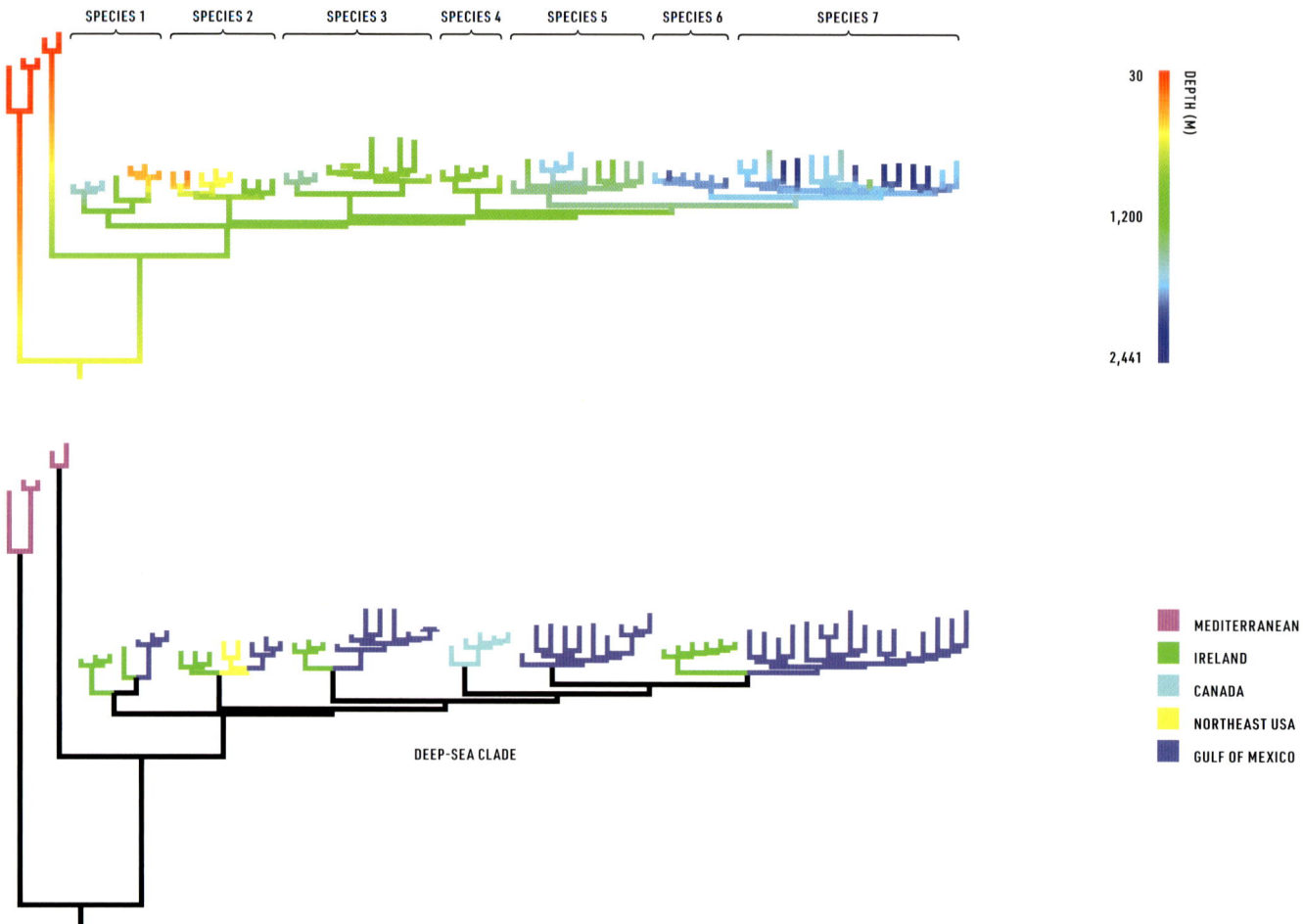

Depth as a driver of speciation

Studies of closely related deep-sea species often show congeners (species in the same genus) inhabiting the same geographic area, but separated by depth. In pelagic species such as copepods (small crustaceans), each member of a closely related species pair may be restricted to either the mesopelagic or bathypelagic zone. In demersal fishes such as grenadiers, a bathyal species may be the closest relative to an abyssal species. In deep-sea corals such as *Paramuricea*, multiple cryptic species (hard to distinguish except genetically) exist in discrete depth bands. What drives this speciation?

In some cases, the driver may not be depth per se but water masses that don't mix with other water masses and thus inhibit the movement of larvae, with the result that each species is directly associated with a water mass. This pattern is seen in *Paramuricea* spp. at a local scale in the highly stratified Gulf of Mexico and on a broader scale across the whole North Atlantic. Currents can similarly prevent mixing of populations; they may be acting as barriers to dispersal now, or they may have been stronger or had a slightly different course historically, such that their potential influence is not obvious at present.

But selection may also play an important role. Environmental variability is high in the mesopelagic zone and on the continental slope, where depth-related gradients are steepest. Temperature, light, oxygenation, and food resources all vary by depth. Traits that make an animal better adapted to the environmental conditions at shallower or deeper depths are likely to be selected for, and species at different depths can slowly separate into divergent populations and from there into sister species. If selection relies on environmental variability, we might expect abyssal species, which live in a potentially less variable environment, to have broader ranges, and this has been shown to be the case for some grenadier species.

Speciation by depth

Above: these phylogenetic trees of seven *Paramuricea* species show (top) the depth of different lineages when they diverged, and (below) the location where those lineages are found today. This demonstrates that the speciation into seven *Paramuricea* species appears to have occurred first by depth, with some of those lineages subsequently splitting geographically. The result is five species within the Gulf of Mexico, four in Ireland, and single species in Canada, Northeast USA, and the Mediterranean. (Source: Quattrini et al., 2022.)

Fan coral
Paramuricea sp.

Right: *Paramuricea* species in the deep sea off Ireland; in both images, the coral is hosting a brittle star associate, the long arms of which are wound around the coral branches.

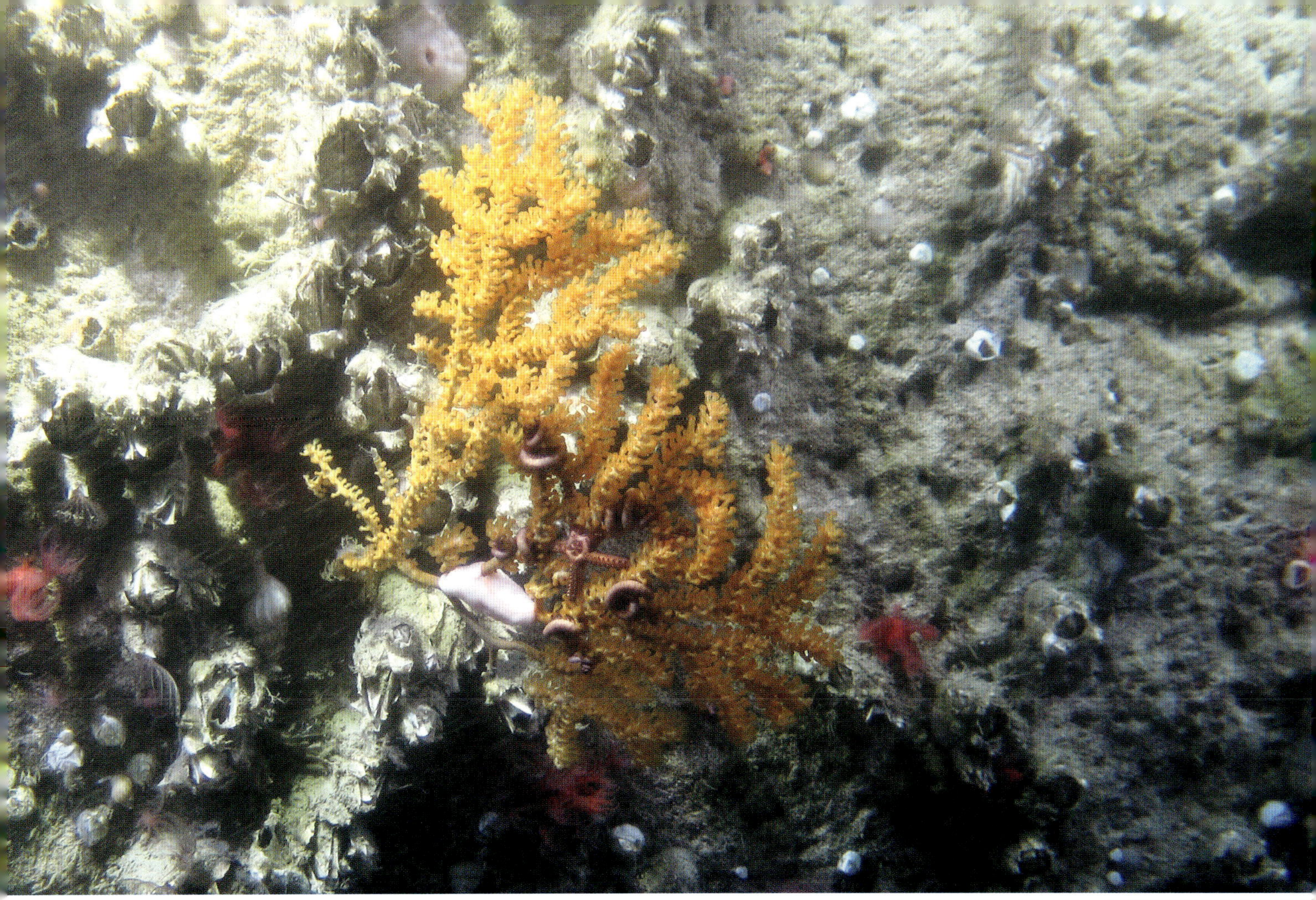

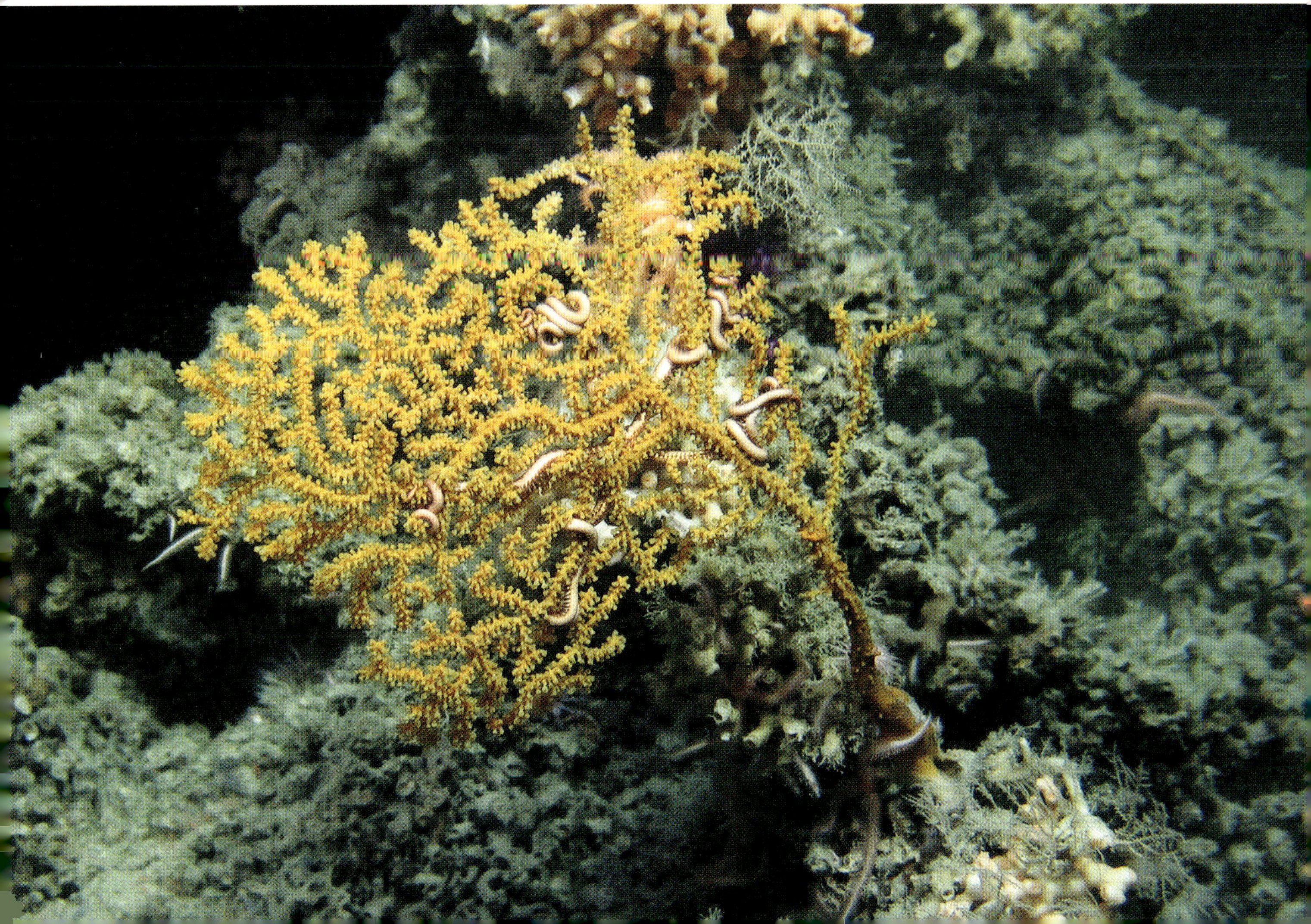

Oxygen-minimum zones as drivers of speciation

Oxygen-minimum zones (OMZs) occur where productivity is high and water circulation is poor, with the result that oxygen is used up and is not replenished. OMZs occur in the Eastern Tropical Pacific ecoregion and the Arabian Sea among other places (see pages 172–173). Only a limited number of species can tolerate low oxygen concentrations, so species diversity is reduced in OMZs, and a few tolerant species dominate the community. How, then, can OMZs drive speciation?

Speciation usually occurs through one of two mechanisms. Allopatric speciation occurs through geographic separation of populations, whereby each population becomes locally adapted to environmental conditions, and the populations diverge until they are two different species. Sympatric speciation, in contrast, occurs without geographic separation and is often the result of selection for particular traits, as is the case with depth (pages 224–225). OMZs can potentially influence speciation through both mechanisms—allopatrically through geographically separating populations of species that cannot tolerate low oxygen concentrations, and sympatrically through selection for tolerance to hypoxic (low-oxygen) conditions.

Furthermore, OMZs have likely been acting through geological time. Periods of global warming lead to an increase in OMZs, promoting both selection and allopatric speciation, while periods of global cooling lead to their decrease, allowing species to invade the once again oxygen-rich abyss and other newly oxygenated areas. Those species evolving through selection probably contribute to the endemic faunas that we recognize to be associated with some well-known low-oxygen regions (e.g., the Eastern Tropical Pacific). Selection for particular character traits, such as dwelling structures, is common in these areas. For example, polychaetes are often found in tubes, much as they are at hydrothermal vents, an adaptation that allows them to regulate their immediate surroundings and thus limit the impact of the challenging environmental conditions.

Lollipop Catshark
Cephalurus cephalus
The Lollipop Catshark is found in the eastern central Pacific off Mexico and California, a region well known for OMZs. It is thought the enlarged head is an adaptation to accommodate the oversized gills that extract oxygen from the low concentrations available in the OMZ.

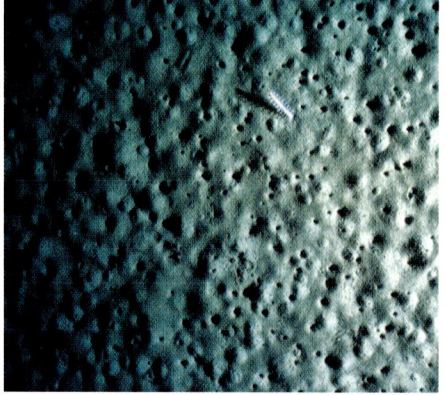

OMZ environments
Evidence of dwelling structures on the seafloor in the Arabian Sea OMZ.

Allopatric and sympatric speciation

In allopatric speciation, a species (below left) is separated into two populations by an OMZ, where oxygen concentrations are too low for the species to survive. Over time, each population diverges into a distinct species, both of which are intolerant to hypoxia. In sympatric speciation (below right), some fishes are more tolerant to low oxygen than others. Selection on this trait eventually leads to the evolution of a new (blue) hypoxia-tolerant species.

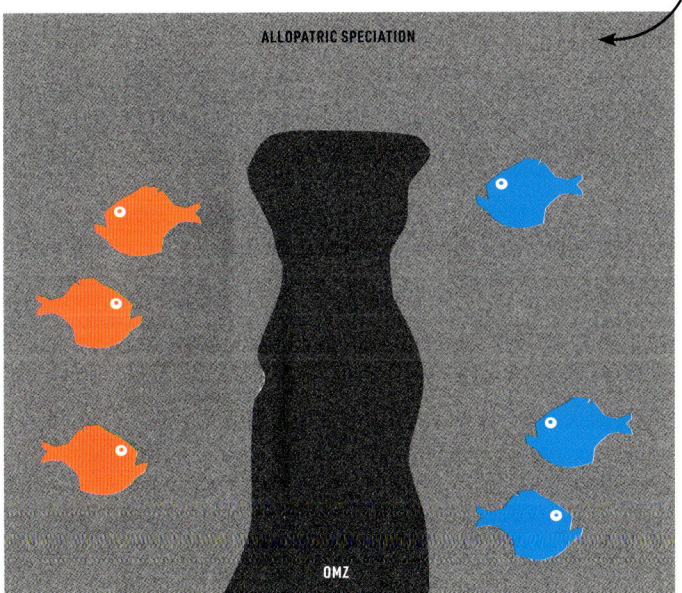

ALLOPATRIC SPECIATION

OMZ

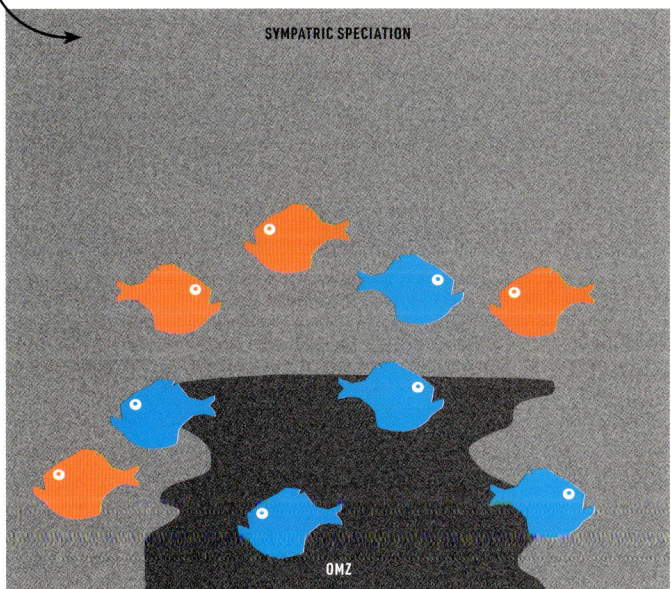

SYMPATRIC SPECIATION

OMZ

"OMZs can potentially influence speciation through two mechanisms—allopatrically through geographically separating populations of species that cannot tolerate low oxygen concentrations, and sympatrically through selection for tolerance to hypoxic (low-oxygen) conditions."

Bioluminescence as a driver of speciation

Deep-sea animals use bioluminescence to lure prey and to defend against predation (see pages 118–119), but it also has an important role in inter- and intraspecific communication. In the large Humboldt Squid (*Dosidicus gigas*), bioluminescence is used to backlight chromatophore patterns in the skin. Chromatophores are muscular pigment organs that allow cephalopods in shallow sunlit waters to rapidly change color. Backlighting them with bioluminescence allows changing skin patterns to be seen in the darkness by other squid in the school, and it is thought the squids use this to communicate their movement, enabling schools to move fast and change direction without collisions. Bioluminescence is also used in species recognition. Closely related midwater micronekton species differ visually from their relatives only in the patterns of the light-producing organs, or photophores, on their body. Additionally, the patterns of photophores in many species of fishes and cephalopods differ between males and females, presumably for identifying mates in the vast darkness. It is the use of bioluminescence to signal to potential mates that appears to drive evolution.

There are several hugely abundant groups of bioluminescent fishes. The order Stomiiformes includes the hatchetfishes (family Sternoptychidae) and bristlemouths (family Gonostomatidae), which have only downward-facing photophores; and the dragonfishes (family Stomiidae), which have downward-facing photophores as well as bioluminescent chin barbels and photophores on the head that distinguish each species. Dragonfishes also often have enlarged photophores behind the eye in males, which may be attractive to females and possibly indicate male fitness. The order Myctophiformes includes blackchins (family Neoscopelidae), which have only downward-facing photophores; and lanternfishes (family Myctophidae), which have downward-facing photophores and photophores on the head and sides of the body. These head and side photophores are arranged in species-specific patterns that sometimes differ between males and females. For example, the male *Notoscopelus bolini* has a large patch of luminous tissue behind the eye, and the male *Dasyscopelus asper* (Prickly Lanternfish) has a photophore above the tail, neither of which are present in the females of those species.

Downward-facing photophores are probably just used for camouflage, counter-illuminating hatchetfishes, bristlemouths, and blackchins to protect them from predators (see pages 118–119). But the greater variety of photophores in dragonfishes and lanternfishes likely aids in species recognition and mate choice. Supporting this hypothesis, lanternfish species that have sex differences in their photophore arrangement also have sex differences in the retina, presumably as an adaptation facilitating the detection of luminous mating signals. If selection of a particular photophore pattern is driving evolution and hence speciation, then we might expect to find many more species of dragonfishes and lanternfishes because of their use of bioluminescence for recognition. Indeed, this is the case: there are fewer than 100 hatchetfish species, 21 bristlemouths, and just 6 blackchins, compared with nearly 300 species each of lanternfishes and dragonfishes—indicating bioluminescence can drive speciation in the deep sea.

Barbeled dragonfish
Eustomias sp.

Top left: a male dragonfish with a bioluminescent chin barbel and a large photophore behind the eye, which may indicate fitness to potential female partners.

Hatchetfish
Argyropelecus hemigymnus

Top right: downward-facing photophores on hatchetfishes help camouflage the fishes from upward-looking predators. The photophores break up the silhouette against a moonlit surface. They don't play a role in speciation.

Lanternfish
Lepidophanes guentheri

Right: the pattern of photophores on the sides of this lanternfish are specific to this species and help in mate recognition. Selection for specific patterns of lateral photophores can drive sympatric speciation.

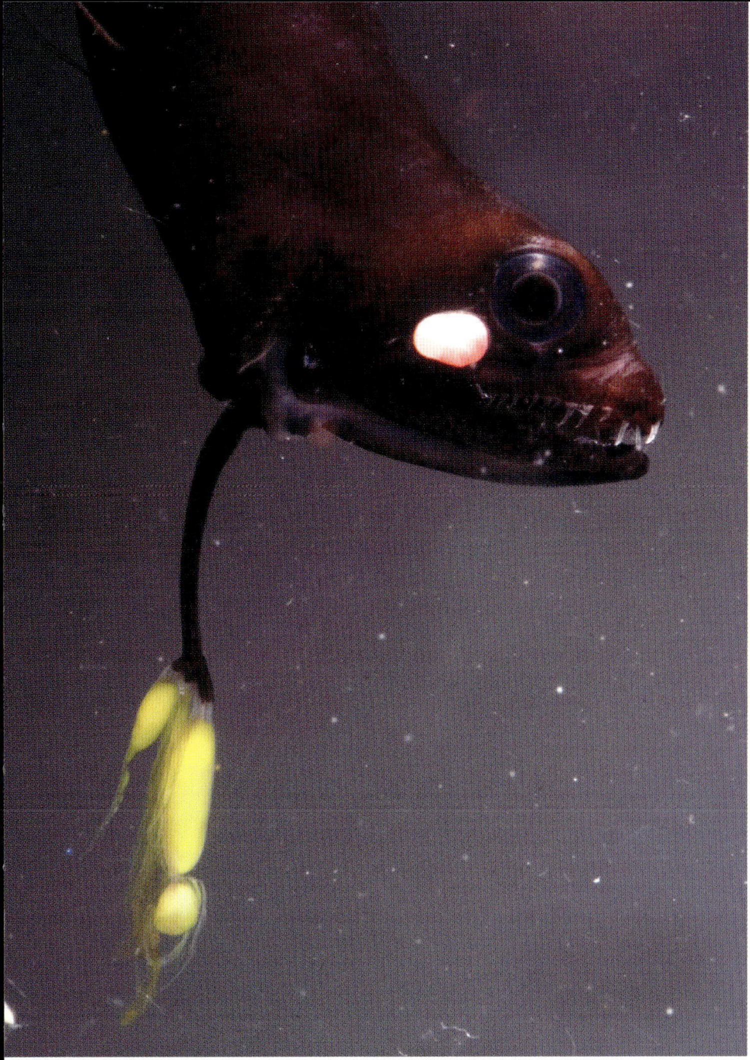
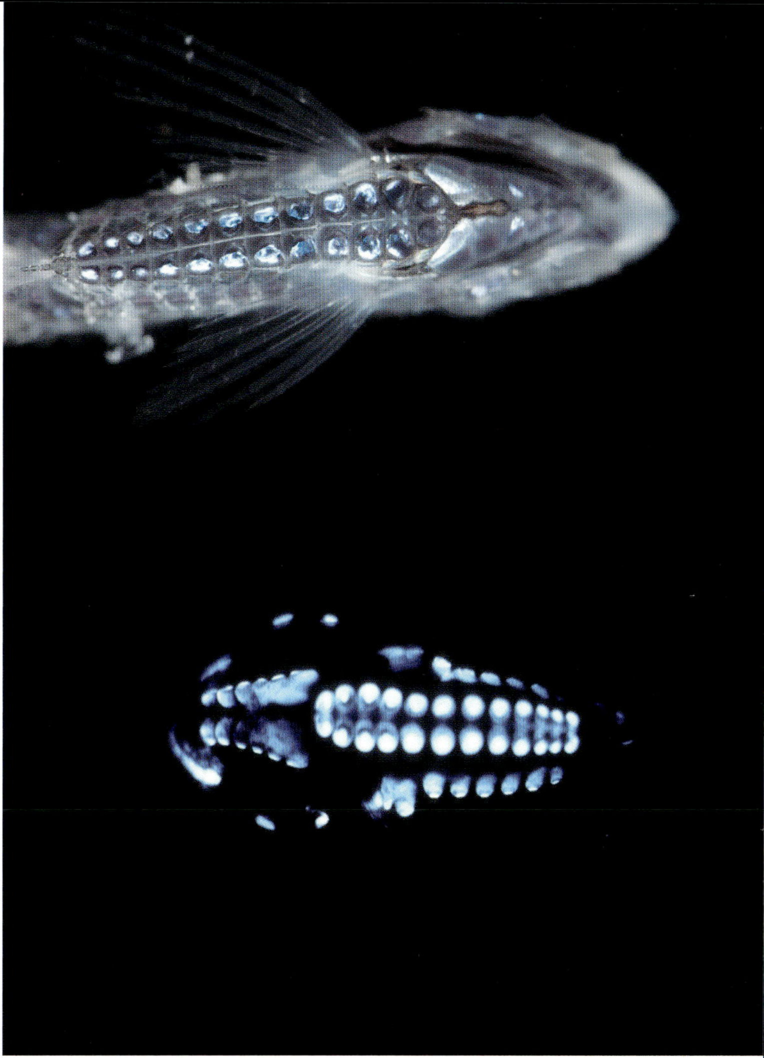
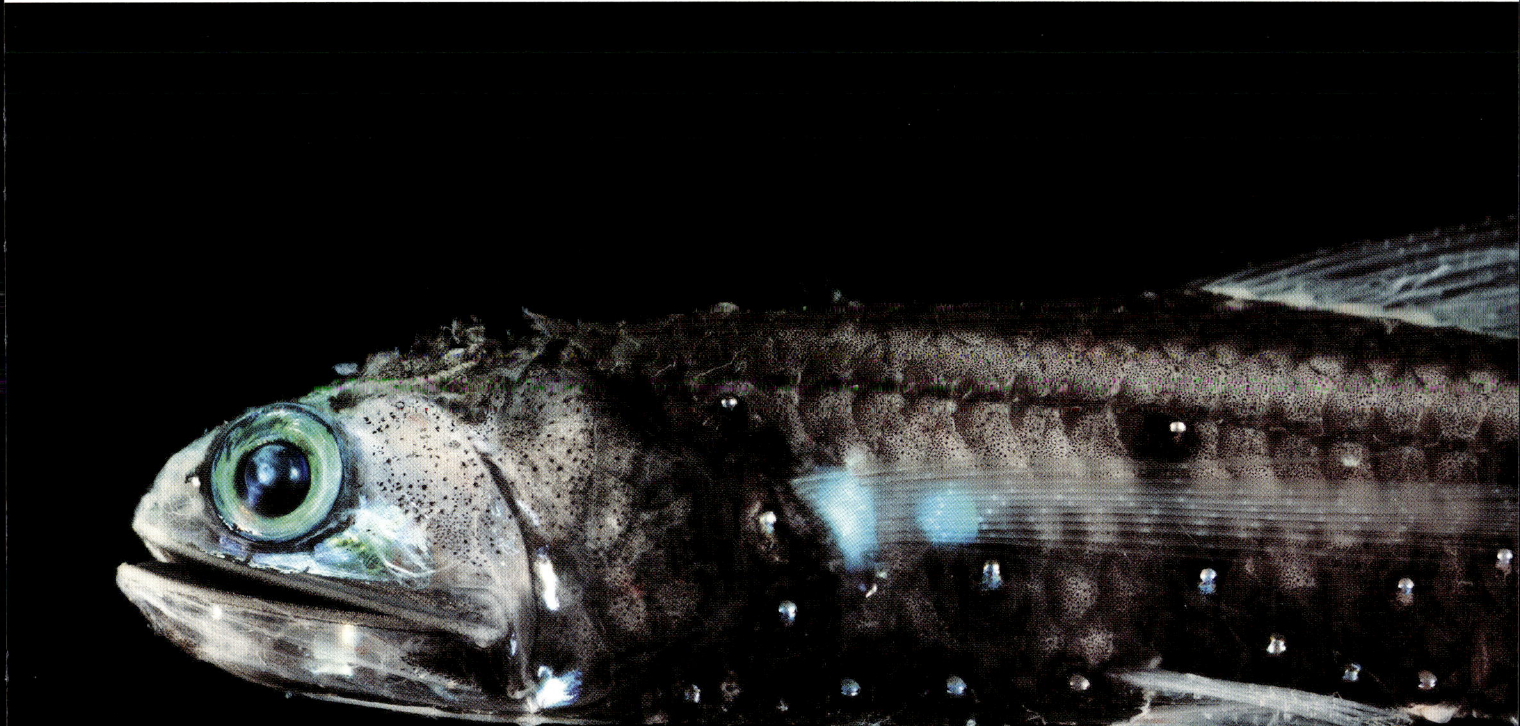

HUMANITY
AND THE
DEEP OCEAN

FISHERIES AND WHALING

Human exploitation of deep-sea fishes and whales

Previous pages: **Deep-water coral**
Desmophyllum pertusum
Reef in Trondheim Fjord, North Atlantic Ocean, where this species is impacted by mining practices.

Deep-sea fisheries

There is a long unwritten history of artisanal deep-sea fishing from oceanic islands and coastlines where the bottom drops off steeply, allowing fishermen using small boats to lower lines to great depths within sight of land. South Pacific Polynesian islanders traditionally captured Oilfish (*Ruvettus pretiosus*), a giant mackerel that grows up to 10 ft. (3 m) long, from depths of 660–1,640 ft. (200–500 m) using baited hooks on long lines. Around the islands of the Azores and Madeira in the North Atlantic, similar methods were used to target the Black Scabbardfish (*Aphanopus carbo*), a much-prized fish found at depths of 2,300–4,265 ft. (700–1,300 m). The Inuit of western Greenland captured Greenland Halibut (*Reinhardtius hippoglossoides*) at 1,150–2,300 ft. (350–700 m) depth in fjords by lowering lines through holes cut in the ice during midwinter. Occasionally Greenland Shark (*Somniosus microcephalus*), which grows to over 13 ft. (4 m) long, was also caught. Off the coast of Portugal at Setúbal, south of Lisbon, a long-established deep-sea shark fishery captured species such as the Portuguese Dogfish (*Centroscymnus coelolepis*) at 1,800–2,400 ft. (550–730 m) depth. These and other fisheries, pursued since time immemorial, were sustainable, providing regular sustenance for coastal and island communities.

Fishing changed fundamentally in the decades after World War II, becoming industrialized with the advent of large, high-powered boats that could work far from their home port. Highly productive fisheries on continental shelves in such places as the North Sea, the Grand Banks, and the North Pacific had reached capacity, and the Soviet Union, Japan, Poland, Spain, Taiwan, and South Korea developed distant-water fleets that roamed the global ocean searching for fishing opportunities. New technologies, including synthetic fibers for nets and lines; freezers for storing fish; navigation aids such as radar, sonars, loran (long-range navigation), and Decca radio navigation; and, from the 1980s, Global Positioning System (GPS) satellites, all greatly increased the fishing effectiveness of these fleets. From the 1960s, they began to target deep-water resources. There followed a pattern of discovery, exploitation, depletion, and abandonment of one fish stock after another.

Traditional fishing

Left: Greenland Inuits in kayaks fishing for halibut (*Hippoglossus hippoglossus* and *Reinhardtius hippoglossoides*). Fish are caught in deep water over 1,500 ft (460 m) close inshore.

Large fishing trawlers

Above: deep water trawling for hake (*Merluccius capensis* and *M. paradoxus*) off Cape Town, South Africa. Scavenging seabirds follow the ship as the net is hauled to the surface.

Marine technology

Right: wheelhouse on a small modern trawler with two radars, two GPS antennas, two VHF radio antennas, MF radio antenna, satellite internet, and cell phone. The trawler has hydraulic cranes for handling nets and equipment.

Deep-sea exploitation

One of the first species to be exploited in this way was the Pacific Ocean Perch (*Sebastes alutus*), also known as Pacific Rockfish or Rock Cod, at depths down to 1,640 ft. (500 m) on the slopes of the Gulf of Alaska and around the North Pacific. Catches by the Soviet Union and Japan peaked at 480,000 tons in 1965, followed by a rapid decline until only 8,000 tons were landed in 1984. This slow-growing species reaches sexual maturity at 10 years and lives to a maximum age of 80 years or more. The females produce live young. Since 2000, the Gulf of Alaska stock has been rebuilt to a possibly sustainable yield of 25,000 tons per year under a management regime imposed by the United States.

Just as the Pacific Ocean Perch fishery passed its peak, Soviet trawlers in 1967 discovered large aggregations of Slender Armorhead (*Pentaceros wheeleri*) around the Emperor Seamount chain in the northwest Pacific. The international fleet, including Japan, caught 50,000–200,000 tons of armorhead per year. By 1978 the catch had declined to only 900 tons; fished to commercial extinction, the population never recovered.

During the 1960s, the Roundnose Grenadier (*Coryphaenoides rupestris*) was discovered on the slopes of the North Atlantic and the Mid-Atlantic Ridge. The first commercial landings were recorded in 1967, and a peak catch of 84,000 tons was attained within four years, followed by a sharp fall. By 1995 the species was over 99% depleted and critically endangered in the northwest Atlantic. Fishing shifted to other parts of the North Atlantic, culminating in a catch of 50,000 tons in 2000 before stocks collapsed to a catch of less than 4,000 tons in 2014.

To avoid the problems of unregulated offshore fisheries, in 1977 most countries declared an exclusive economic zone out to 200 nautical miles (230 mi./370 km) from their coasts. This began to curtail the activities of the itinerant deep-sea fishing fleets. In the early 1970s the Soviet Union discovered throngs of Orange Roughy (*Hoplostethus atlanticus*) on seamounts and plateaus around New Zealand. Under New Zealand management from 1977, the fishery grew to 55,000 tons by 1989, the most valuable fishery in these waters. Within a decade the stock was depleted to 3% of its original size. Off Australia, a roughy fishery began in 1985 and peaked at 40,000 tons in 1990. Roughy stocks were discovered around the world, exploited, and depleted, such as the fishery off the west coast of Scotland, which began in 1990, peaked at 3,500 tons in 1991, and was exhausted by 1995. Other species followed, with peaks of the Patagonian Toothfish (*Dissostichus eleginoides*) at 44,000 tons in 1995, Oilfish at 43,000 tons in 2007, and Boarfish (*Capros aper*) at 138,000 tons in 2010. None of these species now figures in the list of top deep-sea commercial species.

Pacific Ocean Perch
Sebastes alutus

Top left: formerly abundant on the upper continental slopes of the North Pacific Ocean but severely overfished in the 1960s. Slow-growing, Pacific Ocean Perch can live to over 100 years.

Splendid Alfonsino
Beryx splendens

Bottom left: captured in a deep-water longline fishery off the Azores islands in the mid-Atlantic Ocean. A traditional method with small boats, but modern technology can be managed sustainably.

Orange Roughy
Hoplostethus atlanticus

Top right: trawl cod end full of Orange Roughy being emptied onto the deck of a trawler. Abundant catches were made during the 1980s. Most stocks have been overfished. Orange Roughy can live to over 140 years if not captured.

Patagonian Toothfish
Dissostichus eleginoides

Bottom right: a Southern Ocean species found in the Atlantic Ocean, Indian Ocean, and southwest Pacific sectors. Caught mainly by bottom set longlines in a fishery managed by international agreement.

Deep-sea fishing today

Today, about 300 species of deep-sea finfishes are exploited. As more species became targeted, the total world deep-sea catch increased from 170,000 tons in 1960 to a peak of 3.6 million tons in 2005. About half the official deep-sea catch is blue whitings (*Micromesistius* spp.), which are shoaling pelagic fishes found in great abundance at 985–1,310 ft. (300–400 m) depth in the North Atlantic and off New Zealand and Argentina. They are exploited by large midwater trawlers that switch between these and other pelagic species, such as herrings and mackerels. From 2000 to 2021 the blue whiting catch has fluctuated between 2.5 and 0.5 million tons, whereas the rest of the deep-sea fishes have amounted to around 1 million tons annually, or roughly 1% of the global fish catch. About 20% of this is southern hakes (*Merluccius* spp.) off South Africa and Chile, 20% is hoki (*Macruronus* spp.) off New Zealand and South America, 8% is Greenland Halibut, and 6% is ocean perch or redfish (*Sebastes* spp.) in the North Pacific and North Atlantic. The rest of the world catch is made up of a diverse assemblage of species caught by bottom trawls, longlines, set nets, or traps. An important additional component is deep-water shrimps or prawns that form the basis of immensely valuable bottom-trawl fisheries. These include Blue and Red Shrimp (*Aristeus antennatus*) in the Mediterranean, Northern Shrimp (*Pandalus borealis*) in the North Atlantic and North Pacific, nylon shrimps (*Heterocarpus* spp.) in the tropics, and Knife Shrimp (*Haliporoides triarthrus*) in the southwest Indian Ocean.

Motivated by economic, social, political, and food-security considerations, national governments tend to subsidize fishing activity. A study showed that cumulative subsidies in 2014 accounted for about 60% of the total cost of global high-seas fishery operations that would otherwise be far from profitable. So far it has proved very difficult to reduce this amount of subsidy.

> *"Today, about 300 species of deep-sea finfishes are exploited. About half the official deep-sea catch is blue whitings (Micromesistius spp.). From 2000 to 2021 the blue whiting catch has fluctuated between 2.5 and 0.5 million tons."*

Deep-sea bottom trawl

A mixed catch being hauled on board from about 3,280 ft. (1,000 m) depth in the Tasman Sea west of New Zealand. The catch is primarily Common Mora (*Mora moro*—locally known as Ribaldo), Birdbeak or Shovelnose Dogfish (*Deania calcea*), and Orange Roughy (*Hoplostethus atlanticus*). Bulging eyes and stomachs are caused by decompression while the *Mora moro* are brought to the surface.

Blue whiting
Micromesistius spp.

There are two species: *Micromesistius poutassou*, in the North Atlantic and Western Mediterranean and *Micromesistius australis* in the Southern Hemisphere. Fluctuations in abundance are poorly understood but well-managed fisheries can be very productive.

Super trawlers

FV *Margiris*, the world's second-biggest fishing vessel, a trawler and factory ship that catches various species around the world. In February 2022 she controversially dumped (possibly accidentally) a large quantity of blue whiting in the Atlantic Ocean off France.

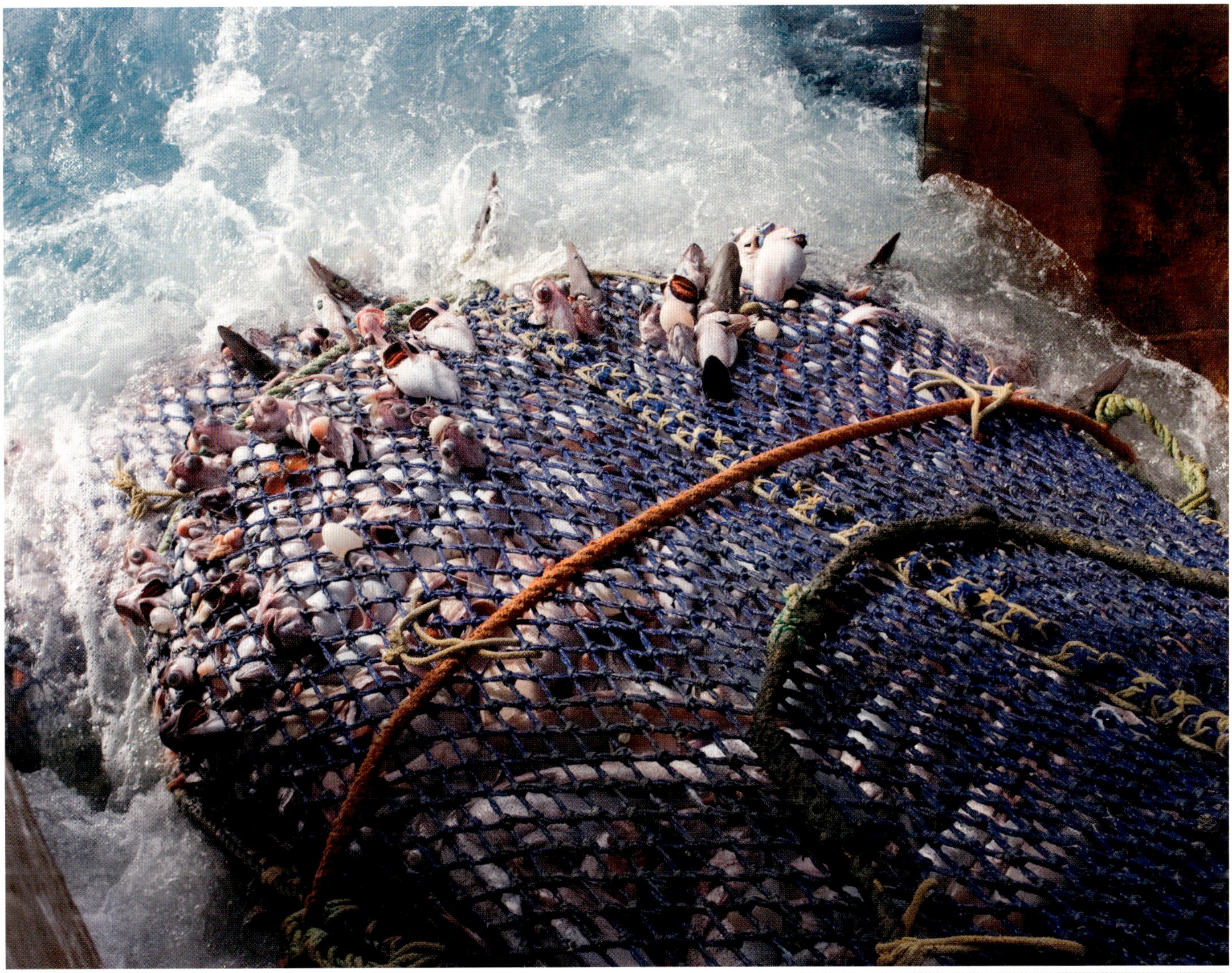

LYRV
MARGIRIS
LYRV
KL855

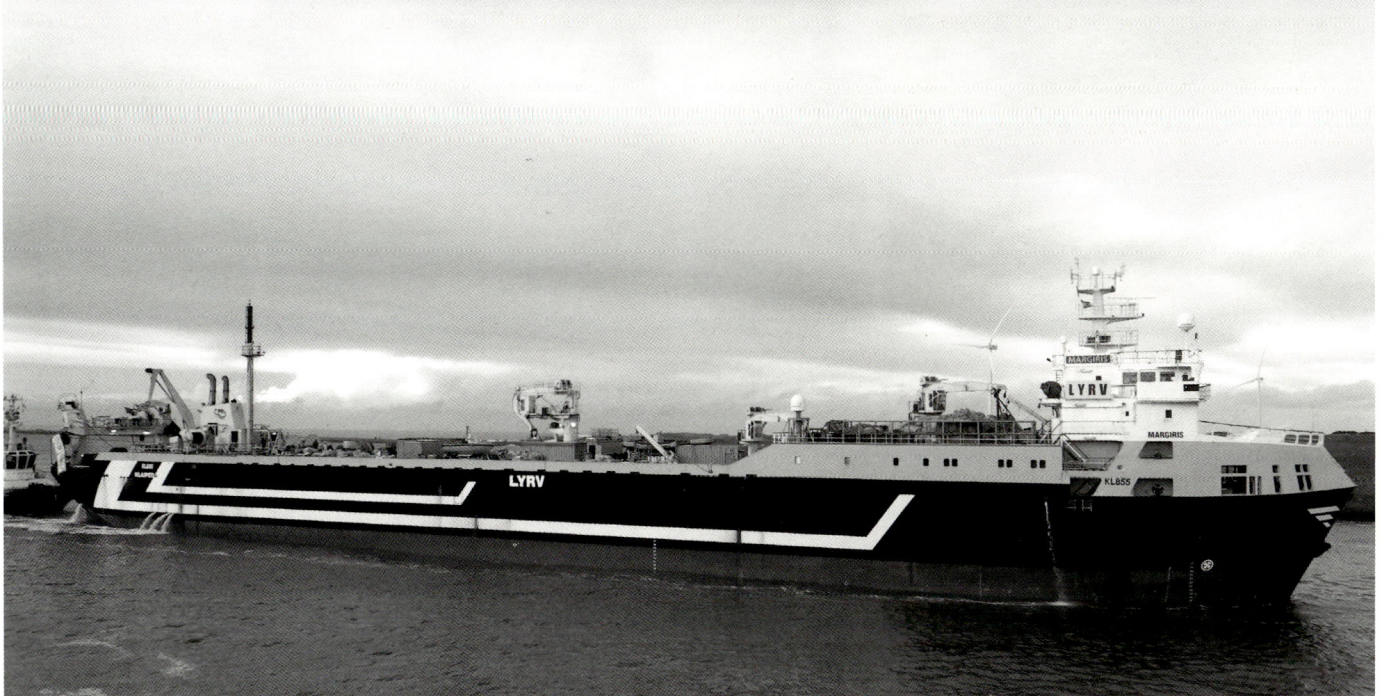

Trawling methods

Pelagic trawling, such as for shoals of blue whitings, provides a clean catch of one species, and because the net does not touch the seafloor, there is no collateral damage to the environment. Bottom trawling creates a major disturbance on the deep seafloor. Parts of the trawl weighing many tons, including otter boards, rollers, wires, and chains, scrape along the seabed stirring up sediment and cutting down and smashing corals, sponges, and stalked fauna in the path of the trawl. Some of these animal-created structures may have taken centuries to grow. This is quite unlike trawling on the continental shelf, where strong currents regularly shift bottom sediments and trawling may be a small additional disturbance. Deep-sea bottom trawling destroys designated Vulnerable Marine Ecosystems (VMEs; see pages 128–133). On soft sediments, trawling accelerates the rate of erosion, smooths the bottom contours, and reduces the amount of organic matter and the biodiversity of fauna living in the sediment. The environment is impoverished in the way that plowing destroys natural habitat on land. However, farmers on land typically plow once per year, whereas deep-sea trawlers plow over their favored fishing grounds on an almost daily basis.

Trawls often bring broken corals to the surface, and indeed during the armorhead fishery on the Emperor Seamounts, a targeted fishery was developed to capture precious corals (*Corallium* spp.) in special tangle nets down to 4,920 ft. (1,500 m) depth. This peaked at 300 tons in 1981, and by 1990 the coral resource was exhausted. Bottom trawling also catches nontargeted species, which are discarded, returned to the sea, and unlikely to survive. In the northeast Atlantic, the fishery for Roundnose Grenadier and Orange Roughy has resulted in depletion of nine other fish species never consumed by humans. Since fish are mobile, the depletion effect can extend far beyond the limits of the fishing area. Commercial deep-sea vessels fish to a maximum depth of 4,920 ft. (1,500 m), but reductions in fish abundance can be detected down to 9,840 ft. (3,000 m).

Pelagic and bottom trawl
The pelagic trawl (upper image) does not touch the seafloor. Large mesh panels herd fish into the cod end. In the bottom trawl, clouds of sediment and noise of the otter boards scare fishes into the path of the trawl, which drags along the seafloor.

Trawling damage
Underwater image of the lead rope and foot rope of a bottom trawl moving from right to left. The foot rope is reinforced with chains and digs into the seafloor.

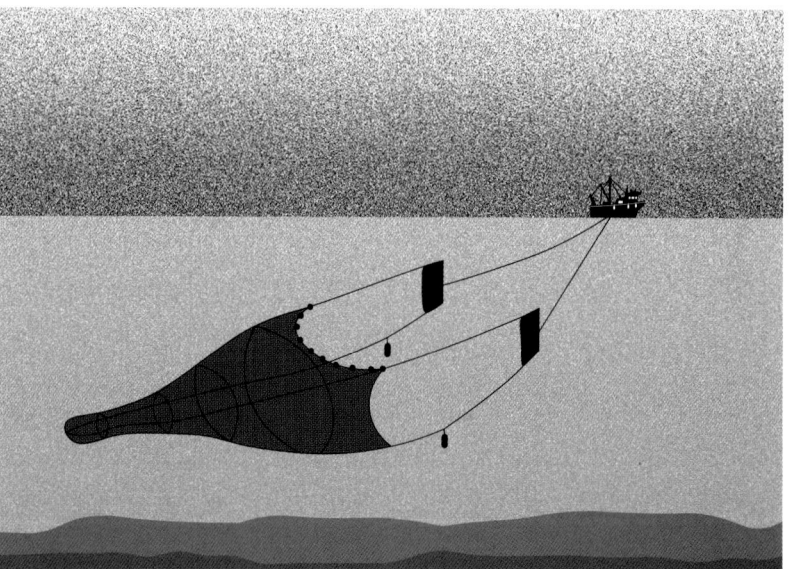

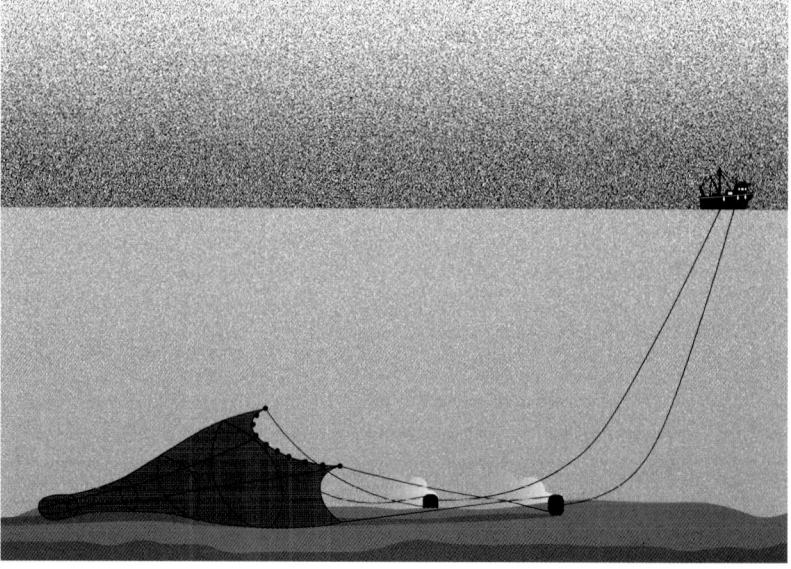

Traditional whale hunting

A Bowhead Whale (*Balaena mysticetus*) being speared with hand harpoons in the Arctic c. 1850. Engraving by A. M. Fournier after E. Traviès.

Whaling

Although they are air-breathing mammals, whales have a dual significance for the deep sea, firstly as predators and secondly as a rich food source when they die. Toothed whales, especially the beaked whales and the Sperm Whale (*Physeter macrocephalus*), dive to 3,280 ft. (1,000 m) or more along slopes of continents, islands, and seamounts, consuming vast quantities of deep-sea squids and mesopelagic fishes and playing an important role as apex predators in the marine food web. When a whale dies in the open ocean of old age, injury, or disease, it may float on the surface but is likely to sink eventually to the seafloor. This whale fall, which becomes a small chemosynthetic ecosystem, provides an important habitat for over 100 species that consume and colonize the carcass (see pages 166–167). Some, such as the bone-boring worm *Osedax*, are so specialized that they are in danger of extinction in the absence of whale falls. Fossil whale-fall communities have been found from up to 30 million years ago, and their lineages may be even older, exploiting giant reptile carcasses such as ichthyosaurs and plesiosaurs that lived 150 million years ago in the Jurassic and Cretaceous ages of dinosaurs.

Like deep-sea fishing, artisanal or aboriginal hunting of whales has a long history, exploiting whales stranded on the coast, herding them into confined bays, or hand-harpooning individuals from small boats. Whaling developed into a major industry of coastal regions of Spain and France by the 17th century. Then companies in England, Amsterdam, Spain, and Denmark began to send whaling expeditions to the Arctic off Spitsbergen and Greenland and in the Davis Strait to hunt the Bowhead Whale (*Balaena mysticetus*). In the Pacific Ocean, Japan exploited the North Pacific Right (*Eubalaena japonica*), Humpback (*Megaptera novaeangliae*), Fin (*Balaenoptera physalus*), and Gray Whales (*Eschrichtius robustus*). From around 1750, New England became a center for whaling, initially specializing in the North Atlantic Right Whale (*E. glacialis*) and the Humpback but then moving farther afield to exploit the more valuable Sperm Whale. After gaining independence from Britain in 1776, the United States grew to become the preeminent whaling nation in the Pacific as well as the Atlantic region.

Decline and effects of whaling

By 1900 the whaling industry had declined because existing resources—the Bowhead, Gray, northern Humpback, and right whales—were almost extinct. Yet, the 20th century saw massive expansion of whaling with steam and diesel-powered vessels, improved harpoon guns and, from 1920, large factory ships that could support a fleet of catching vessels and process the whales on board. In contrast to shore-based whaling stations that could economically catch whales only within a radius of 200 mi. (320 km), the factory ship pelagic fleet could operate far offshore, pursuing whale populations throughout the world. Whaling shifted south in the Pacific, Atlantic, and Indian Oceans and to the Southern Ocean around Antarctica. Despite hiatuses during the world wars, catches steadily increased until over 80,000 whales were killed in 1970. It was evident to all concerned that this was unsustainable, and from 1986 a moratorium on all commercial whaling was implemented. Some stocks have recovered, and small-scale harvesting is permitted for aboriginal, scientific, and commercial whaling, amounting to about 4,000 whales per year.

Overall, humans have reduced the abundance of great whales by 66–90%. For the deep-sea fauna specialized on whale falls, this has several consequences. There is a proportional reduction in the food supply reaching the seafloor. Whaling has also removed the largest species and individuals from the seas, and small carcasses are unlikely to persist as deep-sea reef habitat, unlike the bones of large whales, which could last for over 50 years before they are finally eroded away. Fewer carcasses also means that the average distance between whale falls increases. For organisms dependent on the regular supply of whale falls from the surface, their habitat has become much reduced, ephemeral, and more difficult to colonize. There is concern that whaling may have inadvertently caused the extinction of some of these specialized deep-sea species.

With the decline of the large-scale pelagic whaling operations, investment and some of the equipment was redirected to the expanding deep-sea fishing industry, thus continuing the human propensity to utilize the natural resources of the planet up to and beyond sustainable limits.

Whale harvesting

Top: a Blue Whale (*Balaenoptera musculus*) hauled onto a coastal slipway for processing (1924). Bottom: flensing Southern Minke Whales (*Balaenoptera bonaerensis*) on board a modern Japanese factory ship in the Antarctic. Since the 1930s, factory ships have allowed whaling to become global after coastal stocks had been depleted.

Recovery of stocks

Left: a breaching Humpback Whale (*Megaptera novaeangliae*). This species was hunted almost to extinction. Only 5,000 remained in 1966, but following the end of commercial exploitation, the world population has recovered to around 150,000 individuals.

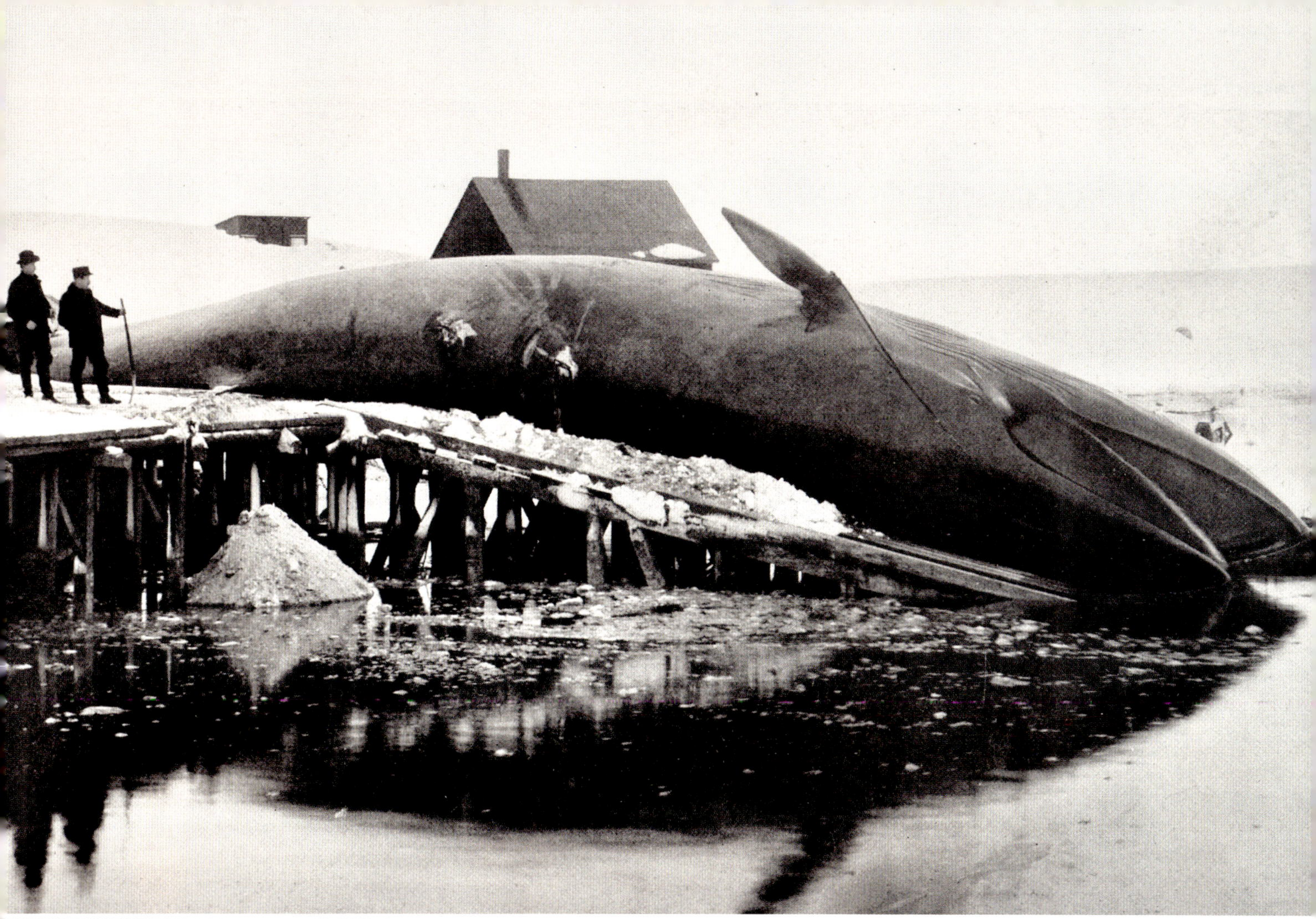

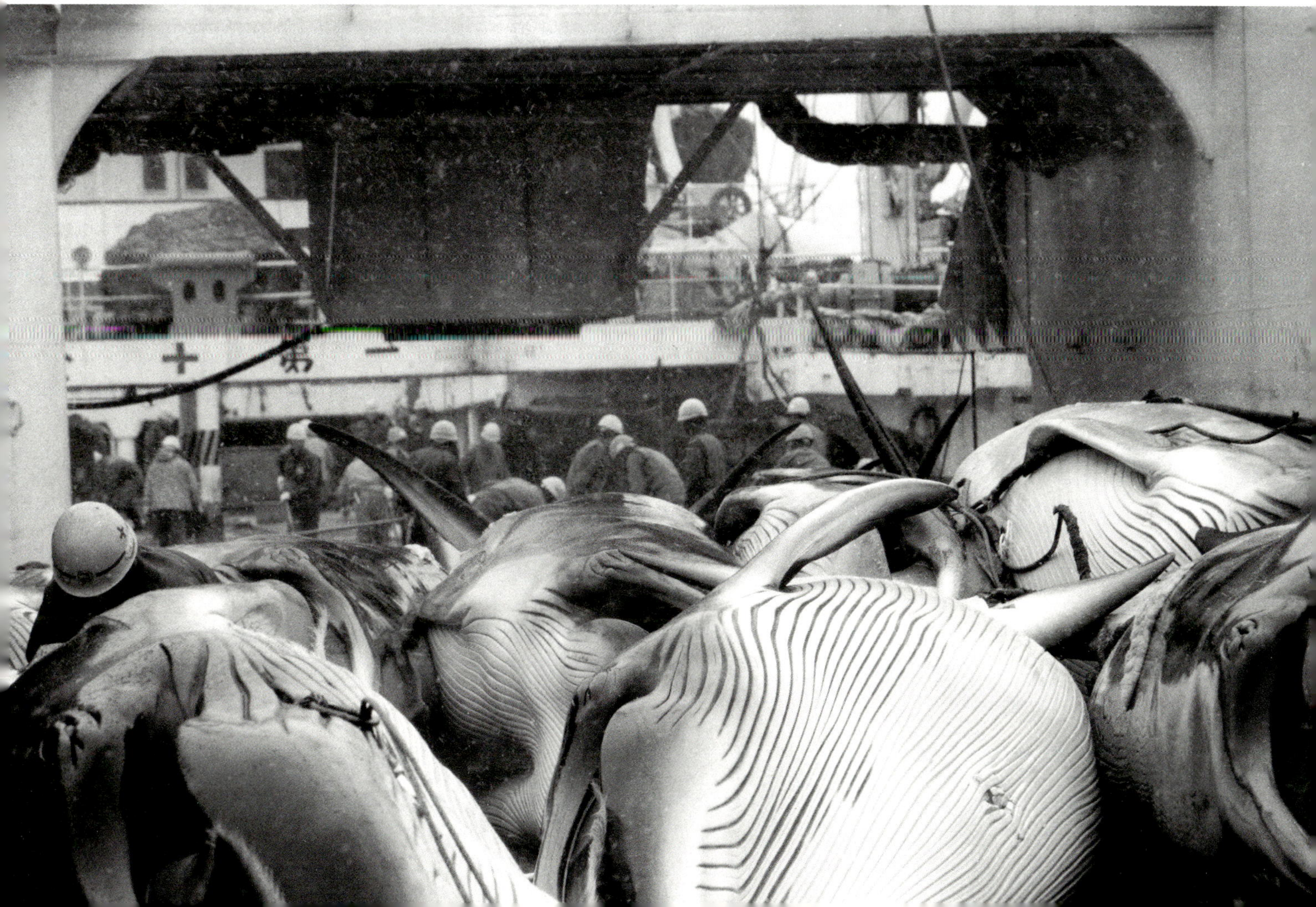

PETROCHEMICALS AND MINING

Extraction of nonliving resources

In addition to widespread harvest of many types of animals, humans also extract nonliving resources from the deep sea. This includes drilling and pumping oil and gas from increasing depths for the petrochemical industry and ongoing development of mining for various minerals needed for modern technology. Similar to fisheries, as onshore and nearshore sources of petroleum and minerals are either used up or rendered inaccessible by international conflict, the search for replacement sources has extended into increasingly deep waters in spite of the technical difficulties and extreme expense required for deep-sea exploitation. Increasing prices of these resources have made the expense of deep-water exploitation economically feasible.

Seismic survey
Powerful low-frequency sound waves produced by air guns or explosives are reflected off the bottom and rock layers beneath the seabed to build up a geological picture showing potential oil and gas reservoirs.

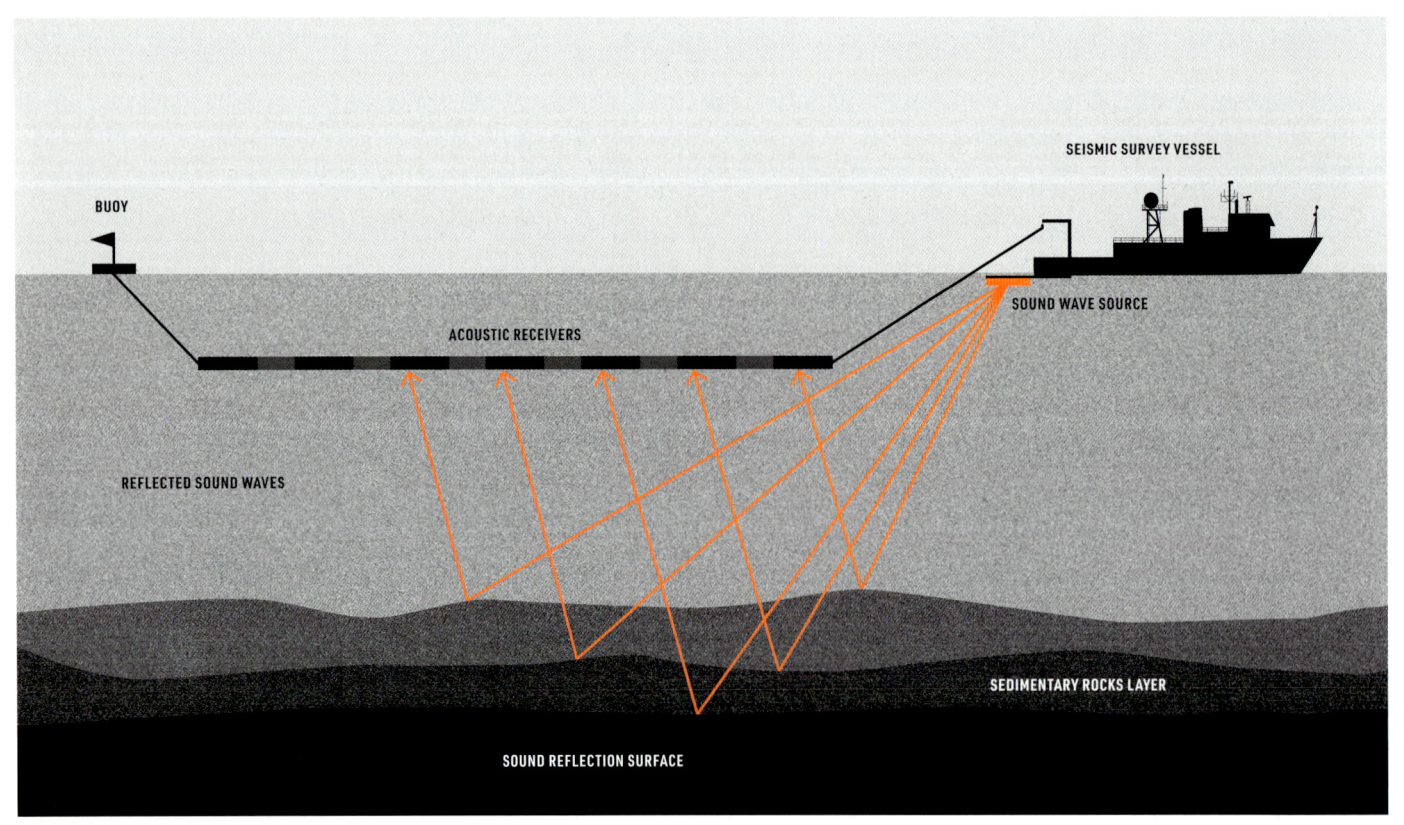

BUOY

SEISMIC SURVEY VESSEL

SOUND WAVE SOURCE

ACOUSTIC RECEIVERS

REFLECTED SOUND WAVES

SEDIMENTARY ROCKS LAYER

SOUND REFLECTION SURFACE

Prestige oil tanker
The 26-year-old oil tanker *Prestige* broke in half during a storm in November 2002 off the Iberian Peninsula. Carrying 60–70,000 metric tons of oil, the two parts sank in water over 9,840 ft. (3,000 m) deep. In addition to the much-publicized impacts at the surface and along the coast, this disaster affected the deep benthos and water column as well.

Oil and petroleum gas

Getting petroleum out of the ocean floor and to refineries is a multistage process, with each stage subject to its own problems and the potential for environmental damage. Oil and gas drilling is now common down to 6,560 ft. (2,000 m) depth and is extending to deeper than 11,480 ft. (3,500 m). Deep-water stages in the process include: (1) seismic exploration; (2) drilling; (3) production; and (4) transport.

Exploitable pockets of oil and gas are typically very deep under the surface of the seafloor sediment. To map potential locations of exploitable petroleum pockets, a very loud type of sonar is used to probe the geological structure under the bottom. Whereas high sound frequencies provide more detailed resolution, lower frequencies and more intense transmission are necessary for better penetration. For deep sub-bottom profiling, exploration ships tow devices called sparkers, boomers, or air-gun arrays, which have replaced the earlier practice of detonating explosives with bubble generation either by electrical pulse or compressed air to create signals loud enough to result in echoes from deep in the sedimentary rocks. Signals from acoustic exploration are among the loudest sounds in the ocean, detectable thousands of miles away. As with the explosives that preceded them, these sounds can harm nearby animals. How nearby? Oil exploration operations north of Spain were regularly followed within days by discoveries of dead Giant Squid (*Architeuthis dux*) floating at the surface in the explored area. The squid, which live at about 1,970–3,280 ft. (600–1,000 m) depth, consistently had damage to their organ system that senses gravity and acceleration, as well as to muscles and other tissues. This species may be an indicator of broader biological damage from the acoustic exploration for oil. It is large enough not to be scavenged quickly after death, and because its tissues accumulate ammonia (which is lighter than seawater) to prevent sinking, the corpses can float to the ocean surface, where they may be encountered by seafarers. It seems quite likely that other species are damaged or killed, but those impacts have gone undetected.

**Semisubmersible
drilling rig**

A typical semisubmersible
deep-water oil rig. The
rig is supported by the
large, floating, vertical
cylinders and is held in
place by multiple anchors.
Approaching the rig is one
of the large support ships
used by the oil companies
to meet the many logistical
needs of an offshore oil
operation.

Large-scale drilling

Once a promising geological formation is located by the
acoustic profiling, large-scale drilling operations are
initiated to access the hoped-for deposit. Unlike shallow-
water drill rigs that stand on the bottom with solid legs,
deep-water drilling uses floating drill ships or so-called
semisubmersible platforms. Both options must be held
in place during drilling operations, either by multiple
anchors buried in the mud below or by a system called
dynamic positioning, in which computer controls link the
desired location, determined by the satellite-based GPS,
with multiple thrusters that keep the drill rig over its hole.
While the drilling continues, large volumes of drill mud,
a fluid mixture of clay with water or oil and chemical
additives (some of which are toxic), are used to lubricate
the drill bit and equipment within the hole. Disposal
of the used fluid, along with sediment and rock from
the hole, can contaminate the water column and
smother the bottom.

Once the drilling strikes oil, the drilling rig is replaced by
a production operation. Whereas shallow-water rigs often
are connected to shore by pipelines, farther offshore the
production rigs usually are connected to a storage facility
that can transfer oil to a tanker for transport. Storage
may be on the rig or several production wellheads may
be connected by pipelines to an offshore terminal with
storage capacity. Oil spills can occur in any of these
components. Also, in addition to chemical contamination,
these rigs are noisy and brightly lit 24 hours per day,
potentially affecting behavior such as vertical migration
of nearby animals.

Once the oil is transferred to a ship, it becomes a marine
transportation problem. An example of the potential of
tanker shipping to impact the deep sea is the wreck of
the *Prestige*. In November 2002, the tanker broke in half
and sank near the edge of Galicia Bank off northwestern
Spain. The two halves sank in depths of 11,700 and
12,570 ft. (3,565 and 3,830 m). In spite of efforts to
recover the cargo from the wreck, 20 million gallons
of oil were spilled into the deep sea.

The *Deepwater Horizon* disaster

Deep-water oil drilling exploded
into the public consciousness
worldwide in April 2010, when a well
being drilled by the dynamically
positioned semisubmersible drilling
platform *Deepwater Horizon* blew
up, killing 11 people and releasing
hundreds of millions of gallons of
oil over several months at about
4,920 ft. (1,500 m) depth. Whereas
news coverage showed the visible
impacts at the surface and in coastal
ecosystems, only a portion of the
oil and gas reached the surface.
Some of it precipitated onto the
deep-sea bottom, contaminating
and smothering coral, sponge, and
sedimentary communities. However,
all of the discharged oil and gas,
together with chemical dispersants
injected at the wellhead, passed
through the pelagic environment,
forming a great plume at about
3,280 ft. (1,000 m) depth, through
which passed the vertical migrators
and nonmigrating animals that live
at those depths. When it happened,
scientists knew very little about the
deep pelagic community in that area
and had no baseline data to allow
comparison of the environment
before versus after the accident.
Sufficient information from
post-spill studies now indicates
dramatic impacts, especially on
small swimmers (micronekton)
and coral communities. Plankton
populations crashed immediately but
recovered quickly. Above: deep-sea
coral smothered by oil that sank from
the plume created by the *Deepwater
Horizon* disaster. In addition to the
effects on sessile animals such as
corals, other animals associated with
the coral ecosystem, such as this
serpent star, were impacted as well.

MASSIVE SULFIDE DEPOSITS
AT HYDROTHERMAL VENTS

POLYMETALLIC MANGANESE NODULES
ON ABYSSAL PLAINS

COBALT-RICH FERROMANGANESE CRUSTS
ON SEAMOUNTS

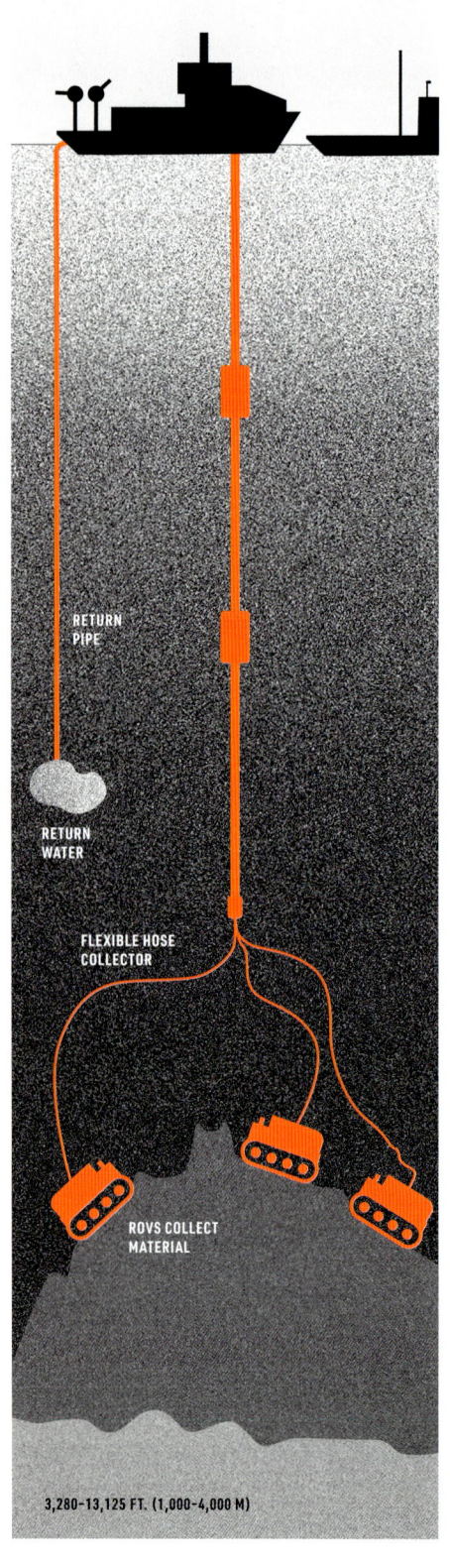

RETURN
PIPE

RETURN
WATER

FLEXIBLE HOSE
COLLECTOR

ROVS COLLECT
MATERIAL

3,280-13,125 FT. (1,000-4,000 M)

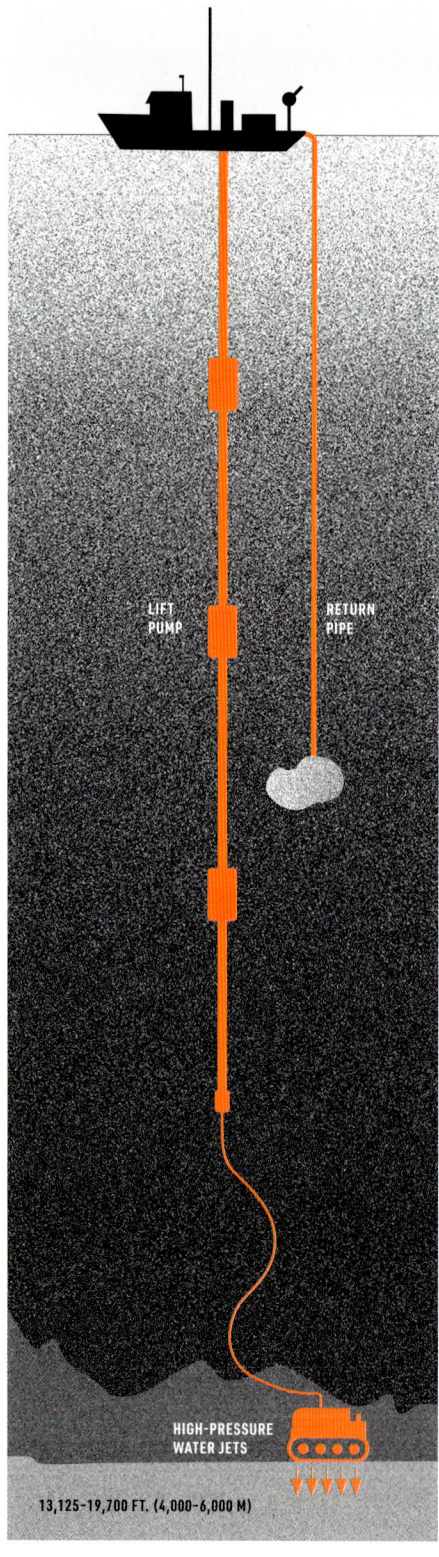

LIFT
PUMP

RETURN
PIPE

HIGH-PRESSURE
WATER JETS

13,125-19,700 FT. (4,000-6,000 M)

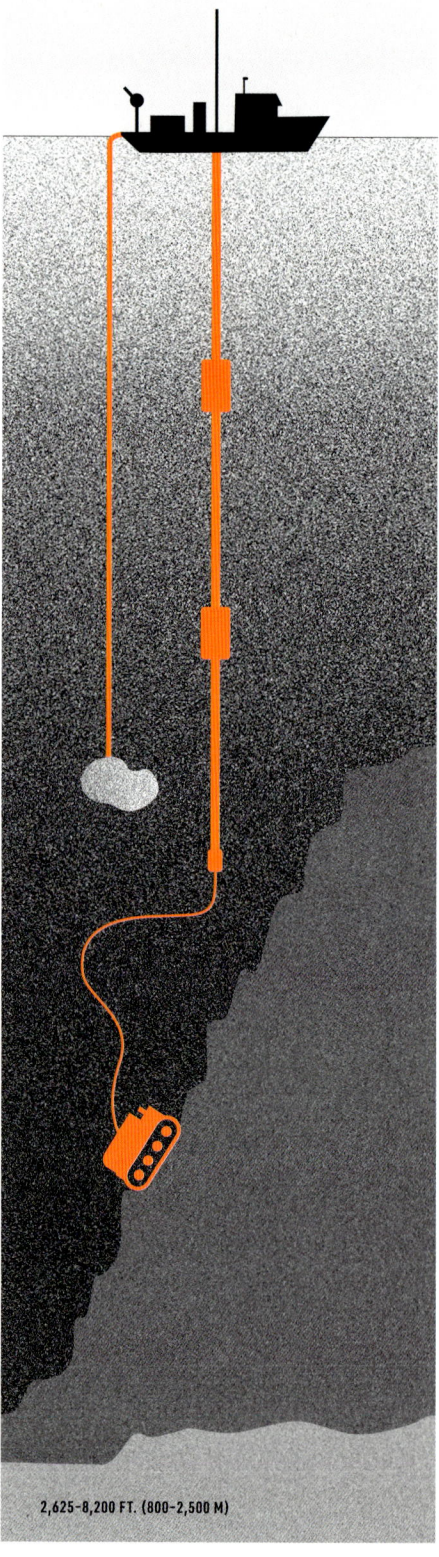

2,625-8,200 FT. (800-2,500 M)

Mining

Over the past few decades, great interest has developed in the possibility of mining valuable minerals from the deep seafloor. This is a controversial expansion of human activities in the deep sea. The minerals are primarily those needed for manufacture of such technology as alternative energy production (e.g., wind and solar power) as well as the increasing number of computer-controlled devices for modern life. Three types of geological targets in the deep sea have been proposed for extraction of these minerals: (1) massive sulfide deposits at hydrothermal vents; (2) polymetallic manganese nodules found in some areas on abyssal plains; and (3) cobalt-rich ferromanganese crusts on seamounts. The controversy in deep-sea mining results from the proposed methods, all of which are very destructive, for mining the ores containing the target minerals. Limited experimental mining operations have begun, and large-scale commercial operations are very likely to commence soon. We note here that nodules and mineral crusts require very long times to form and the resources therefore cannot be considered renewable.

Manganese nodules

Rocks similar in size and shape to a potato, manganese nodules form by gradual precipitation from near-bottom seawater of concentric layers of minerals. Although it includes nickel, copper, and so-called rare-earth elements (yttrium, molybdenum, tellurium, niobium, zirconium), the composition is primarily oxides of manganese and iron. Precipitation begins around a nucleus, which may be a shark tooth, fragment of shell, or squid beak. The nodules form very slowly at the sediment surface on the abyssal plains in areas where sediment accumulation is not rapid enough to bury the growing nodule. In such places the bottom may be almost covered with a pavement of nodules extending for hundreds or thousands of miles. The fauna in nodule fields is a patchwork of animals using the hard substrate on the nodules and soft-substrate animals in the sediment between nodules. As a result, biodiversity in nodule fields may be quite high.

Planned recovery of nodules uses hydraulic dredging: high-pressure water jets are used to stir up the bottom, and then the resulting slurry, including nodules, sediment, and entrained organisms, is sucked up through a very long pipe, called a riser, to the ship at the surface, thousands of feet above. There the nodules are separated from the slurry, which then must be discharged. If this discharge occurs at the surface, then a plume forms and affects the entire water column as it sinks and spreads. The alternative is to pipe the discharge back down for release close to the bottom, where it forms a near-bottom sediment plume that covers the bottom when it settles. In other words, the mining operation removes the entire top layer of the bottom, and discharge of unwanted material may affect the water column and will certainly cover the benthos, possibly in areas not yet disturbed by dredging. An experiment in the southeast Pacific Ocean that simulated these operations has been monitored for decades, and very little sign of recovery of the benthic community has been detected in the disturbed areas

Mining minerals

Left: plans for mining various mineral-bearing substrates are variations on the theme of digging up the benthic environment using deep-sea robotic machinery, piping the excavated material up to a ship at the surface for processing, and then disposing of the unwanted fraction back into the ocean. Details of the spread, sinking, and impact of the discharged material will depend on characteristics of the material, how it is released, and, especially, at what depth.

Manganese nodules

Far right: a polymetallic nodule from 16,400 ft. (5,000 m) depth in the Pacific Ocean. Right: a *Parapagurus* sp. crab with an *Epizoanthus* coral on its back makes its way across a spectacular and unexpectedly densely packed field of ferromanganese nodules blanketing the seafloor of Gosnold Seamount in the North Atlantic.

Ferromanganese crust
Cobalt-rich ferromanganese crust from the central Pacific Ocean with the knobby appearance characteristic of older, thicker deposits. Note the mushroom coral, representative of the hard-substrate benthic fauna living on this crust.

Ferromanganese crusts and massive sulfides

We combine these two categories because, although the minerals, their deposition, and their geography differ, the mining methods are quite similar. Crusts form on the rocks of seamounts in a manner similar to the precipitation of manganese nodules. The crusts are similar in composition to the nodules (cobalt-rich ferromanganese containing nickel, platinum, and rare-earth elements), but they are not in convenient little packages waiting to be sucked up. They are a crust on the basalt rocks left by the volcanic action that formed the seamounts. The massive sulfide deposits at hydrothermal vents precipitate when the superheated hydrothermal fluids, saturated with minerals, mix with cold seawater and cannot hold the minerals in solution because of the cooling. The precipitate, in addition to sulfide minerals, contains copper, zinc, gold, and silver. We have discussed elsewhere the fauna that lives on the rocks in these two types of ecosystems (see pages 138–145 and 158–165).

To be mined, the crusts and the sulfide deposits first must be ground into small pieces that can be sucked up a riser pipe to the mining ship. This grinding is accomplished by a very large, bottom-crawling, remotely operated vehicle with a toothed rotating drum on its front. The drum chews up the surface layer of rock, either on the seamount or on the hydrothermal vent. The resulting rubble, along with the fauna that lived there (or its remains), is sucked up the riser to the ship, where the target ore is removed and remaining tailings are separated for disposal. As with nodule mining, the unwanted material can be disposed either at the surface (more convenient for the mining operation) or via return pipe to near the bottom (less disruptive of the pelagic environment, with perhaps somewhat less spreading across the benthic environment).

Gas hydrates

We mention gas hydrates here briefly because they have been discussed as a potential target for future mining. Very extensive deposits are known along continental slopes and in the Arctic. Methane could be separated from the hydrate and used as an abundant form of fossil fuel. Of course, burning of fossil fuels is a primary contributor to global climate change, and environmental advocates are firmly opposed to development of methane hydrates, since they are a contributor of greenhouse gases. Methods of safe extraction of the volatile hydrates have not yet been developed, so this form of mining is not likely to develop as quickly as the mineral mining mentioned previously.

Mining the deep ocean
Assembly workers provide a visual scale in this photo of a huge robotic deep-sea mining machine being built to dig copper and gold from the seabed off Papua New Guinea.

Jurisdiction and regulatory management

Because of the geological history of the deposits, petroleum extraction occurs almost entirely within the exclusive economic zones of coastal nations. The United Nations Convention on the Law of the Sea (UNCLOS) has extended this from 200 nautical miles (230 mi./370 km) offshore to the "base of the extended continental shelf" (i.e., where the continental rise meets the abyssal plain). This has resulted in a flurry of deep-sea mapping activities in order that coastal nations can claim as much territory as possible for mineral (especially petroleum) exploitation.

Conversely, most hydrothermal vents and seamounts and almost all nodule fields are in areas beyond national jurisdiction. Under UNCLOS, that places them within the jurisdiction of the UN International Seabed Authority (ISA) for those nations that have signed and ratified UNCLOS. ISA has authority to assign geographic areas for exploration and exploitation of mineral resources as well as to reserve areas for environmental protection and conservation. We hope that such areas, off-limits to mining, will be ecologically representative of the areas leased for mining.

RADIOACTIVE CONTAMINATION OF THE DEEP SEA

How our nuclear activity affects the oceans and ocean studies

A radionuclide is an atom (element) with an unstable nucleus (core). The nucleus of the atom has excess energy that is released by different types of radioactive decay. Radionuclides (also known as radioisotopes) in our environment are produced by minerals in the earth's crust, by cosmic rays hitting atoms in the atmosphere, and by human activities. Anthropogenic sources include shore-based leaks from nuclear power plants, nuclear fallout and runoff, nuclear bomb testing (with the largest effects from explosions in the atmosphere), sunk nuclear ships and reactors, and deliberate waste disposal. Potential problems of anthropogenic radionuclides in ecosystems include their potential harmful effect on living organisms due to long-term radioactive decay. The burial of nuclear waste in the seafloor, the opposite of deep-sea mining in some respects, will directly affect the slowly growing deep-sea species and their habitat via the suspension of sediment from the digging. The long-term effects of radionuclides are unknown, as the ocean waters move slowly in the deep sea but are not stagnant, and turbulent events sporadically occur throughout. The corrosive effects on the very long time scale (greater than 1,000 years) are not well studied.

Radionuclide tracers for circulation

Human-caused radioactive contamination over and in the deep sea can be turned to scientific purposes. As deep-sea and near-surface circulation patterns are difficult to capture by observational means, accidental radioisotope releases have been used to advantage. For example, nuclear fuel reprocessing plants in the United Kingdom (Sellafield, which borders the Irish Sea in northern England) and France (La Hague, which borders the English Channel in Normandy) leaked radioisotope material, notably cesium-137 and plutonium-239, mostly in the 1970s. After tracking the isotope distributions,

scientists confirmed the North Sea's surface circulation, establishing a major southward flow around the UK, along their east coast, past the European continent, and northward into the Norwegian Sea. The North Pacific Ocean surface circulation from west to east was revealed by a similar, more recent release of radioactive isotopes on a larger scale, from the 2011 nuclear power plant accident in Fukushima, Japan, which released large amounts of cesium-137 and cesium-134 in a relatively brief period of time. These isotopes were traced to the North American west coast. The tracing results from oceanographic sampling were in general accordance with previous observational and modeling work on the North Pacific surface circulation.

As may be expected, large-scale deep-sea basin circulations are even more difficult to establish and quantify, and purposeful attempts cost years of preparation. Examples of studies are sulfur-hexafluoride tracer release experiments that have been performed in the Brazil Basin and over its adjacent Mid-Atlantic Ridge topography. This nontoxic gas was artificially released via deep-towed instrumentation from research vessels. The experiments demonstrated and confirmed the notion that turbulent mixing is larger over sloping topography than it is in the basin interior or over a relatively flat, topographically featureless deep seafloor.

Radioactive material deposited from the atmosphere after the Chernobyl (Ukraine) nuclear power plant accident in 1986 enabled ad hoc studies of deep mixing in the Eurasian Black Sea. From the cloud of outfall of radioisotopes, again the cesium components were used to establish the mixing between oxygenated and anoxic waters in the Black Sea, mainly in the upper 330 ft. (100 m) from the surface.

Nuclear fuel reprocessing plants

Workmen building a waste release pipe outside La Hague nuclear fuel reprocessing plant, France (top), and aerial view of Sellafield, United Kingdom (bottom). In the 1970s these plants accidentally leaked radioisotope material into the sea. Although damaging to the environment, this enabled scientists to track water transport circulation patterns.

Radiocarbon dating

On a global scale, deep-sea waters have been dated via radiocarbon (carbon-14). Although radiocarbon is naturally formed in the atmosphere, humankind has added substantial levels of it via nuclear bomb testing since the 1950s. Most atmospheric bomb testing, in terms of number of tests and radiocarbon production, occurred between 1958 and 1963. This is a short time interval relative to the time scales of many ocean-circulation processes, such as ocean-basin turnover, which typically lasts 1,000 years. This radiocarbon time history, coupled with the level of contamination and the fact that radiocarbon becomes intimately involved in the oceanic carbon cycle, allows bomb-produced radiocarbon, albeit unintentionally, to be valuable as a tracer for several ocean processes. These include biological activity, air–sea gas exchange, thermocline ventilation, basin-wide ocean circulation, and upwelling. This radiocarbon has also aided in age determination of large deep-sea sharks and coral skeletons. Naturally formed radiocarbon has been used to confirm the large-scale, deep-ocean circulation that was previously established via distributions of deep-sea measurements of salinity, temperature, and oxygen.

A new development with radiocarbon determination of the large-scale ocean circulation has been the relative dating of the waters, which reveals that deep waters in the Atlantic Ocean are generally newer, or younger, than those in other ocean basins.

Nuclear bomb testing

A nuclear weapon test at Bikini Atoll, Micronesia, 1946. Radiocarbon produced from these tests proved useful for oceanographers to track ocean circulation patterns.

Fukushima

The release of radioactive isotopes following the 2011 nuclear power plant accident in Fukushima, Japan, helped oceanographers track the North Pacific Ocean surface circulation.

Radio isotope studies

Researchers lift a water sample bag on deck to be used for radioisotope testing.

OUR FOOTPRINT IN THE DEEP

Waste disposal and noise pollution in the deep sea

The ocean has long been viewed as a cheap and easy place to dispose of unpleasant things. For over a hundred years the residues of burned coal were tossed over the sides of steamships. This "clinker" can be found on the seafloor of abyssal plains, slopes, and canyons, along with broken crockery and other trash from the ships. The ocean was assumed to have limitless capacity to absorb humanity's discards. This was incidental waste—generated on the ocean and troublesome and costly to store, carry home, and dispose of more carefully—but other waste was deliberately towed out to sea and dumped.

Out of sight, out of mind

Deep Water Dump Site 106, named for its distance of 106 nautical miles (123 mi./196 km) from the former Ambrose Light off New York, is the largest sludge dump site in the world. Located just a few miles east of the edge of the continental shelf, it received waste from 1961 to 1992, dumped at the surface and left to percolate to the seafloor 1.5 mi. (2.5 km) below. Acids, unspecified chemicals, munitions, and over 30 million tons of sewage sludge from the New York–New Jersey metropolitan area were dumped.

The first disposal of radioactive waste in the ocean was off California in 1946, but 95% of global radioactive waste has been dumped in the Arctic and northeast Atlantic Oceans. The Arctic dumps tend to be shallower sites, although some are as deep as 985 ft. (300 m), but those in the northeast Atlantic are at depths of 4,920–16,400 ft. (1,500–5,000 m). Dumping of radioactive waste has been prohibited since 1994, but releases from nuclear plants still occur—for example, Japan intends to release treated wastewater containing tritium from the Fukushima Daiichi Nuclear Power Plant starting in 2023.

One of the biggest chemical dump sites was the Puerto Rico Dump Site, just north of Puerto Rico over the Puerto Rico Trench, which reaches depths of 19,700–26,250 ft. (6,000–8,000 m). From 1973 to 1981, almost 2.5 million metric tons of pharmaceutical waste were dumped. It is not clear what the impacts of that dumping were. Laboratory studies showed the waste to be toxic to a variety of marine organisms, but the response depended on the dilution—and the intention, of course, of disposal

at sea was that the chemicals would be highly diluted. Although disposal on this scale no longer takes place, there is current concern that other widely used pharmaceuticals, such as livestock antibiotics, human birth-control hormones, and antidepressants, enter the ocean through sewage and river systems. Again, these may be widely diluted, but persistent organic pollutants (which do not break down) ultimately end up in the deep ocean, as do heavy metals.

Following World War II, large numbers of munitions were also dumped in the ocean. Some were placed in quite shallow sites, and these have potentially serious consequences for fishers and for coastal communities. Over 1 million tons of munitions have been dumped in Beaufort Dyke, a submarine depression around 985 ft. (300 m) at its deepest part, between Northern Ireland and Scotland, and in recent years munitions have washed up on nearby beaches. Some ocean munition dumps contain chemical weapons too, and many are poorly documented.

The dumping described here is mostly historical. Although cleanup is almost impossible, the worst practices have at least been prohibited by national laws or international treaties. In the following pages, ongoing waste disposal is explored.

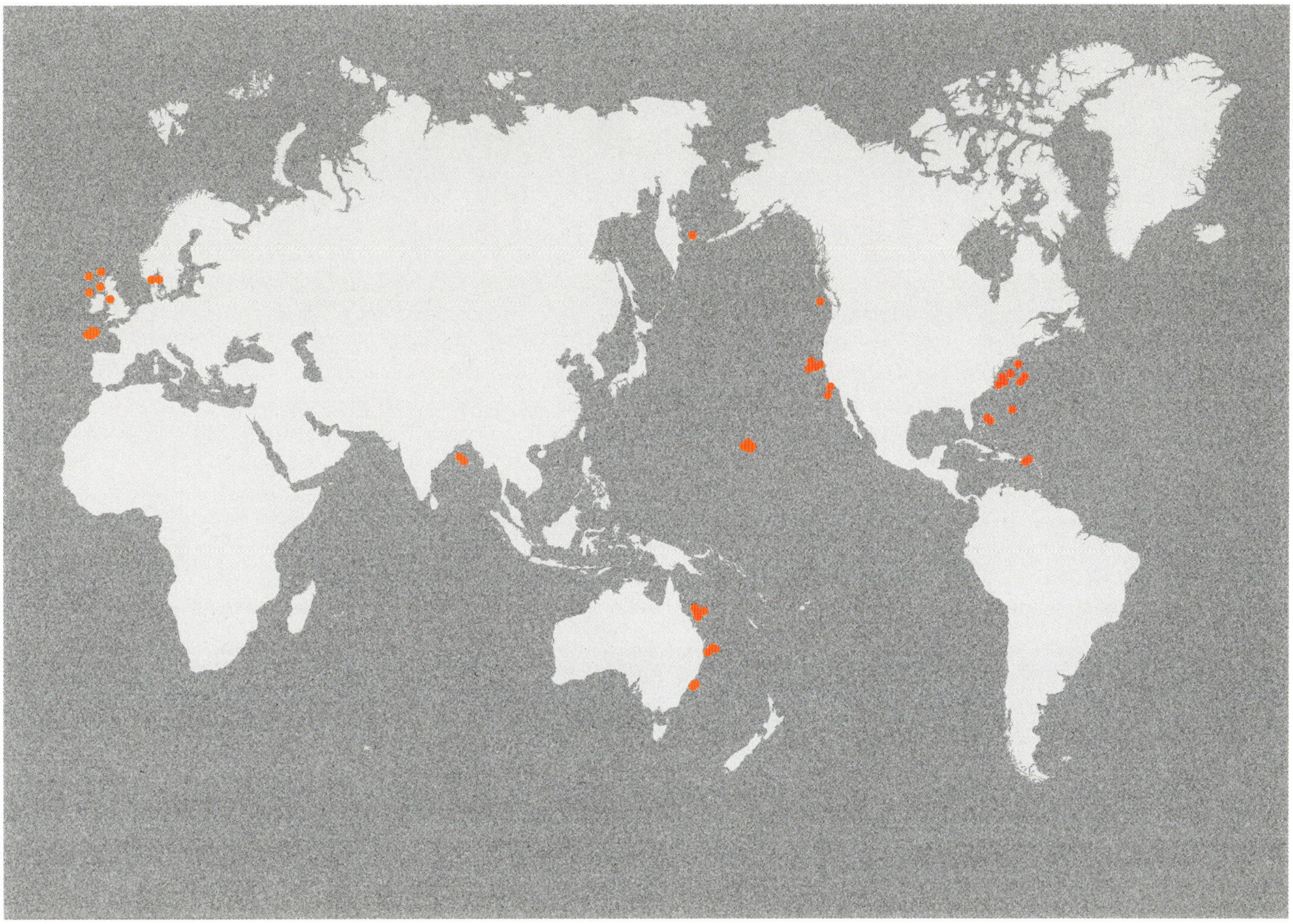

Oceanic dumping grounds

Location of deep-water chemical weapon dump sites in our oceans. (Source: The James Martin Center for Nonproliferation Studies, 2017.)

Deepwater dump sites

Bottom, left to right: Beaufort Dyke, dump site of World War II munitions; Deep Water Dump Site 106, dump site of sewage sludge and other waste; Puerto Rico Dump Site for pharmaceutical waste.

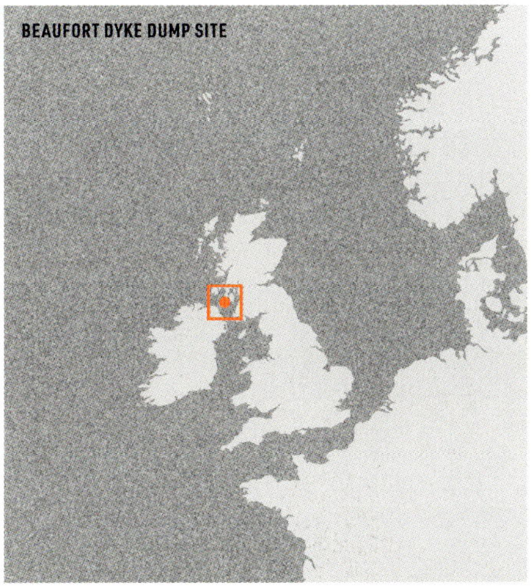
BEAUFORT DYKE DUMP SITE

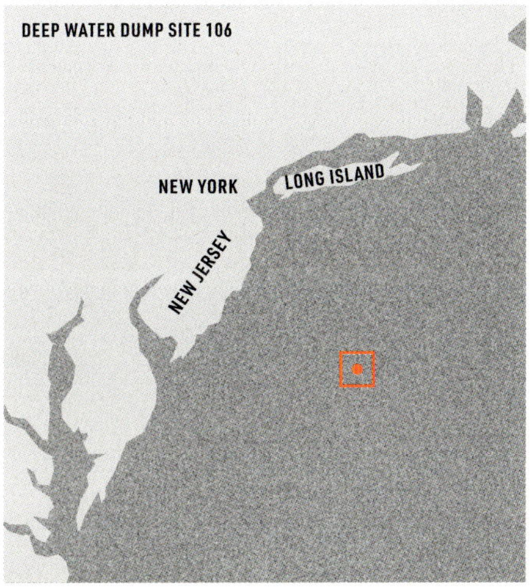
DEEP WATER DUMP SITE 106
NEW YORK
LONG ISLAND
NEW JERSEY

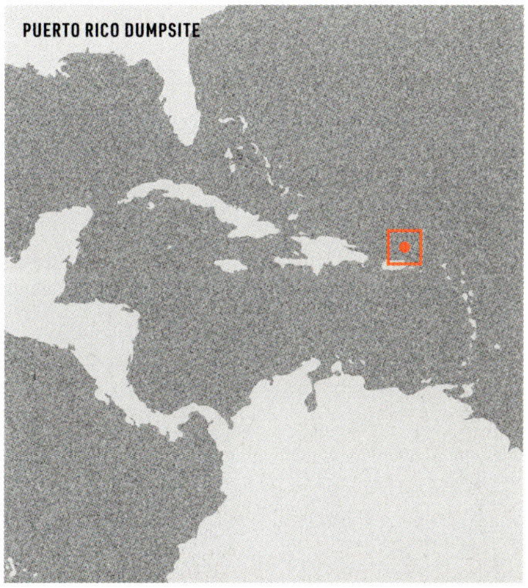
PUERTO RICO DUMPSITE

Mine tailings disposal

As the need increases for batteries and components for electronic devices and other technology, so does the need for metals, and mining is a dirty, destructive process that generates a lot of waste. Metals occur in mineral-rich rocks called ores. The metals are just a small proportion of the ore; a typical copper ore contains less than 1.0% copper, a typical gold ore less than 0.1% gold. Mine tailings are the waste from the extraction process. Tailings consist of a slurry of finely crushed rock polluted with heavy metals such as cadmium, lead, and mercury, and chemicals used in processing such as sodium cyanide. Mine tailings are usually pumped into land-based ponds, separated from watercourses by dams, to allow sedimentation.

Occasionally, large-scale collapse of tailings dams has occurred, causing human catastrophes, through immersion of villages in slurry, and massive environmental disasters. In 2015, a tailings dam collapsed at an iron ore mine near Mariana, Brazil, and released 40 million tons of tailings that killed entire fish populations. The fine sediment can clog gills and other sensitive surfaces of aquatic animals and may be more damaging than the chemicals in the tailings. Sedimentation resulting from the disaster was reported in Atlantic Ocean coastal waters more than 400 mi. (650 km) downstream. Because tailings dams are inherently unsafe, a condition exacerbated in some countries by climate or seismic activity, some mining operations dispose of their tailings directly in the ocean instead. Many such outfalls are shallow, but have the potential to create environmental hazards. Therefore, some operations prefer deep-sea tailings disposal (DSTD), discharging the slurry at depths greater than 330 ft. (100 m), usually at the edge of a drop-off, where the outflowing tailings can form a turbid current accelerating downward, speeding their path to deeper waters. Meeting these physical requirements, submarine canyons are often considered suitable sites; for example, a nickel mine discharges tailings into the head of the Basamuk Canyon in Papua New Guinea, and a copper and gold mine discharges tailings into the Senunu Canyon in Indonesia. This raises concern about the Vulnerable Marine Ecosystems (VMEs) found in submarine canyons, which are rich in corals and sponges (see pages 128–133), animal groups that rely on delicate, easily-clogged structures to feed. In the Basamuk Canyon, cold seeps (see pages 164–165) have been found in the vicinity of the Ramu refinery, and impacts on the environment were recorded a short time after discharge first commenced.

Only a few countries use DSTD, but it could be utilized in the future by nations comprising small islands of volcanic nature that have drop-offs close to shore, particularly as demand for metals rises to support increased production of microchips and batteries for electric vehicles.

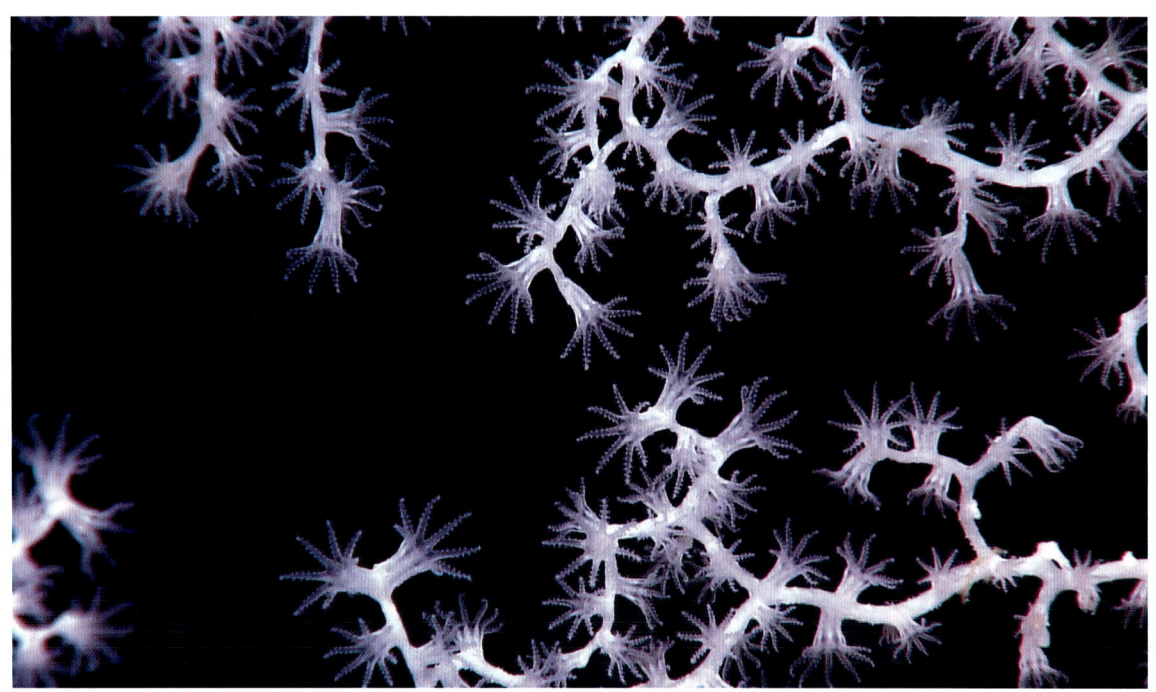

Vulnerable ecosystems

Left: the fine sediment associated with deep-sea tailings disposal can smother the delicate feeding tentacles of suspension feeders such as this deep-sea octocoral.

Collapse of tailings dams

Right: brown sludge spreading through the environment following the collapse of a land-based tailings dam in Brumadinho, Brazil.

Deep-sea tailings disposal (DSTD)

Tailings pipes are designed such that discharges flow downward from the end of the pipe to deeper water.

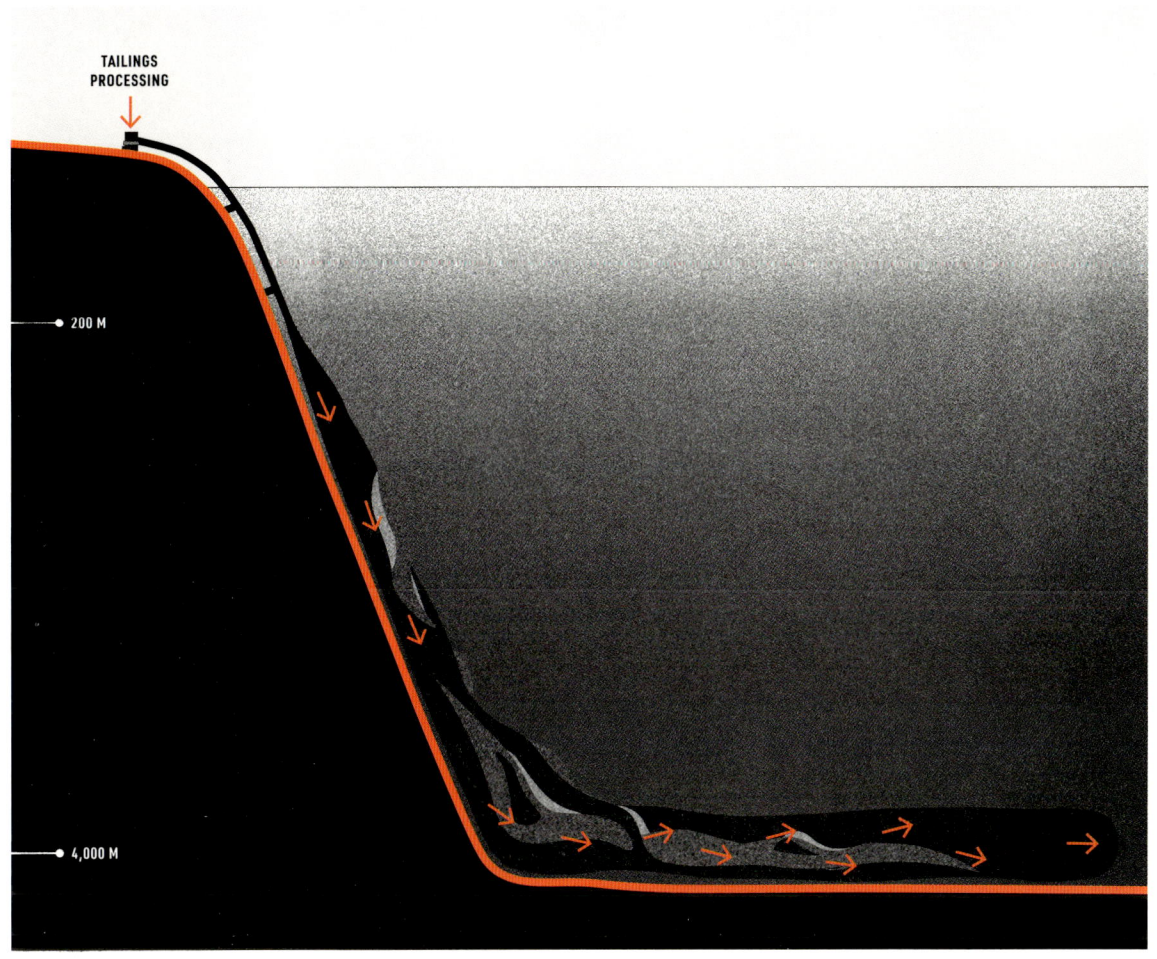

TAILINGS PROCESSING

200 M

4,000 M

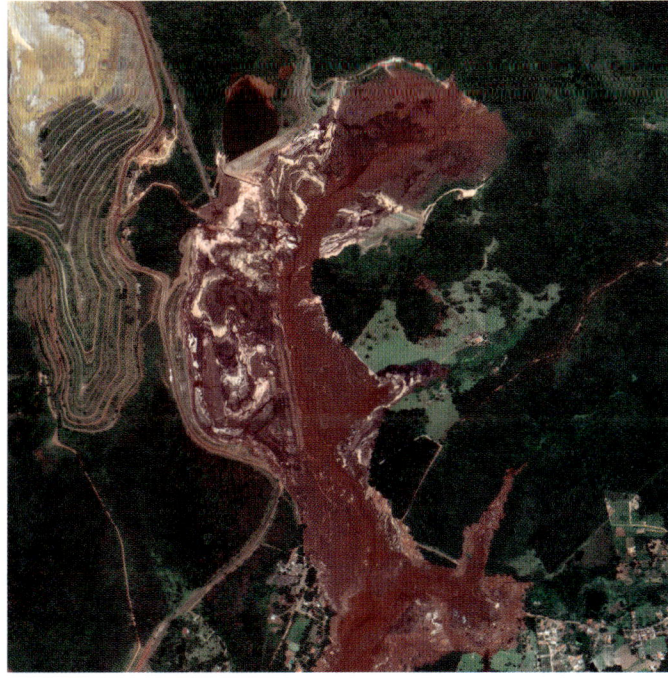

Marine litter

Marine litter includes plastics, metal, and glass. Estimates of how much litter enters the ocean varies, but it is thought to exceed 6 million tons per year. Some is deliberately dumped (either legally or illegally), and some is accidentally lost at sea (e.g., fishing gear, shipping containers). In 1992, a shipping container with nearly 30,000 plastic bath toys (ducks, beavers, turtles, and frogs) was lost in the North Pacific, and plastic toys were recovered around the globe over the next 15 years, at least one even reaching as far as Scotland in the northeast Atlantic, showing the remarkable persistence of plastic litter in our oceans. Bath toys are designed to float, but not all items that end up in our ocean are so buoyant. About 80% of ocean litter comes from land—the majority via rivers, including massive (accidental) input to rivers from terrestrial dump sites. Much of this, maybe 2 million tons per year, is plastic waste. It may float—indeed there are thought to be 3 million tons of floating plastic waste in the North Pacific—but the majority of it eventually sinks to a final destination on the deep seafloor.

There have been various efforts to remove waste from the ocean, but these mostly fail. Efforts should be focused on stopping litter entering the oceans by legislating against open dump sites and by reducing production and use of plastics. In March 2022, the United Nations Environment Assembly (UNEA) began negotiating a global agreement to end plastic pollution.

About 225 million tons of plastic are produced each year, and the cost of cleanup for coastal communities is enormous, so there is economic incentive for governments to act. A project to clean up plastic debris from Aldabra—an atoll in the Indian Ocean that is a World Heritage Site—cost US$225,000 to remove 25 tons. It is estimated that 500 tons remain. While much of the waste was lost fishing gear, there were nearly 2 tons of discarded flip-flops.

Scientists studying marine litter in European seas found that the amount of litter on the continental slope, in submarine canyons, on seamounts, and on mid-ocean ridges was higher than on the continental shelf. They found differences among these locations too. Plastic dominated in submarine canyons, suggesting the canyons accumulate litter from land, whereas seamounts were littered with lost or discarded fishing gear, indicative of the high fishing pressure in these rich habitats.

Plastic waste is also prevalent at the deepest depths, making up over 50% of all litter found below 19,700 ft. (6,000 m) in the Pacific and Indian Oceans. Worse, almost all of this debris is single-use plastic. Government policies preventing the manufacture and sale of single-use plastic could protect our deep oceans and the animals that inhabit them from this plastic waste. The deepest litter known is a plastic bag observed at the bottom of the Mariana Trench at 35,755 ft. (10,898 m). It will probably remain there forever.

Litter composition

The litter present in deep-sea habitats reflects human activities in the waters above. A large proportion of the litter items found at submarine canyons are plastic waste, whereas at seamounts and ocean ridges, there is a higher percentage of lost or discarded fishing gear. The large proportion of clinker found on continental slopes and in deep basins is the discarded burned coal remains from the engines of old steam ships. (Source: Pham et al., 2014.)

CONTINENTAL SHELVES   CONTINENTAL SLOPES   SUBMARINE CANYONS

SEAMOUNTS, BANKS, AND MOUNDS   OCEAN RIDGES   DEEP BASINS

- Plastics
- Fishing gear
- Metal
- Glass
- Other
- Clinker

Deep-sea plastic

Plastic is increasingly prevalent in the deep sea, and may be colonized by animals such as these anemones (top) and this echinothuriid urchin (bottom).

Microplastics

As the name suggests, microplastics are very small pieces of plastic, from 1μm (a thousandth of a millimeter) to around a millimeter. There are many different types of plastics—for example, PVC (polyvinyl chloride), and polypropylene—and hence many different types of microplastics, and their chemical composition affects their potential toxicity.

Microplastics can be primary or secondary. Primary microplastics are manufactured small—for example, the microplastic beads in some facial scrubs and toothpastes. Secondary microplastics are degraded from plastic waste. This slowly happens to all plastic litter in the sea, but microplastics can also be transported on wind currents for as far as 60 mi. (100 km) and so can enter the sea from land, as well as via rivers. Sources include fine debris from car tires and fibers from our synthetic clothes flushed into watercourses from washing machines. The deep sea is considered to be a microplastic sink. Like suspended sediment and marine snow, the ultimate fate of microplastics is downward. Although studies on deep-sea microplastics are in their infancy, high levels of microplastics have been found in deep-sea sediments across the globe.

Microplastics have been found in the guts of amphipods from the deepest ocean trenches in the Pacific Ocean, at depths of 22,970–35,730 ft. (7,000–10,890 m). Microplastics are also ingested by deep-sea corals. It is likely that corals aren't able to detect the difference between a plastic particle and a food particle, and they suffer two immediate consequences: (1) the ingested plastic particle has no nutritional value; and (2) they expended energy obtaining it. Experimental work has shown that the deep-sea reef-building coral *Desmophyllum pertusum* ingests less food, grows less, and makes less calcium carbonate skeleton when exposed to microplastics. There is also concern that microplastics can provide an adsorption surface for heavy metals, chemicals, and bacteria, which might have further detrimental effects on any organism that consumes them. There are already many pressures (e.g., warming temperatures, ocean acidification) on deep-sea corals, and there is concern that this extra pressure may further weaken their resilience to climate change.

Contaminated by microplastics

Microplastic threads are visible in this plankton sample from the South Atlantic Ocean. Their ultimate resting place will likely be deep benthic sediments.

Deep-sea amphipod
Eurythenes plasticus

The deep-sea amphipod species from the Mariana Trench so named because it was found with plastic in its guts.

Breakdown of microplastics

Microplastics are estimated to be up to 10,000 times more abundant in deep-sea sediments from the Atlantic and Indian Oceans than in surface waters. Some will result from in situ degradation of plastic waste, but much will settle from the water column.

Ocean noise

Anthropogenic noise is increasingly being recognized as a concern in the oceans, and it is not limited to shallow or coastal waters. Animals such as whales use sound to find prey and to communicate, and noise produced by human activity can hinder these behaviors. Extreme noise also generates pressure waves that can be damaging to internal organs. Giant Squid (*Architeuthis dux*) with damaged internal organs and shattered ear bones (statoliths) were found washed up on the shore of northern Spain following seismic surveys (see also page 245). In the North Atlantic, analysis of more than 10 years' worth of seismic survey data and whale-stranding records suggested that the number of strandings of Long-finned Pilot Whales (*Globicephala melas*) was increased by offshore seismic surveys. Similarly, analyses of acoustic data sets and known US military training exercises around the Mariana archipelago in the Pacific Ocean have found that strandings of Cuvier's Beaked Whale (*Ziphius cavirostris*)—the deepest-diving whale, which can reach depths of almost 2 mi. (3 km)—were preceded by mid-frequency sonar, used in antisubmarine warfare.

Seismic noise is intended to reach the seafloor and hence the deepest parts of the ocean, but other anthropogenic noise reaches these depths too. In 2015, scientists put a hydrophone array in the Challenger Deep, an area in the Mariana Trench more than 6.2 mi. (10 km) below the surface, for 24 days. They heard natural sounds from baleen and toothed whales and from sea-surface winds and waves, including from a passing typhoon, as well as from earthquakes. But they also heard ship propellors, despite the depth of the hydrophone, air guns (from seismic surveys), and sonar. If these noises can be heard in the Challenger Deep, they are likely penetrating the entire ocean.

Scientists are increasingly calling for "quiet-seas" reserves as a form of marine protected area. Such refuges are essential to protect species that rely on sound.

Seismic damage

Giant Squid (*Architeuthis* sp.) stranded on a beach in Norway. Not all squid strandings are related to seismic noise, but there is increasing evidence that seismic noise can cause strandings in multiple marine species.

Cuvier's Beaked Whale
Ziphius cavirostris

Cuvier's Beaked Whales that strand following military sonar use often have chronic and acute tissue damage. This likely results from gas bubble formation as they surface too quickly as a stress response to the sound waves—akin to decompression sickness in divers.

Seismic surveys

There is increasing evidence that offshore seismic activity causes some animals to strand.

GLOBAL CLIMATE CHANGE AND OCEAN ACIDIFICATION

Effects of burning fossil fuels

It is clear that human actions warm the earth, which includes warming of the ocean (see the UN Intergovernmental Panel on Climate Change, or IPCC, 2021 report). This effect is known as global climate change, and it follows from the burning of fossil fuels. The burning leads to human-driven increased levels of carbon dioxide (CO_2) in the atmosphere. Because of the coupling between atmosphere and ocean via surface exchange, this also leads to more CO_2 dissolving into the ocean. The ocean's average pH value (a measure of acidity) is now around 8.1 near its surface. Values of pH greater than 7 indicate basic, or alkaline (low acidity) conditions. As the ocean continues to absorb more CO_2, the pH decreases and the ocean becomes more acidic, currently by about 0.5% in 15 years. The seawater will not actually become acidic (pH below 7), even with all the CO_2 that is dissolving into the ocean, but the changes toward increasing acidity will impact ocean life.

When CO_2 dissolves in seawater, the lowering of the pH binds carbonate ions to hydrogen ions and makes them less available for organisms. Corals, oysters, mussels, and many other organisms need carbonate ions to build shells and skeletons by binding carbonate with calcium ions. Acidification thus results in less strong shells and skeletons. Carbon dioxide is naturally in the air; plants need it to grow, and animals exhale it when they breathe. But, thanks to humankind burning fuels, there is now more CO_2 in the atmosphere than at any time in the past 15 million years. Most of this CO_2 collects in the atmosphere, and because it absorbs heat from the sun, creates a blanket around the planet, warming its temperature. But some 30% of this CO_2 dissolves into seawater, where it doesn't remain as floating CO_2 molecules. A series of chemical changes break down the CO_2 molecules and recombine them with others, affecting the pH.

So far, ocean-surface pH has dropped from 8.2 to 8.1 since the industrial revolution. If we continue to add CO_2 at current rates, the ocean-surface pH is expected to fall another 0.3–0.4 pH units by the end of the 21st century. A drop in pH of 0.1 may not seem like a lot, but the pH scale is logarithmic, like the Richter scale for measuring earthquakes. The expected ocean-surface pH values of 7.8 or 7.7 are already found in the deep sea because of natural processes. The CO_2 that dissolves in the ocean surface descends into the ocean depths with dead biotic materials and accumulates there. Thus, ocean acidification naturally occurs from the deep sea upward.

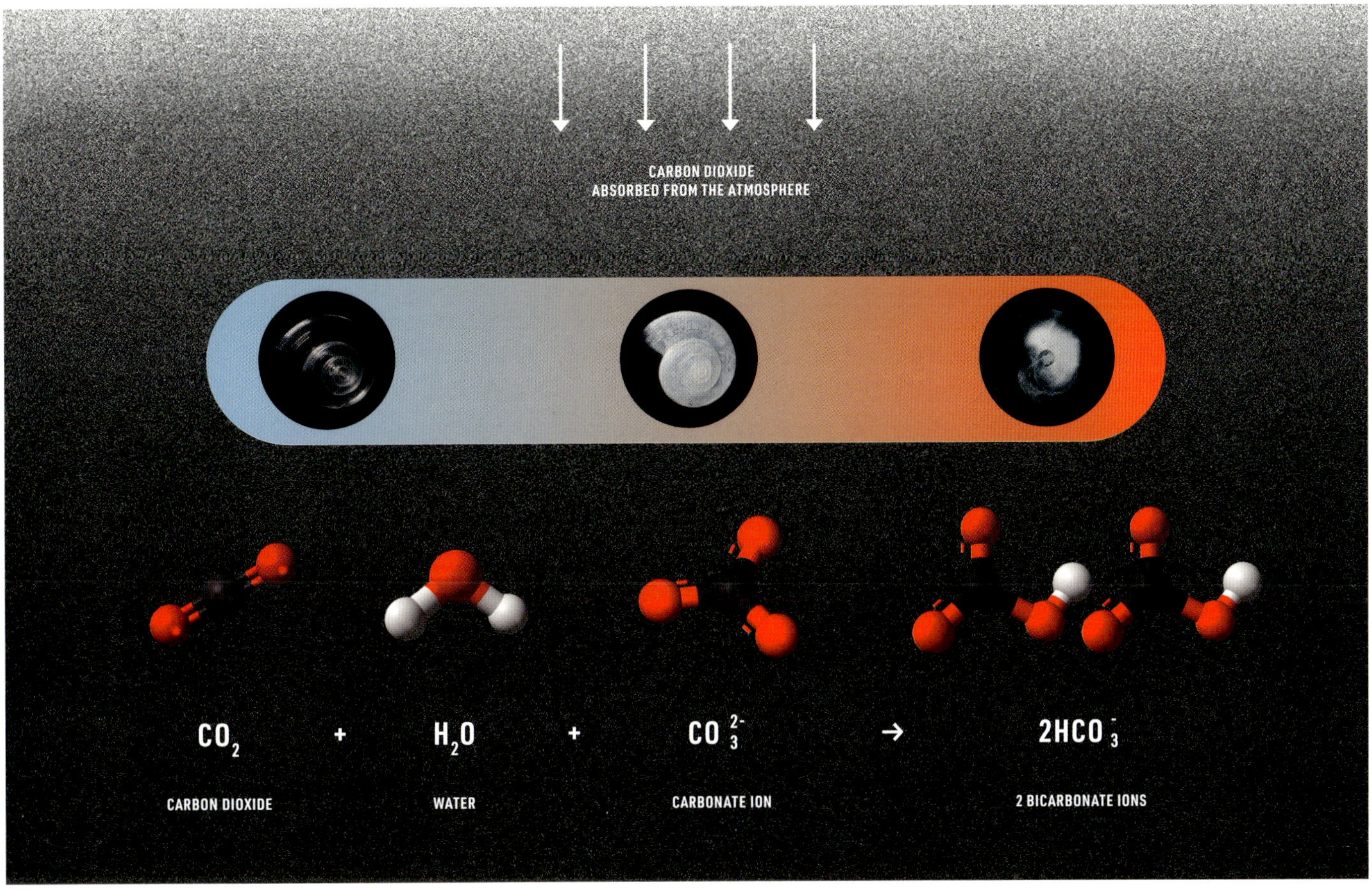

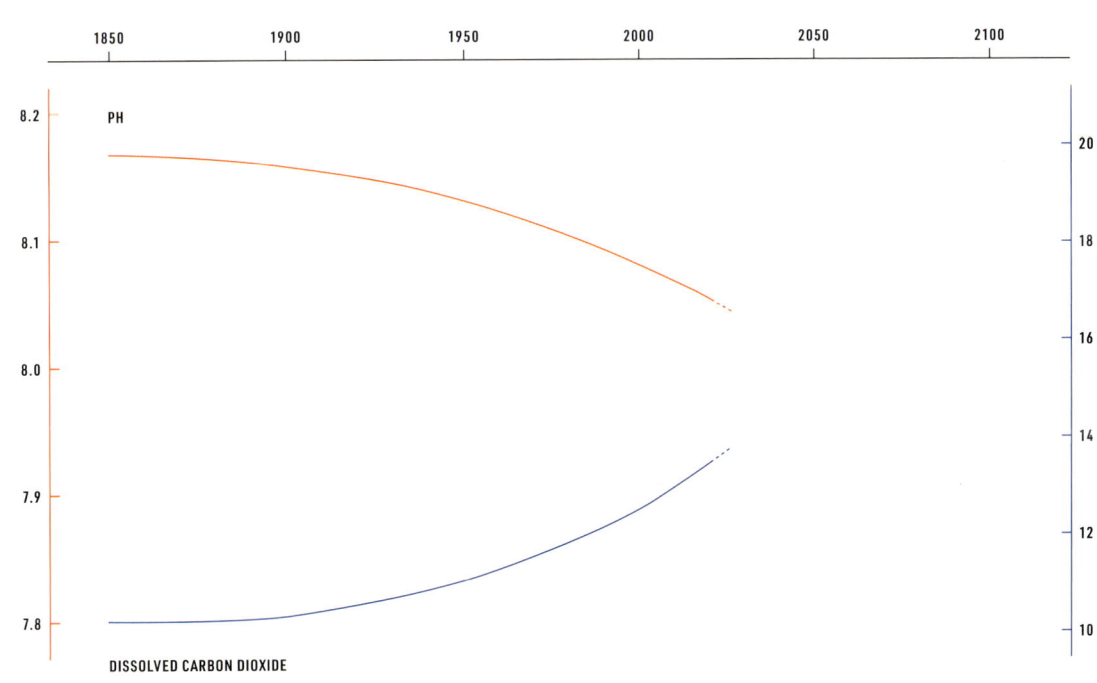

**CARBON DIOXIDE
ABSORBED FROM THE ATMOSPHERE**

$$CO_2 \quad + \quad H_2O \quad + \quad CO_3^{2-} \quad \rightarrow \quad 2HCO_3^{-}$$

CARBON DIOXIDE   WATER   CARBONATE ION   2 BICARBONATE IONS

Effects of ocean acidification

Sketch of potential effects of ocean acidification. Extra uptake of carbon dioxide by ocean waters involves carbonate ions that become less available for shell building.

Ocean surface pH

Due to human actions the ocean has become increasingly more acidic (a lower pH) since the 1850s. This is set to continue, impacting many ocean species.

PH

DISSOLVED CARBON DIOXIDE

Consequences of acidification

Human-caused acidification will occur more slowly in the deep sea than in surface seawater. But its ecological effects may nonetheless be severe because of the assumed greater sensitivity of the biota. Deep-sea organisms live in a cold, dark environment with low nutrient inputs and reduced reliance on visual interactions between predator and prey. These organisms generally grow slowly and have lower metabolic rates than comparable taxa living in warmer surface waters. In animals, slow metabolism typically corresponds to a low capacity for gas exchange (i.e., oxygen transport and CO_2 release) and reduced enzyme function, including those linked to acid–base (pH) regulation. The environmental stability of the deep sea over long time scales is also postulated to have reduced the tolerance of deep-sea species to environmental extremes through the loss of more tolerant genotypes, thereby decreasing the potential for adaptation to future ocean acidification.

Some likely consequences of future ocean acidification in deep-sea waters can be inferred from organisms inhabiting hydrothermal vent and cold seep environments (see pages 158–167), which often, but not always, have low pH levels. Echinoderms (animals such as sea cucumbers, sea stars, and urchins; see pages 84–85), and some other calcium-carbonate forming taxa are often absent near hydrothermal vents and cold seeps. This is presumably a result of the low ambient pH or other stressful environmental factors—including high concentrations of toxic metals such as cadmium, silver, and others—that may limit distribution of some fauna. Other vent and seep taxa thrive, in spite of high CO_2 levels, and in some cases exploit the energy-rich conditions in these environments to sustain anomalously high rates of growth. Adaptations promoting success for some animals at vent and seep habitats are likely to have evolved over long periods. It remains unknown whether more typical deep-sea animals are capable of adapting to future changes in deep-sea chemistry caused by acidification.

Dissolving shells

Left: in laboratory experiments, this pteropod shell dissolved over the course of 45 days in seawater adjusted to an ocean chemistry projected for the year 2100.

Unhealthy pteropod

Right: an unhealthy pteropod shows the effects of ocean acidification—the white lines on its upper surface, the cloudiness, and abrasions are all effects of dissolution.

Effects on deep-water corals

Marine biological studies under laboratory conditions have demonstrated that calcification rates in the cold-water coral species *Desmophyllum pertusum* were reduced by an average of 30% and 56%, respectively, when the pH was lowered by 0.15 and 0.3 units relative to ambient conditions. Despite this response, calcification rates in this species did not stop completely. However, few studies have existed until recently, and more scientific work is needed. As has been shown for shallow-water corals, genetic variability is an important factor in understanding effects of acidification on deep-sea corals.

Compared to other oceans in the world, the North Atlantic possesses a natural balancing or feedback physical mechanism that has helped it to slow acidification in the deep sea in the past. There, deep, dense-water formation causes the less acidic surface water to sink and mix with the more acidic deep water. This process has kept the deep Atlantic Ocean far less acidic than the North Pacific Ocean, where surface and deep waters have lower pH values than Atlantic waters at similar depths. The less acidic North Atlantic may seem a safe haven for calcifying marine organisms such as shellfish and, particularly, the deep, cold-water hard corals. However, as the acidic waters potentially corrode their shells and skeletons, North Atlantic waters deeper than a critical depth also may become too acidic for these organisms to survive. Nevertheless, cold-water hard corals have been found as deep as 6,560 ft. (2,000 m), well below the ocean's depth zone of minimum pH value. By comparison, the same corals struggle to survive at 1,640 ft. (500 m) in the Pacific Ocean.

The effects of acidification are compounded by other types of human-induced damage (discussed in depth throughout this chapter). The primary human impact on deep cold-water corals to date is from deep-water trawling. Trawlers drag nets across the ocean floor, disturbing sediments and breaking and destroying deep cold-water corals. Another harmful activity is longline fishing. Oil and gas exploration also damages deep cold-water corals. A study conducted in 2015 found that coral injury observed in the Mississippi Canyon in the Gulf of Mexico increased from 4–9% of the population before the *Deepwater Horizon* oil spill to 38–50% after the spill.

Deep cold-water corals grow slowly, so recovery takes much longer than in shallow waters where nutrients are far more abundant. Deep-sea cold-water coral longevity estimates range from decades (at a minimum) to centuries, with at least some species living more than 1,000 years. The slow growth and long recovery time of coral mound communities put them at great risk for lasting damage from human activities.

Deep-water corals
Desmophyllum pertusum
Deep-sea coral ecosystems are threatened by the effects of climate change. Above, healthy *Desmophyllum pertusum* in southern California's Channel Islands National Marine Sanctuary. Left: *Desmophyllum pertusum* interspersed with *D. pertusum* rubble, Gulf of Mexico. Deep-water fishing only adds to the impacts already affecting delicate coral reefs.

Indirect effects on the deep sea

Whereas the main human contributor to climate change is the high demand for energy, the wide variety of indirect effects in the deep sea includes tidal energy reduction. Tidal energy is considered by some to be a potential alternative to fossil fuels for power. Although there is relatively very little tidal energy (it is estimated to cover considerably less than 1% of present-day humankind's energy consumption demand), the extraction of a substantial portion of it, for example by submerged turbines, could strongly affect deep-sea life. A reduction of tidal motions implies a reduction of turbulent energy, as most of the deep-sea turbulence is generated via the breaking of internal waves (see pages 60–65). Capture of tidal energy would result in a reduction of turbulent mixing and exchange across the vertical density stratification and hence a reduction of the transport of nutrients and suspended matter. The reduction of turbulent mixing also results in a reduction of heat transport downward from the surface into the deep sea. Extraction of the ocean tides would thus eventually lead to destruction of the deep-sea stratification, turning the deep sea into a lifeless, stagnant pool of cold water.

SHIPWRECKS AND OTHER SINKINGS

Wrecking the marine ecosystem

The sinking of the tanker *Prestige*, mentioned in "Petrochemicals and Mining" (pages 245 and 247), is an example of a surprisingly common impact of human commerce on the deep sea. It is difficult to determine an annual average number of ships sunk in deep waters (or even on the high seas—i.e., areas beyond national jurisdiction), since disappearances and the reasons for them may go unrecorded, and piracy and other practices confuse matters. Estimated average sinkings range from about 10 to over 100 per year. An average year, of course, doesn't include wartime, when countries are intentionally sinking each other's ships. A lot of ships were sunk in the deep during both world wars, and such sinkings continue with smaller maritime conflicts.

One published estimate, based on an assumption that just 10% of the ships lost on the high seas due to accidents or weather sink into the deep sea, is that about 18 large ships impact the deep in an average year. Why don't we hear more about these catastrophes? Many of these ships are registered with "flags of convenience" (countries with no real connection to the ship but where safety requirements are lax) and manned by crews from developing nations, and their activities often fly under the radar. In addition to accidental sinkings, some ships are discarded by scuttling at sea or are used for military target practice. These target ships—unlike those lost by accident—may have been cleaned of fuel and lubricants prior to sinking.

Shipwrecks and the ecosystem

Regardless of whether we have heard about it, when a sinking ship reaches the bottom of the ocean it does several things to the local ecosystem, both short-term and long-term. The area is torn up and crushed by the mass settling on it. Because most of the deep-sea bottom is sedimentary, the wreck suddenly adds hard substrate, which can be colonized by a completely different set of organisms. The addition of different fauna can cause shifts in the food web.

Ships are treated with special paints to prevent corrosion as well as toxic antifouling paint on the bottom. As the paint flakes off a shipwreck and all that metal slowly corrodes, it changes the local chemistry of the water and sediment. Whereas wooden shipwrecks are, or used to be, prime habitat for xylophagous (wood-eating) bivalves and associated fauna, iron and steel hulls develop "rusticles," dangling structures that form from microbe-mediated corrosion and then flow down the sides. The wreck of RMS *Titanic* is famously disintegrating into rusticles. In addition to its construction materials (e.g., steel, wood, composites), every modern ship contains a large amount of fuel, along with lubricants, insulators (e.g., PCBs), supplies for the crew, and so on. Several nuclear-powered ships, mostly submarines, have been lost in accidents, and more have been sunk for disposal. These wrecks would include their radioactive propulsion units as well as nuclear fuel and construction materials that became radioactive during the use of the propulsion system.

Wreck of the RMS
Titanic

Sponges, corals, and
feather stars take
advantage of rare hard
surfaces in the abyssal
North Atlantic. Iron is
removed from the steel
by bacteria creating the
brown-colored growths
or "rusticles."

WWII bomber

In addition to shipwrecks, aircraft also crash at sea and sink into the deep. This wreck of a B-29 Superfortress from World War II, resting upside down on the seafloor near Tinian Island (part of the Northern Mariana Islands) is investigated by an ROV.

Gun carriage

A cannon resting on top of a pile of artifacts at the site of the wreck of an early-19th-century, wooden-hulled sailing vessel deep in the Gulf of Mexico.

Shipping container

The common use of shipping containers in modern transportation has resulted in many containers that are lost and sunk at sea, along with their varied contents, even when the carrying ship does not sink. After seven years on the deep seafloor, this sunken shipping container had been colonized by a variety of deep-sea animals.

Cargo sinkings

Cargo ships contain all the typical operational materials and may also be carrying very large quantities of almost anything. An estimated 90% of all goods in worldwide commerce are transported by ships. The variety of potential cargoes is matched by their potential impacts to local ecosystems. Whereas much cargo is packaged into shipping containers (see below), bulk materials, both liquid and solid, are generally transported in large cargo holds within the ships. If the holds are not immediately breached by the sinking, they may open many years later, due to the slow action of corrosion, releasing their contents. When a large volume of organic material, such as wood pulp for paper production, is enclosed in a flooded cargo hold, it may become anoxic and establish a local environment functionally similar to a whale fall. Observations exist of such cargoes being colonized by chemosynthetic organisms similar to those that form the communities found on cold seeps or the chemosynthetic phase of community succession on a whale fall (see pages 166–167).

Some warships are cargo vessels, but others are fighting ships, which, when sunk, carry down their load of munitions, whether nuclear, conventional, or even chemical or biological weapons. If their weapons are rocket propelled, as is now common, each projectile contains its own toxic fuel as well as explosive warheads. This, again, is in addition to fuel, lubricants, and other substances.

Shipping containers

Much of the volume and most of the variety of cargo is now transported in shipping containers. These standardized steel boxes are stacked and secured on the tops, as well as the insides, of ships. However, the securing system does not always hold. Approximately 1,000, give or take a few hundred, shipping containers are lost at sea each year, usually in heavy weather. These lost containers may float around for a while (endangering passing vessels) but eventually sink. Very many sink into the deep sea. Each then becomes, functionally, a little wrecked cargo vessel, although without fuel and other substances involved in propulsion. Much of the stuff inside your home, office, or garage may have been carried in shipping containers. Similar stuff may be sitting at the bottom of the ocean in corroding containers.

Other deposits in the deep

Although most of the human-made deposits in the deep sea are related to ships, transoceanic aircraft also occasionally crash into the open ocean. These accidents are less common than shipwrecks but more likely to receive public attention. Like ships, airplanes sink into the deep sea with, in addition to their main painted-aluminum structures, fuel, electronics, fabrics, luggage, and cargo.

An old song states that "many brave hearts are asleep in the deep." It may not be very pleasant to consider, but sinking ships and air crashes also result in the deposition of human bodies in the deep. This, again, is much more common during naval warfare. For the scavenging communities of the abyss, a human body would be essentially like any other moderate-size food fall, which the scavengers have evolved to find and exploit. These human bodies would become part of the deep-sea food web and the global transfer of carbon into the deep sea.

APPENDICES

CLASSIFICATION OF DEEP-SEA SPECIES

The classification of organisms in this book follows the World Register of Marine Species (www.marinespecies. org). Common names are given where available, but many deep-sea species lack common names. Throughout we follow the standard convention of droping the "ae" from "-idae" for shortening family names for English use, such that we refer to family Histioteuthidae as "histioteuthids" and family Moridae as "morids."

KINGDOM CHROMISTA

Phylum Foraminifera	foraminifera or "forams"
· · · · · *Superfamily* Xenophyophoroidea	xenophyophores

KINGDOM ANIMALIA

Phylum Porifera	sponges
· *Class* Demospongiae	demosponges
· · *Family* Cladorhizidae	carnivorous sponges
· *Class* Hexactinellida	glass sponges or hexactinellid sponges
Phylum Cnidaria	
· *Class* Hydrozoa	hydrozoans (includes jellyfishes and polyps)
· · *Order* Siphonophorae	siphonophores (colonial jellies)
· · · *Suborder* Physonectae	physonect siphonophores
· *Class* Cubozoa	box jellies
· *Class* Scyphozoa	true jellyfishes
· *Class* Anthozoa	
· · *Subclass* Ceriantharia	tube anemones
· · *Subclass* Hexacorallia	six-arm corals
· · · *Order* Actinaria	true anemones
· · · *Order* Zoantharia	zoantharians
· · · *Order* Scleractinia	stony corals
· · · *Order* Antipatharia	black corals
· · *Subclass* Octocorallia	octocorals (eight-arm corals)
· · · *Order* Alcyonacea	
· · · · *Suborder* Alcyoniina	soft corals
· · · · · *Family* Alcyoniidae	includes mushroom corals
· · · · · *Family* Nephtheidae	includes cauliflower corals
· · · · *Suborder* Calcaxonia	
· · · · · *Family* Keratoisididae	bamboo corals
· · · · *Suborder* Holaxonia	
· · · · · *Family* Gorgoniidae	whip corals
· · · *Order* Pennatulacea	sea pens
Phylum Ctenophora	ctenophores or comb jellies
Phylum Chaetognatha	arrow worms
Phylum Annelida	
· *Class* Polychaeta	polychaetes or bristle worms
· · *Subclass* Errantia	mobile bristle worms
· · · *Order* Phyllodocida	phyllodocids
· · · · *Family* Aphroditidae	aphroditid worms, includes sea mice
· · · · *Family* Nereididae	nereid worms or ragworms
· · *Subclass* Sedentaria	sessile bristle worms
· · · *Order* Sabellida	includes tube worms
· · · · *Family* Siboglinidae	highly specialized family of tube worms found at cold seeps and hydrothermal vents (includes species of *Riftia*)
· · · *Order* Terebellida	includes tube worms
· · *Subclass* Echiura	spoon worms
· *Class* Sipuncula	peanut worms

Phylum Mollusca	
· *Class* Gastropoda	gastropods or sea snails (includes limpets and shell-less snails)
· · · *Order* Pteropoda	shelled and naked pteropods (includes sea butterflies and sea angels)
· · · *Order* Nudibranchia	naked-gill snails (a group of shell-less snails)
· · · *Order* Littorinimorpha	
· · · · · *Superfamily* Pterotracheoidea	sea elephants or heteropods
· *Class* Bivalvia	bivalves (includes clams, mussels, oysters, and relatives)
· *Class* Cephalopoda	cephalopods (includes octopods, squids, bobtails, *Spirula*, and relatives)
· · · *Order* Oegopsida	oceanic squids
· · · · · *Family* Chiroteuthidae	chiroteuthid squids
· · · · · *Family* Histioteuthidae	histioteuthid squids (jewel squids)
· · · · · *Family* Mastigoteuthidae	mastigoteuthid squids (filamentous squids)
· · · · · *Family* Ommastrephidae	ommastrephid squids (flying squids, includes shortfin squids)
· · · *Order* Sepiolida	bobtails
· · · *Order* Octopoda	eight-arm cephalopods
· · · · *Suborder* Cirrata	cirrate, finned, or dumbo octopods
· · · · *Suborder* Incirrata	finless (incirrate) octopods
Phylum Arthropoda	
Subphylum Crustacea	
· *Class* Copepoda	
· · · *Order* Harpacticoida	harpacticoid copepods
· *Class* Malacostraca	
· · *Order* Decapoda	shrimps, crabs, and lobsters
· · · · · *Family* Sergestidae	sergestid shrimps
· · · · · *Family* Galatheidae	squat lobsters
· · · *Order* Euphausiacea	
· · · · · *Family* Euphausiidae	shrimp-like krill
· · · *Order* Amphipoda	amphipods (scuds or sideswimmers)
· · · *Order* Cumacea	hooded shrimps
· · · *Order* Isopoda	isopods (pillbugs)
· · · · *Family* Munnopsidae	long-legged isopods
· · · *Order* Lophogastrida	lophogastrids
· · · *Order* Mysida	opossum shrimps
· · · *Order* Tanaidacea	tanaids
· *Class* Thecostraca	
· · · *Order* Balanomorpha	acorn barnacles
· · · *Order* Scalpellomorpha	stalked barnacles
· *Class* Ostracoda	seed shrimps or ostracods
Subphylum Chelicerata	
· *Class* Pycnogonida	sea spiders
Phylum Tardigrada	water bears or tardigrades
Phylum Echinodermata	
· *Class* Holothuroidea	sea cucumbers
· *Class* Asteroidea	sea stars
· · · *Order* Brisingida	brisingids
· · · · · *Family* Brisingidae	
· · · · · *Family* Freyellidae	
· *Class* Ophiuroidea	brittle stars
· *Class* Crinoidea	feather stars and sea lilies
· *Class* Echinoidea	sea urchins
· · · *Order* Spatangoidea	heart urchins
· · · *Order* Echinothurioida	
· · · · · *Family* Echinothuriidae	echinothuriids

Phylum Hemichordata
· *Class* Enteropneusta — acorn worms

Phylum Chordata
Subphylum Vertebrata
· *Infraphylum* Agnatha — jawless fishes
· · · · · *Class* Myxini
· · · · · · *Order* Myxiniformes
· · · · · · · *Family* Myxinidae — hagfishes
· *Infraphylum* Gnathostomata
· · *Parvphylum* Chondrichthyes
· · · · *Class* Elasmobranchii — cartilaginous fishes
· · · · · · *Infraclass* Selachii
· · · · · · · *Superorder* Galeomorphi
· · · · · · · · *Order* Carcharhiniformes — includes catsharks
· · · · · · · *Superorder* Squalomorphi
· · · · · · · · *Order* Hexanchiformes — cow sharks, six-gilled sharks, frilled sharks

· · · · · · · · *Order* Squaliformes — true sharks
· · · · · · · · · *Family* Dalatiidae — kitefin sharks
· · · · · · · · · *Family* Etmopteridae — lantern sharks
· · · · · · · · · *Family* Somniosidae — sleeper sharks
· · · · · · · *Infraclass* Batoidea
· · · · · · · · *Order* Rajiformes — deep-sea rays and skates
· · · · · *Class* Holocephali
· · · · · · *Order* Chimaeriformes — chimaeras (also known as rabbitfishes or ghost sharks)

· · · · · · · *Family* Chimaeridae — short-nosed chimaeras
· · *Parvphylum* Osteichthyes — bony fishes
· · · *Gigaclass* Actinopterygii — ray-finned fishes
· · · · · · · · · *Order* Cetomimiformes — whalefishes
· · · · · *Subclass* Teleostei
· · · · · · · · *Order* Alepocephaliformes
· · · · · · · · · *Family* Alepocephalidae — slickheads or smoothheads
· · · · · · · · *Order* Argentiniformes
· · · · · · · · · · *Family* Bathylagidae — deep-sea smelts
· · · · · · · · · · *Family* Opisthoproctidae — barreleyes and spookfishes
· · · · · · · · *Order* Aulopiformes
· · · · · · · · · *Family* Ipnopidae — deep-sea tripod fishes
· · · · · · · · · *Family* Bathysauridae — deep-sea lizardfishes
· · · · · · · · · *Family* Evermannellidae — sabertooth fishes
· · · · · · · · · *Family* Giganturidae — telescopefishes
· · · · · · · · · *Family* Scopelarchidae — pearleyes
· · · · · · · · *Order* Gadiformes — cods
· · · · · · · · · *Family* Gadidae — North Atlantic codlings
· · · · · · · · · *Family* Macrouridae — grenadiers, rattails, and whiptails
· · · · · · · · · *Family* Moridae — morid cods or morids
· · · · · · · · *Order* Lophiiformes
· · · · · · · · · *Family* Ceratiidae — anglerfishes
· · · · · · · · *Order* Myctophiformes
· · · · · · · · · *Family* Myctophidae — lanternfishes
· · · · · · · · · *Family* Neoscopelidae — blackchins
· · · · · · · · *Order* Notacanthiformes — halosaurs and spiny eels
· · · · · · · · *Order* Ophidiiformes
· · · · · · · · · *Family* Aphyonidae — blind cusk eels
· · · · · · · · · *Family* Ophidiidae — cusk eels
· · · · · · · · *Order* Perciformes
· · · · · · · · · *Family* Cottidae — sculpins
· · · · · · · · · *Family* Liparidae — snailfishes
· · · · · · · · · *Family* Sebastidae — rockfishes
· · · · · · · · · *Family* Zoarcidae — eelpouts
· · · · · · · · · *Suborder* Notothenioidei — notothenioids
· · · · · · · · · *Family* Channichthyidae — Antarctic icefishes

· · · · · · · · · · · *Order* Pleuronectiformes — flatfishes
· · · · · · · · · · *Order* Stomiiformes
· · · · · · · · · · · *Family* Gonostomatidae — bristlemouths
· · · · · · · · · · · *Family* Phosichthyidae — lightfishes
· · · · · · · · · · · *Family* Sternoptychidae — hatchetfishes
· · · · · · · · · · · *Family* Stomiidae — dragonfishes
· · · · · · · · · · *Order* Trachichthyiformes
· · · · · · · · · · · *Family* Trachichthyidae — roughies and redfishes
· · · · · · · · · · *Order* Zeiformes
· · · · · · · · · · · *Family* Oreosomatidae — oreos
· · · · · · · · · *Superorder* Elopomorpha
· · · · · · · · · · *Order* Anguilliformes — true eels
· · · · · · · · · · · *Family* Congridae — conger eels
· · · · · · · · · · · *Family* Nemichthyidae — snipe eels
· · · · · · · · · · · *Family* Serrivomeridae — sawtooth eels
· · · · · · · · · · · *Family* Synaphobranchidae — cutthroat eels
· · · · · · · · · · *Order* Saccopharyngiformes
· · · · · · · · · · · *Family* Eurypharyngidae — gulpers
· · · · · · · · · · · *Family* Saccopharyngidae — gulpers
· · · · · *Megaclass* Tetrapoda
· · · · · · *Class* Mammalia
· · · · · · · *Infraorder* Cetacea
· · · · · · · · *Superfamily* Mysticeti — baleen whales
· · · · · · · · *Superfamily* Odontoceti — toothed whales
· · · · · · · · · *Family* Ziphiidae — beaked whales
· · · · · · *Class* Reptilia
· · · · · · · *Order* Testudines — turtles
Subphylum Tunicata — tunicates or sea squirts
· · · · · *Class* Appendicularia — larvaceans
· · · · · *Class* Ascidiacea — tunicates and sea squirts
· · · · · *Class* Thaliacea — salps, dololids, and pyrosomes

GLOSSARY

abyss/abyssal
Seafloor beyond the continental slope at depths of 9,840–19,700 ft. (3,000–6,000 m).

abyssopelagic
Pelagic waters at abyssal depths.

amphipod
Order of crustaceans with a sideways flattened curved body (e.g. sandhoppers).

anoxic
Absence of oxygen.

barbel
Fleshy appendage on the jaws of fishes with sensory or luring function.

bathyal
Seafloor on slopes at 660–9,840 ft. (200–3,000 m) depth.

bathymetry
Mapping the depth of the seafloor.

bathypelagic
Pelagic waters at depths greater than 3,280 ft. (1,000 m) where there is no sunlight.

benthic
Pertaining to the seafloor.

benthic boundary layer
Layer of seawater in contact with the seafloor.

benthos
Collective term for organisms living on the seafloor.

bioluminescence
Light generated by living organisms.

biota
Collective term for all living organisms, including microbes, plants, and animals.

bivalve
Mollusk, such as a clam, with two shells joined by a hinge.

box corer
Device for taking precise square, undisturbed samples of seafloor sediment.

calcium carbonate
CaCO3, the main component of shells and skeletons of many marine organisms.

carbonate binding
The process of combining carbonate and calcium from seawater to grow hard calcium carbonate skeletons as in corals and mollusks.

cephalopod
Class of mollusks that includes squids, octopus, cuttlefish, and nautilus.

chemosynthesis
Conversion of inorganic compounds into organic matter such as sugars and amino acids using chemical energy.

chromatophore
Colored cells in the skin that produce changing patterns on the surface of certain animals.

cold seep
Emission of natural gas, oil, or hydrogen sulfide from the seafloor at low temperatures.

continental drift
Motion of continents through geological time.

continental shelf
Seabed less than 660 ft. (200 m) deep around continents.

continental slope
Steep edge of continents between the shelf and the abyss.

copepod
Subclass of planktonic crustaceans usually 0.04–0.08 in. (1–2 mm) long.

Coriolis effect
Deflection of currents and winds toward the right in the Northern Hemisphere and the left in the south caused by the earth's rotation.

counter-illumination
Underwater camouflage using lights on the underside of the body.

cryptic
Hidden or indistinguishable.

decapod
Order of crustaceans with a carapace such as crab, lobster, or shrimp.

deep scattering layer (DSL)
Concentration of mesopelagic animals at a particular depth detected by echosounder.

demersal
Animals living close to the seafloor.

density stratification
Layers of water separating out according to their density.

deposit feeder
Animal that feeds on detritus particles on the seafloor.

diel
Twenty-four hour periodicity.

dorsal
Upper side of an organism.

echinoderm
Animal phylum that includes sea stars, sea urchins, sea cucumbers, and feather stars.

eddy
Rotating mass of water.

epipelagic
The upper 660 ft. (200 m) of the open ocean.

exoskeleton
Jointed external skeleton of crustaceans.

filter feed
Extracting small food particles from water using a sieve or comblike apparatus.

Foraminifera
Phylum of single-celled amoebas with a hard external shell or test.

gastropod
Mollusk, such as a snail or slug, with a single large foot.

guyot
Undersea mountain with a flat top.

gyre
Large system of circulating ocean currents, e.g. North Pacific Gyre.

hadal zone
Depths greater than 19,700 ft. (6,000 m) in ocean trenches.

hadopelagic
The water column at hadal zone depths.

hot-spot volcanism
Volcanoes caused by molten rock ascending from hot parts of the mantle.

human-occupied vehicle (HOV)
A deep-diving submersible with a human crew on board.

hydrocarbon
Compound made up of hydrogen and carbon (e.g. natural gas, petroleum).

hydrothermal vent
Mineral-rich hot-water spring on the seafloor.

hypoxic/hypoxia
Low oxygen concentration, less than 30% saturation, in which most animals cannot survive.

igneous rock
Rock formed by cooling and solidifaction of molten material.

inertial
Motion tending to continue at a constant speed and direction but, owing to rotation of the earth, susceptible to deflection by Coriolis effects.

isopod
Order of crustacea with a flattened (wider than deep) segmented body.

isopycnals
Contour lines joining points of equal seawater density.

kinetic energy
Energy of a body due to its motion, proportional to mass times velocity squared.

laminar flow
Smooth fluid flow with no mixing between layers.

mantle plume
Area under the earth's crust where hot molten rock ascends by convection, creating volcanoes at the surface.

marine snow
Particles originating in surface water, rich in organic matter, that fall toward the deep seafloor.

megafauna
Sea floor animals larger than 1 in. (2.5 cm).

meiofauna
Animals smaller than 0.04 in. (1 mm) living in sediments.

mesopelagic
Depths of 660–3,280 ft. (200–1,000 m) throughout the open ocean—the twilight zone with insufficient sunlight for photosynthesis.

mesoscale eddy
Rotating mass of water around 60 mi. (100 km) in diameter.

microbe
Small living organisms including viruses, archaea, bacteria, and protozoa.

micronekton
Small actively swimming pelagic organisms 0.8–4.0 in. (2–10 cm) in size.

nekton
Pelagic animals large and strong enough to move independent of water currents.

oceanographic
Pertaining to study of the oceans.

oxygen-minimum zone (OMZ)
Depth range in the ocean with low oxygen concentration.

pelagic
Environment of the open ocean water column distant from the coast and seafloor.

pH
Scale of acidity 0–14; values less than 7 are acid; the ocean is slightly alkali at ph 8.1.

photophore
Light-producing organ.

photoreceptor
Light-detecting cell (e.g. in the retina).

phytoplankton
Planktonic organisms that use energy from sunlight for photosynthesis.

plankton
Organisms, mainly small, that drift with ocean currents.

plate tectonics
The theory that the earth's crust is composed of rigid plates that move relative to one another over geological time scales.

polychaete
Class of segmented marine bristle worm, with paired limbs bearing bristles.

polyp
Attached individual in a coral colony or anemone of the phylum Cnidaria.

primary production
The creation of new organic matter by photosynthesis or chemosynthesis.

pteropod
Order of pelagic gastropod mollusks, sea snails, and sea slugs. Often transparent with winged feet.

radiocarbon
Radioactive isotope of carbon (14C) used for determining the age of biological remains.

radionuclide
Unstable radioactive isotopes occurring naturally or produced by nuclear bombs and reactors.

remotely operated vehicle (ROV)
Deep-sea submersible vehicle with no human occupants.

resonance
Increase in wave amplitude at a particular frequency determined by the size of an ocean basin or entity.

sampling gear
Any device for taking samples of water, biota, sediments, particles, or rocks.

seamount
Submarine mountain.

seismic survey/noise
Use of underwater shock waves to map rock layers beneath the seafloor.

sessile
Attached, immobile.

shear stress or force
Forces resulting from layers of fluid moving relative to one another (e.g. wind over the ocean).

sonar
Underwater detection and location using sound waves.

spicule
Small needlelike structures making up the skeletons of sponges and other animals.

subduct/subduction zone
Collision between tectonic plates where one plate sinks beneath the other.

submersible
Deep-sea underwater vehicle.

substrate
Surface on which an animal lives.

suspension feeder/ing
Feeding on small particles in the water column.

symbiont
The smaller member of a symbiotic relationship.

symbiotic/sis
Two or more species of organism living together.

taxon/taxa
A scientifically recognized group of organisms: species, genus, family etc.; plural taxa.

tectonic plate
One of 15–20 massive slabs of rock comprising the hard surface of Earth: African plate, Pacific plate, and so on.

thermocline
Transition between warm surface and cold deep waters that can be permanent, persisting throughout the year, or seasonal.

thermohaline circulation
Water movement caused by differences in salinity and temperature.

thiotrophic (sulfur-eating) microbes
Microbes that live by oxidation of sulfide (e.g. at hydrothermal vents).

topography
The shape and properties of the seafloor.

trade wind
Permanent easterly winds blowing toward the equator in tropical regions.

trench
Deep pockets of seafloor at the edge of tectonic plates where subduction occurs.

turbulence
Chaotic flow in a fluid creating mixing as opposed to laminar flow.

ventral
Lower or underside of an organism.

volcanic arc
Curved chain of volcanoes above the edge of a subducting plate.

westerlies
Prevailing west to east winds at mid-latitudes (30–60 degrees latitude).

zooplankton
Animal plankton.

RESOURCES

Further reading

Baker, M., Ramirez-Llodra, E., and Tyler, P. *Natural Capital and Exploitation of the Deep Ocean.* Oxford University Press, 2020.

Boyle, P. *Life in the Mid Atlantic.* Bergen Museum Press, 2009.

Clarke, M.R., Consalvey, M., and Rowden A. A. *Biological Sampling in the Deep Sea.* Wiley-Blackwell, 2016.

Copley, J. *Ask an Ocean Explorer.* Hodder & Stoughton, 2019.

Gage, J.D. and Tyler, P. *Deep-Sea Biology: A Natural History of Organisms at the Deep-Sea Floor.* Cambridge University Press, 1991.

Herring, P. *The Biology of the Deep Ocean.* Oxford University Press, 2002.

Jamieson, A.J. *The Hadal Zone: Life in the Deepest Oceans.* Cambridge University Press, 2015.

Koslow J.A. *The Silent Deep: The discovery, ecology, and conservation of the deep sea.* University of Chicago Press, 2007.

Marshall, N.B. *Developments in Deep-Sea Biology.* Blandford Press, 1979.

McIntyre, A. *Life in the World's Oceans.* Wiley-Blackwell, 2010.

Murray, J. and Hjort, J. *The Depths of the Ocean.* Macmillan, 1912.

National Research Council. *Ocean Acidification: A National Strategy to Meet the Challenges of a Changing Ocean.* The National Academies Press, 2010.

Priede, I.G. *Deep-Sea Fishes.* Cambridge University Press, 2017.

Pinet, P.R. *Invitation to Oceanography,* 8th edn. Jones & Bartlett Learning, 2019.

Roberts, J.M., Wheeler, A., Freiwald, A., et al. *Cold-Water Corals: The Biology and Geology of Deep-Sea Coral Habitats.* Cambridge University Press, 2009.

Scales, H. *The Brilliant Abyss.* Atlantic Monthly Press, 2021.

Thomson, C. Wyville and Murray J. (eds). *Report on the scientific results of the voyage of H.M.S. Challenger during the years 1873–76 under the command of Captain George S. Nares R.N., F.R.S and the late Captain Frank Tourle Thomson, R.N.* 32 volumes plus narratives and summaries, 1880–1891. Available online:
https://www.biodiversitylibrary.org/bibliography/6513
https://doi.org/10.5962/bhl.title.6513

Thorpe, S.A. *The Turbulent Ocean.* Cambridge University Press, 2005.

Tyler, P.A. (ed) *Ecosystems of the Deep Oceans,* Volume 28. Elsevier Science, 2003.

Widder, E. *Below the Edge of Darkness.* Penguin Random House, 2021.

Wunsch, C. *Modern Observational Physical Oceanography: Understanding the Global Ocean.* Princeton University Press, 2015.

Presentations

Ingels, J., Clarke, M.R., Vecchione M., et al. "Open Ocean Deep Sea." Chapter 36F in: Inniss, L. and A. Simcock (joint coordinators). *First Global Integrated Marine Assessment, World Ocean Assessment I.* New York: United Nations (2016): p.37.

Ramirez-Llodra, E., Brandt, A., Danovaro R., et al. "Deep, diverse and definitely different: unique attributes of the world's largest ecosystem." *Biogeosciences 7* (2010): 2851–99.

Useful websites

Deep Sea Biology Society
http://dsbsoc.org/

Deep Sea Conservation Coalition
https://www.savethehighseas.org/deep-sea-fishing/

The Deep Sea – Smithsonian Ocean
https://ocean.si.edu/ecosystems/deep-sea/deep-sea

FishBase – A global information system on fishes
https://fishbase.net.br/home.htm

Monterey Bay Aquarium Research Institute (MBARI)
https://www.mbari.org/

National Oceanic and Atmospheric Administration (NOAA)
https://www.noaa.gov/

World Register of Marine Species (WoRMs)
https://www.marinespecies.org/

NOTES ON CONTRIBUTORS

Michael Vecchione went to sea as cabin boy on a three-masted schooner at the age of 16. After earning a B.S. in biology, he spent several years in the army. He has been studying cephalopods and deep-sea biology since earning a Ph.D. in 1979. As a National Oceanic and Atmospheric Administration (NOAA) employee at the US National Museum of Natural History, Vecchione is Curator of Cephalopods and Curator of the Sant Ocean Hall. He was Director of the NOAA laboratory for 18 years. He teaches deep-sea biology and has participated in over 80 research expeditions, many of them as chief scientist.

Hans van Haren is a senior scientist in physical oceanography at the NIOZ, the Royal Netherlands Institute for Sea Research in Texel. His major research topics are tidal motions, internal waves, and turbulent exchange in seas and oceans, which he researches via in-situ observations using custom-made instrumentation. He applies his knowledge of the physical oceanographic environment to research on the redistribution of nutrients and suspended matter by participating in multidisciplinary international research programs.

Imants (Monty) Priede is Professor Emeritus of Zoology at the University of Aberdeen, Scotland. He began his studies on deep-sea fish longlining and trawling on the Royal Research Ship *Challenger* in the 1970s and has since participated and led expeditions to the Pacific Ocean, Atlantic Ocean, Mediterranean Sea, and the Mid-Atlantic Ridge. He founded Oceanlab in Aberdeen, which pioneered robotic landers for investigating deep-sea life down to full ocean depth. He is Editor-in-Chief of the journal *Deep-Sea Research Part 1* and recipient of the Beverton Medal of the Fisheries Society of the British Isles.

Louise Allcock is Professor of Zoology at the University of Galway, where she teaches zoology and marine science. She began her career working on octopuses in the Southern Ocean and expanded to the deep sea worldwide. She has explored the North Atlantic with ROVs (remotely operated vehicles), leading many expeditions as chief scientist, and she has dived in submersibles in the Indian Ocean. Her heart lies in molecular systematics—using DNA sequencing to define species and to understand their evolution, and she has published over a hundred scientific papers on this and related topics. She is passionate about protecting the oceans, and contributes to conservation initiatives led by the Irish government, the EU, and the International Union for Conservation of Nature (IUCN).

INDEX

A

abyssal circulation 39
abyssal ecoregions 190
abyssal plains 23, 46, 146–53, 181, 216, 220, 221
abyssopelagic layer 174
acidification 15, 131, 266–71
acidity 22, 50
acorn barnacles 86
acorn worms 86, 103, 152, 190
acoustic exploration 245, 264
adaptations 94, 110–21, 269
Adélie Penguins 203
Agassiz, Alexander 39
Agulhas Current 52, 68
airplanes 275
albatrosses 144
Albert I of Monaco, Prince 39
Aldabra 260
Alicella gigantea 86, 156, 216
allopatric speciation 226
Alvin 40
Alvinella pompejana 162, 192
amphipods 86, 148, 156–7, 216, 222, 263
anemones 82
anglerfishes 109, 118
animal forests 127, 132, 136
Annelida 79, 86
anoxic events 222
Antarctic circumpolar basin 70
Antarctic Circumpolar Current 21, 52, 54, 61, 68, 186, 196, 207
Antarctic glaciation 222
Antarctic Herring 200, 203
Antarctic icefishes 113
Antarctic invertebrates 197
Antarctic megafauna 203
Antarctic shelf 196
antifreeze glycoproteins 200
Aphanopus carbo 232
Aptenodytes 101
Aptenodytes forsteri 203
Aptenodytes patagonicus 203
archaea 50, 158, 164
Architeuthis dux 24, 144, 245, 264
Arctic Ocean 206–11, 256
Arctic Ocean Diversity Project (ArcOD) 210
Arnoux's Beaked Whale 203
arrow worms 99, 140
Arthropoda 79, 86, 103
Atlantic Footballfish 104
Atlantic Ocean
· acidification 270
· basin 67
· floor 44
· hydrothermal vents 162
· radioactive waste disposal 256
· radiocarbon dating 254
Atlantification 210
autonomous underwater vehicles (AUVs) 35
azoic hypothesis 37

B

back-arcs 23
Backus, Richard 40
bacteria 50
· mats 127, 158
· photophores 118
· see also chemosynthetic ecosystems; microbes
baits 32
Balaena mysticetus 210, 241
Ballard, Bob 40
Baltimore Canyon 133
bamboo corals 82, 212
barnacles 192
barreleyes 106

Barton, Otis 39
Basamuk Canyon 258
basins 55, 66–75, 206, 207
bathyal ecoregions 190, 216, 220
Bathymodiolus 164, 182, 192
Bathynomus 86, 110
bathypelagic layer 174
bathysphere 39, 40
Bay of Fundy 70
beaked whales 35, 101, 144, 203, 241, 264
Beaufort Dyke 256
Beebe, William 39
Beluga 210
benthic boundary layer 147, 176
benthic ecoregions 190–1
benthic zone 16, 26
· Arctic 210
· fish 88–91
· invertebrates 78–87
· sampling 32
· vertical zonation 212–17
Berardius arnuxii 203
Bering Strait 207
bicarbonate 50
biodiversity 218–21
bioluminescence 12, 32, 50, 104, 113, 118–121, 170, 212, 228–9
birds 144
bivalves 86, 132, 148, 156, 164
blackchins 228
black corals 82
Blackmouth Catsharks 128
Black Scabbardfish 232
Black Sea 68, 73, 252
bluefin tunas 170
blue whitings 236, 238
Boarfish 234
bobtails 100
bony fishes 90–1
bores 63
Bosporus Strait 73
bottom trawling 131, 238, 270
boundary flows 55, 71, 72
Bowhead Whale 210, 241–2
box jellies 97, 114
brisingid sea stars 84
bristlemouths 106, 228
bristle worms 148, 154
brittle stars 84, 103, 156, 222
Broecker, Wallace 39
Brunn, Anton 39
bubble streams 30
buoyancy 12, 88, 94, 110, 144, 150, 182, 200, 214
buoyancy chambers 110, 214

C

calcium 50
Calyptogena 164
cameras 32
camouflage 94, 104, 228
Canadian Basin 207
Cancer borealis 124
canyons 55, 61, 71–2, 80, 124, 126–7, 131, 132–7, 181, 184, 258, 260
Capros aper 234
carbonate compensation 214
carbon flux 64
Carcharodon carcharias 170
cargo ships 275
cartilaginous fishes 88
catsharks 88
Census of Marine Life (CoML) 40
Central Indian Ridge 55
Centrophorus squamosus 139
Centroscymnus coelolepis 232
cephalopods 100, 103, 113, 114, 156, 194
ceriantharians 82

chaetognaths 99
Challenger Deep 35, 40, 46, 154, 264
Challenger, HMS 37
chemoreceptors 117
chemosynthetic ecosystems 29, 40, 86, 127, 158–67, 181, 184, 220, 241, 275
Chernobyl incident 252
chimaeras 88
Chondrichthyes 88
chromatophores 228
Chrysomallon squamiferum 162
Chun, Carl 39
circulation
· abyssal 39
· radiocarbon dating 254
· radionuclide tracers 252
· thermohaline 182, 222
· World Ocean Circulation Experiment 39
Cirrothauma murrayi 113
Cladorhizidae 80
clams 132, 148, 156, 164
climate change 131, 173, 209, 210, 266–71
clinker 152, 256
cnidarians 79, 82–3, 97
cobalt 249, 250
cods 90
coelacanths 220
cold seeps 40, 127, 158, 164, 182, 184, 192, 258, 269
Colossendeis colossea 86
comb jellies 97
continental drift 39, 40
continental plates 44–6, 158
convection 21
Cookie-cutter Shark 104
copepods 148, 224
corals 79, 82–3, 127, 132–3, 136, 139, 184, 212
· acidification 270–1
· bottom trawling 238
· *Deepwater Horizon* disaster 247
· gardens 131, 133, 139
· mounds 128
· NOAA survey 194
· plastic 263
· reefs 128
· speciation 224
· summits 140
· whale fall 167
corers 32
Coriolis effect 21, 52, 54, 60–1
Corliss, Jack 40
Coryphaenoides 90, 216
Coryphaenoides rupestris 124, 140, 234
cow sharks 88
crabs 32, 82, 86, 124, 164, 192
crocodile icefishes 200
crustaceans 80, 86, 99–100, 103, 114, 127, 148, 156, 181, 190, 197, 216
· Arctic 210
· bioluminescence 118
· see also amphipods; copepods; crabs; krill; lobsters; shrimps
ctenophores 97
currents 15, 18, 21, 23, 31, 46, 52–65, 182, 186, 224
cusk eels 84, 90, 136, 150, 157
cutthroat eels 90
Cuvier's Beaked Whale 35, 144, 264
cyanobacteria 50

D

Dasyscopelus asper 228
deep gliders 35
deep scattering layer 39, 40, 106, 124, 136, 140, 170, 181
deep-sea flows 54–5, 57
deep-sea tailings disposal (DSTD) 258
deep sinking 72–3
Deep Water Dump Site 106 256

Deepwater Horizon disaster 247, 270
defense strategies 117, 118
Delphinapterus leucas 210
demosponges 79, 80
dense water-mass formation 56–7, 72
density 48, 52, 54–5, 57, 72–3
depth
· speciation 224
· zones 16
Dermochelys coriacea 101, 170
Desmophyllum dianthus 184
Desmophyllum pertusum 82, 128, 140, 263, 270
digestion 110
dispersal 182, 184, 189
Dissostichus 200
Dissostichus eleginoides 234
dogfish sharks 88
dolphins 101, 104, 144
Dosidicus gigas 189, 228
dragonfishes 106, 118, 144, 228
dredges 32, 37
drill mud 247
dwarfism 29
dynamic positioning 247

E

earthquakes 46, 154, 264
Eastern Tropical Pacific 189, 226
echinoderms 79, 84–5, 103, 114, 127, 156, 222, 269
echinothuriids 84
ecoregion classification 186–95
ecosystems
· connected 180–95
· *see also* chemosynthetic ecosystems
eddies 31, 54, 68, 71, 168
eelpouts 90, 200, 210
eels 39, 84, 90, 104, 108, 136, 150, 157
Ekman, Vagn Walfrid 39
electrical receptors 117
Electrona antarctica 186
elephant seals 35, 101, 203
Emperor Penguins 203
endangered species 162
energy management 110
Eschrichtius robustus 241
Eubalaena 241
Euphausia superba 197
Eurypharynx pelecanoides 108
evaporation 72
evolutionary relationships 29
extinction events 220, 222
eyes 114
eyespots 114

F

faunal data 188–9
feather stars 84, 103
feeding 29
· continental slopes 124
· ecosystems 181
· trenches 157
· zonation 216–17
feeling 117
ferromanganese crusts 23, 249, 250
finned octopods 100
fishes 37, 110, 194
· abyssal plains 150
· Arctic and Antarctic 210
· benthic 88–91
· biodiversity 221
· bioluminescence 228
· canyons 136

· continental slopes 124
· ecoregions 190
· electroreceptors 117
· eyes 114
· mesopelagic 40
· pelagic 104–9
· Southern Ocean 200
· speciation 224
· summits 140, 142
· trenches 157
fishing 131, 232–9, 270
flatfishes 90
flux studies 40
foraminifera 148, 154
Forbes, Edward 37
fracture zones 55, 57, 74, 152
frilled sharks 88
fronts 186
fulmars 144
fungi 24

G

Galathea 39
Galeus melastomus 128
gas hydrates 251
Gazelle 39
ghost sharks 220
Giant Squid 24, 144, 245, 264
giant viruses 24
Gibraltar, Strait 70
gigantism 29, 82, 197
Gigantocypris 114, 156
Glacial Eelpout 210
glass sponges 79, 80
glass squid 194
gliders 35
Globicephala melas 144, 264
Goblin Shark 104
grabs 32
Gray Whales 241–2
great ocean conveyor 39, 182
Great White Sharks 170
Green Bomber Worm 118
Greenland coral gardens 133
Greenland Halibut 232, 236
Greenland Shark 88, 232
grenadiers 90, 124, 136, 140, 150, 157, 194, 200, 216, 224, 234, 238
Guam 35
Gulf of Alaska 234
Gulf of Mexico 189, 218, 224, 270
Gulf Stream 52, 54, 71
guyots 46
gyres 52, 68, 182, 186

H

habitat 26, 122–77
hadal ecoregions 190
hadal trenches 46, 74, 90, 154–7, 176, 190
hadopelagic zone 176
hagfishes 88, 166–7
halibuts 90, 124
halosaurs 90
harpacticoid copepods 148
hatchetfishes 106, 228
Hawaiian Islands 23, 124, 142, 194
hearing 117
heart urchins 84
Heezen, Bruce 40
hemocyanin 113
hemoglobin 158, 167, 200, 203
hexactinellid sponges 79, 80

Himantolophus groenlandicus 104
Hirondellea 157
histioteuthid squids 114
hoki 236
holoplankton 99
Honjo, Susumu 40
Hoplostethus atlanticus 142, 234
hot-spot volcanism 23
Hovis, Warren 40
human-occupied vehicles (HOVs) 35, 40
humans 8, 15, 230–75
Humboldt Squid 189, 228
Humpback Whale 35, 241–2
hydraulic dredging 249
hydraulic-jump mixing 64
hydrothermal vents 40, 158, 162, 184
· acidification 269
· ecoregions 192–3
· jurisdiction 251
· sulfide deposits 249, 250
hydrozoans 82, 97
Hyperoodon ampullatus 136

I

ice algae 209
icebergs 209
ice dynamics 209
incirrate octopods 100
Indian Ocean
· basin 68
· floor 44
"intermediate-disturbance" hypothesis 220
internal waves 18, 21, 23, 57, 60–4, 71, 271
· breaking 61–3
· canyons 127, 136
· kinetic energy 39
marginal seas 71
International Seabed Authority (ISA) 251
invertebrates
· Antarctic 197
· pelagic 94
Ipnops 114
iron 22
Isistius brasiliensis 104
isopods 86, 110, 148, 216
isopycnals 57, 64

J

jawless fishes 88
jellyfishes 97, 99
jet stream 21
Jonah Crab 124

K

Kaikōura Canyon 136
kinetic energy 60–1, 71, 73
King Penguins 203
kitefin sharks 88, 104
Kiwa tyleri 192
krill 100, 101, 106, 108, 136, 140, 144, 186, 197, 203
Kula oceanic ridge 192
Kuroshio Current 52, 54, 61, 71

L

Lamellibrachia 164
landers 32
lanternfishes 106, 144, 186, 194, 200, 228
lantern sharks 104, 200
Laplace, Pierre-Simon 36
larvaceans 98
larvae 182–4, 197
Latimeria 220
Leafscale Gulper Shark 139
Leatherback Sea Turtle 101, 170, 173
lebensspuren (life signs) 86
lecithotrophic larvae 182
Leiogalathea 222
Leptonychotes weddellii 203
lidar 31
light
·   zonation 212
·   *see also* bioluminescence
lightfishes 106
Limacina helicina 210
limpets 192
litter 260–3
lizardfishes 90, 106
lobsters 86, 110, 222
Longfinned Pilot Whales 144, 264
longlines 32
lophogastrids 100
lures 32
Lycodes frigidus 210

M

macrofauna 148
Macropinna microstoma 106, 114
Macruronus 236
magma 46
magnetic fields 117
Malacosteus niger 113
manganese nodules 23, 152, 249, 251
mantle plumes 23
marginal seas
·   basins 68–72
·   subbasins 74
Mariana Snailfish 35, 90
Mariana Trench 35, 40, 46, 90, 154, 260, 264
marine litter 260–3
marine snow 80, 98, 127, 140, 146–7, 181
mate choice 228
Matthews, Drummond 40
mechanoreceptors 117
Mediterranean Sea 68, 70, 72, 73, 74, 189, 215
megafauna 150
megaplankton 99, 101, 110, 210
Megaptera novaeangliae 35, 241
meiofauna 148, 154
Merluccius 236
meroplankton 99
mesopelagic boundary community 124
mesopelagic layer 170, 181, 224
·   ecoregion classification 186–9
·   fish 40
mesoscale eddies 68, 71
Messina, Strait 74
methane 251
microbes 29, 50
·   pelagic 94
·   *see also* bacteria; chemosynthetic ecosystems
Micromesistius 236
micronekton 100, 186, 228
micronutrients 22
microplastic 263
Mid-Atlantic Ridge 44, 46, 55, 67, 74, 139, 162, 234
mid-ocean ridges 40, 46, 57, 67, 74, 138–9, 158
migration 144, 216, 222
·   magnetic fields 117
·   vertical 31, 100, 140, 170, 176, 181, 209, 210, 247
minerals 22, 23, 46, 181
mine tailings disposal 258
mining 244–51
Mirocaris 192

Mirounga 101
Mirounga angustirostris 35
Mirounga leonina 35, 203
Mitsukurina owstoni 104
Mollusca 79, 86, 148, 156
·   *see also* cephalopods; clams; snails
monkfishes 124
Monodon monoceros 210
morids 194, 200
Morley, Lawrence 40
munitions dumps 256
Murray, John 37
mussels 164, 182, 192

N

Narwhal 210
National Oceanic and Atmospheric Administration
     (NOAA) 194
Nautilus 214
nekton 99–101
nematode worms 148, 154
Nemichthys scolopaceus 104
nepheloid layers 136
net sampling 32
noise pollution 264–5
Northern Bottlenose Whales 136
Northern Elephant Seal 35
Notoliparis 157, 200
Notoscopelus bolini 228
notothenioids 200
nuclear bomb testing 254
nudibranchs 99

O

Oarfish 24
oceanic plates 44–6
oceanography 42–75
octocorals 82, 133
octopods 86, 100, 110, 117, 156, 158
Odobenus rosmarus 210
oil *see* petrochemicals
Oilfish 232, 234
Orange Roughy 142, 234, 238
Orcas 210
Orcinus orca 210
oreos 194
Osedax 167, 182, 241
ostracods 114, 148
ostur 131
oxygen 15
·   anoxic conditions 73
·   hydrographic mapping 39
·   metabolism 113
·   saturation 48
·   zonation 215
oxygen-minimum zones (OMZs) 15, 39, 48, 66–7, 127, 173,
     189, 215, 226
oysters 132

P

Pacific Ocean
·   abyssal plains 152
·   acidification 270
·   basin 67
·   floor 44
·   hydraulic dredging 249
·   plate 46
Pacific Ocean Perch 234
Pacific Plate 44
paleogeomagnetic studies 40
Panamanian Seaway 192
Paraliparis 200
Paramuricea 224
Patagonian Toothfish 234

Paull, Charles 40
peanut worms 79, 86
pearleyes 106
pelagic zone 16, 26
·   biodiversity 220, 221
·   canyons 136
·   fish 104–9
·   organisms 94–109
·   sampling 32
·   trawling 238
Pelican Eel 108
penguins 94, 203
Pentaceros wheeleri 142, 234
perch 236
petrels 144
petrochemicals 244–51, 270
pheromones 117
Philippine Trench 39
photophores 118, 228
Physeter macrocephalus 24, 101, 117, 136, 144, 203, 241
phytoplankton 12, 31, 71, 127, 140, 146–7, 181, 209
Picasso Sponge 131
Piccard, Jacques 40
piezolytes 214
pilot whales 101, 144, 264
planktonic larval duration (PLD) 182
planktotrophic larvae 182
plastics 8, 15, 260–3
plateaus 138–45
Polar Bears 209, 210
polar front 186
polar seas 194, 196–211
polychaete worms 79, 86, 114, 148, 154, 226
Pompeii Worm 162, 192
Porifera 79, 80–1
Portuguese Dogfish 232
pressure 21, 32, 52, 54, 174
·   feeling 117
·   metabolism 113
·   zonation 214
Pressure Drop ship 35
Prestige 247, 272
Prickly Lanternfish 228
provinces 192
Pseudoliparis swirei 35, 90, 157
Psychropotes 150
pteropods 99, 210, 269
Pudgy Cusk Eel 90
Puerto Rico Dump Site 256
pyrosomes 98

R

radioactive contamination 252–6, 272
radiocarbon dating 254
radionuclides 252
Ram's Horn Squid 100
ray-finned fishes 90
rays 88, 139, 200
reefs 82
refuges 220
Regalecus glesne 24
Reinhardtius hippoglossoides 232
remotely operated vehicles (ROVs) 35, 40, 128, 132, 194, 250
remote sensing 40
reproduction 182–4
·   larvae 182–4, 197
·   mate choice 228
resonance 70
Reticulammina 150
Ridgeia piscesae 192
ridges 40, 44–6, 57, 67, 74, 138–45, 158
·   Mid-Atlantic Ridge 44, 46, 55, 67, 74, 139, 162, 234
·   turbulence 64
Riftia pachyptila 158, 192
right whales 241–2
Rimicaris 114
Rimicaris exoculata 162, 192
Risso, Antoine 37, 212
robotic vehicles 35
Ross, Captain James Clark 37
Ross, John 37
Ross Sea 68

roughies 194, 234, 238
Roundnose Grenadier 124, 140, 234, 238
rusticles 272
Ruvettus pretiosus 232

S

sabertooth fishes 106
Saccopharyngidae 108
salinity 15, 21, 22, 31, 32, 39, 48, 50, 67
salps 98
sampling 32
Sandalops melancholicus 194
Sandwell, David 40
Sars, Michael 37
satellites 31, 40, 232, 247
sawtooth eels 108
scale worms 118
Scaly-foot Snail 162
Schmidt, Johannes 39
schreckstoff (fright substance) 117
sclerites 82
Scotoplanes 84
sea cucumbers 15, 84, 103, 110, 118, 147, 150, 156, 167, 190, 216, 222, 269
sea elephants 99, 100
sea fans 82
sea ice 206, 209
Sea of Japan 68
sea lilies 84, 110, 156, 222
seals 203, 210
seamounts 23, 46, 66, 80, 138–45, 181, 190
· biodiversity 220
· ferromanganese crusts 249, 250
· jurisdiction 251
· litter 260
· stepping-stones 184
Sea Pangolins 162
sea pens 82, 131, 212
sea pigs 84
sea spiders 79, 86, 103, 197
sea stars 124, 156, 222
sea urchins 84–5, 156
Sebastes 236
Sebastes alutus 234
Secchi disk 203
seed shrimps 114, 148
selection 224, 226
semisubmersible platforms 247
sessile species 29
sharks 35, 88, 104, 110, 128, 139, 166–7, 170, 200, 220, 232
shipping containers 275
shipwrecks 272–5
shrimps 100, 114, 136, 148, 162, 164, 192, 236
Sicily, Strait 74
siphonophores 94, 97, 99, 171–2
size
groupings 29
· zooplankton 99
skates 88, 158
Slender Armorhead 142, 234
Slender Snipe Eel 104
slickheads 90
slopes 62–3, 124–7, 128, 131
smelling 117
smelts 90, 106
Smith, Walter 40
snailfishes 35, 90, 157, 200
snails 84, 99, 110, 114, 148, 156, 162, 192, 210
snipe eels 108
sofar 48
solar energy 18, 56, 60, 212
Solenosmilia variabilis 140, 184
solibores 63
Somniosus microcephalus 88, 232
sonar 30, 39, 132, 170, 232, 245, 264
Southern Elephant Seal 35, 203
southern hakes 236
Southern Ocean 68, 192, 196–203, 207, 209, 210
sparkers 245
speciation 222–9
Spectrunculus grandis 90

speed of sound 48
Sperm Whale 24, 101, 117, 136, 144, 203, 241
spiny eels 90
Spirula spirula 100
sponges 79, 80–1, 110, 127, 131, 139, 140, 197
spookfishes 106, 114, 115
spoon worms 152
squid 24, 100, 110, 114, 118, 124, 136, 144, 156, 189, 194, 197, 203, 228, 245, 264
Stannophyllum zonarium 150
Staurocalyptus 131
"stepping-stone" hypothesis 184
Stommel, Henry 39
stony coral 82, 128
stoplight loosejaws 106, 113
stromatolites 50
subbasins 74
subduction zones 44, 46, 74, 158
submersibles 35
sulfide deposits 249, 250
sulfur-hexafluoride tracer 252
Swima bombaviridis 118
swim bladders 117, 200, 214
symbiosis 162
sympatric speciation 226
Symphurus 90

T

Talisman 39
tanaids 148
Taonius pavo 194
tardigrades 148
tectonic plates 23, 39, 40, 44–6, 74, 222
telegraph cables 39
telescopefishes 106
temperature 15, 31
· hydrographic mapping 39
· oxygen 173
· pressure effect 21
· zonation 215
· *see also* climate change
Tethys Sea 192, 222
Teuthowenia maculata 194
Teuthowenia megalops 194
Tharp, Marie 40
thermohaline circulation 182, 222
thiotrophic microbes 158
Thompson, Charles Wyville 37
Thunnus 170
tides 36, 60–1
· marginal seas 70
· tidal energy 271
Titanic 272
tonguefishes 90
toothed whales 101, 117, 136, 144, 173, 241
toothfishes 200, 234
topography 23, 30, 40, 44–7
· basins 66–75
· slopes 62–3, 124
· trenches 74
toxicity 162
traps 32
Travailleur 39
trenches 7, 23, 30, 46, 64, 74, 154–7
Trieste 40
tripod fishes 90
tube anemones 82
tube worms 79, 86, 158, 164, 182, 192
tunas 170
tunicates 98, 99
turbulent mixing 21, 23, 39, 46, 48, 52–65, 66, 74, 252, 271
typhoon (hurricane) 62, 264

U

United Nations Convention on the Law of the Sea (UNCLOS) 251
Ursus maritimus 209, 210

Vampire Squid 118
Vampyroteuthis infernalis 118
vent shrimps 162, 192
Vine, Frederick 40
viperfishes 106
viruses 24, 50
vision 113–14
Vityaz 39
volcanic arcs 23, 46
volcanic ridges 46
volcanic vents 50
volcanism 23
Vulnerable Marine Ecosystems 127, 128, 131–2, 238

W

Walrus 210
Walsh, Don 40
waste disposal 256–63
water column 168–77
water-mass formation 56–7
water sampling 32
Weddell Sea 68
Weddell Seals 203
Wegener, Alfred 39
whale fall 166–7, 181, 184, 241, 242
whalefishes 109
whales 101, 117, 136, 144, 173, 203, 210, 264
whaling 240–3
whip corals 212
Whittard Canyon 61, 132, 133–4
wind 52, 54, 60–1, 72, 186
Woods Hole Oceanographic Institution (WHOI) 40
World Ocean Circulation Experiment (WOCE) 39

X

xenophyophores 86, 150

Y

yeti crabs 192

Z

Ziphius cavirostris 35, 144, 264
zoantharians 82
zombie worm 101
zonation 212–17
zooplankton 32, 90, 99, 127, 140, 147, 181, 186, 210

ACKNOWLEDGMENTS

The publisher would like to thank the following for permission to reproduce copyright material and for use of reference material for illustrations. All reasonable efforts have been made to contact copyright holders and to obtain their permission for the use of copyright material where necessary. The publisher apologizes for any errors or omissions and will gratefully incorporate any corrections in future reprints if notified.

56: Adapted from Fig 3 in Marshall, J. and F. Schott. "Open-ocean convection: Observations, theory, and models." *Rev. Geophys.* 37(1) (1999), 1–64, doi:10.1029/98RG02739.
68: Adapted with permission from Katherine Hutchinson.
73: Adapted from GRID-Arendal (http://www.grida.no/resources/5885) based on Zavaterelli and Mellor (1995).
81CR: Figure 2 from Dohrmann, M., Kelley, C., Kelly, M. et al. "An integrative systematic framework helps to reconstruct skeletal evolution of glass sponges (Porifera, Hexactinellida)." *Front Zool* 14, 18 (2017). https://doi.org/10.1186/s12983-017-0191-3
133TR: Fig 10 in Long, Stephen & Sparrow-Scinocca, Bridget & Blicher, Martin & Arboe, Nanette & Fuhrmann, Mona & Kemp, Kirsty & Nygaard, Rasmus & Zinglersen, Karl & Yesson, Chris. (2020). "Identification of a Soft Coral Garden Candidate Vulnerable Marine Ecosystem (VME) Using Video Imagery, Davis Strait, West Greenland." *Frontiers in Marine Science.* 7. 10.3389/fmars.2020.00460.
148T: Wiklund, Durden, Drennan, and McQuaid from Fig 3 (E) in Bribiesca-Contreras G., Dahlgren T.G., Amon D.J., Cairns S., Drennan R., Durden J.M., Eléaume M.P., Hosie A.M., Kremenetskaia A., McQuaid K., O'Hara T.D., Rabone M, Simon-Lledó E., Smith C.R., Watling L., Wiklund H., Glover A.G. "Benthic megafauna of the western Clarion-Clipperton Zone, Pacific Ocean." *ZooKeys* 1113: 1–110. (2022), https://doi.org/10.3897/zookeys.1113.82172
149 (all images): plates 1–4 in Bianca Lintner, Michael Lintner, Patrick Bukenberger, Ursula Witte, Petra Heinz. "Living benthic foraminiferal assemblages of a transect in the Rockall Trough (NE Atlantic)." *Deep Sea Research Part I: Oceanographic Research Papers*, Volume 171, 2021, 103509, ISSN 0967-0637, https://doi.org/10.1016/j.dsr.2021.103509.
173: Adapted from Figure 2 in Moffitt S.E., Moffitt R.A., Sauthoff W., Davis C.V., Hewett K., Hill T.M. "Paleoceanographic Insights on Recent Oxygen Minimum Zone Expansion: Lessons for Modern Oceanography." *PLoS ONE* 10(1): e0115246 (2015), https://doi.org/10.1371/journal.pone.0115246
187T & 188: Adapted from Fig. 1 & Fig 4. in Tracey T. Sutton, Malcolm R. Clark, Daniel C. Dunn, Patrick N. Halpin, Alex D. Rogers, John Guinotte, Steven J. Bograd, Martin V. Angel, Jose Angel A. Perez, Karen Wishner, Richard L. Haedrich, Dhugal J. Lindsay, Jeffrey C. Drazen, Alexander Vereshchaka, Uwe Piatkowski, Telmo Morato, Katarzyna Błachowiak-Samołyk, Bruce H. Robison, Kristina M. Gjerde, Annelies Pierrot-Bults, Patricio Bernal, Gabriel Reygondeau, Mikko Heino. "A global biogeographic classification of the mesopelagic zone," *Deep Sea Research Part I: Oceanographic Research Papers*, Volume 126, 2017, Pages 85–102, ISSN 0967-0637, https://doi.org/10.1016/j.dsr.2017.05.006.
206: Jakobsson, M., Mayer, L.A., Bringensparr, C., Castro, C.F., Mohammad, R., Johnson, P., Ketter, T., Accettella, D., Amblas, D., An, L., Arndt, J.E., Canals, M., Casamor, J.L., Chauché, N., Coakley, B., Danielson, S., Demarte, M., Dickson, M.-L., Dorschel, B., Dowdeswell, J.A., Dreutter, S., Fremand, A.C., Gallant, D., Hall, J.K., Hehemann, L., Hodnesdal, H., Hong, J., Ivaldi, R., Kane, E., Klaucke, I., Krawczyk, D.W., Kristoffersen, Y., Kuipers, B.R., Millan, R., Masetti, G., Morlighem, M., Noormets, R., Prescott, M.M., Rebesco, M., Rignot, E., Semiletov, I., Tate, A.J., Travaglini, P., Velicogna, I., Weatherall, P., Weinrebe, W., Willis, J.K., Wood, M., Zarayskaya, Y., Zhang, T., Zimmermann, M., Zinglersen, K.B., 2020. The International Bathymetric Chart of the Arctic Ocean Version 4.0. Scientific Data 7, 176. Doi: 10.1038/s41597-020-0520-9.
221T: Adapted from M. A. Rex, R. J. Etter and C. T. Stuart. "Large scale patterns of species diversity in the deep sea benthos." *Marine Biodiversity Patterns and Processes*, R.F.G. Ormond, J.D. Gage and M.V. Angel (eds), pp94–121, Cambridge University Press (1997).
222: Fig 21 in Rodríguez-Flores, Paula C., Macpherson, Enrique and Machordom, Annie. "Revision of the squat lobsters of the genus *Leiogalathea* Baba, 1969 (Crustacea, Decapoda, Munidopsidae) with the description of 15 new species." pp. 201–256 in *Zootaxa* 4560 (2) 2019, on page 251, DOI: 10.11646/zootaxa.4560.2.1, http://zenodo.org/record/2627514
224: Phylogenetic trees adapted from Figs 2 and 4: Quattrini, A. M., Herrera, S., Adams, J. M., Grinyo, J., and McFadden, C. S. "Phylogeography of Paramuricea: The Role of Depth and Water Mass in the Evolution and Distribution of Deep-Sea Corals." *7th International Symposium on Deep-Sea Corals* (2022, April), Frontiers Media SA.
260: Adapted from Fig 7 in Pham C.K., Ramirez-Llodra E., Alt C.H.S., Amaro T., Bergmann M., Canals M., et al. "Marine Litter Distribution and Density in European Seas, from the Shelves to Deep Basins." *PLoS ONE* 9(4): e95839 (2014), https://doi.org/10.1371/journal.pone.0095839

AFMA: 235TR.
Alamy Stock Photo: /Sipa US 32T; /Chronicle 36; /Sueddeutsche Zeitung Photo 39; /GRANGER – Historical Picture Archive 41; / LWM/NASA/LANDSAT 42–43; EyeEm 51BL; /ASP GeoImaging/ NASA 69T; /World History Archive 76–77; /Nature Picture Library 96TR, 122–123; 201TL, 265B; /Doug Perrine 101B; /David Fleetham 104T, 215T; /Kelvin Aitken /VWPics 105T; /Andrey Nekrasov 125TR, 211BR; /BIOSPHOTO 137CL; /Science History Images 160–161, 193TL; /Mark Conlin 178–179; /Minden Pictures 197T; / Stocktrek Images, Inc. 197TC; /robertharding 211TR; /Solvin Zankl 230–231; /First Collection 232; /Jeff Rotman 235BL; /Alastair J King 235BR; /Diarmuid 237T; /SOTK2011 243T; /REUTERS 245, 251, 259L; /A.P.S. (UK) 253B; /Morgan Trimble 262T.
Louise Allcock: 132T; 191 shapefiles courtesy of Les Watling (research originally published in: Les Watling, John Guinotte, Malcolm R. Clark, Craig R. Smith, "A proposed biogeography of the deep ocean floor." *Progress in Oceanography*, Volume 111, 2013, Pages 91112, ISSN 0079-6611, https://doi.org/10.1016/j.pocean.2012.11.003); 225T; 225B.
Ardea: /Pat Morris 183BR.
Australian Antarctic Division: /© Martin Riddle 197B.
Australian Museum: Michael J. Miller of the University of Tokyo 104B.
British Antarctic Survey expeditions to the Southern Oceans (collated by Dr Huw Griffiths): 198–199.
BluePlanetArchive: /Michael Aw 9; /Steven Kovacs 92–93, 282; / Jeff Milisen 105B; /Solvin Zankl 119TR; /Espen Rekdal 119B; /Doug Allan 204–205; /Franco Banfi 211BL.
CSIRO: /CC BY 3.0 141B, 143B.
Danté Fenolio: 13, 25B, 107T.
GEBCO: 75 (all images), 133BL, 133BR, 142, 194 reproduced from the GEBCO world map 2014, www.gebco.net.
Getty Images: /Enrique Aguirre Aves 202T; /Danita Delimont 235TL; /Danial_Abdullah 246; /Alain Nogues/Sygma/Sygma 253T; /DigitalGlobe 255B; /DigitalGlobe /ScapeWare3d 259R.
Hans van Haren: 60B; 62 (data from: van Haren, H., W.-C. Chi, C.-F. Yang, Y.J. Yang, S. "Deep sea floor observations of typhoon driven enhanced ocean turbulence." *Progr. Oceanogr.*, Jan 2020. 184, 102315, 12 pp); 63R (data from van Haren, H., and L. Gostiaux. "Energy release through internal wave breaking." *Oceanography* 25(2):124–131, 2012, http://dx.doi.org/10.5670/oceanog.2012.47. 64 (data from van Haren, H., G. Duineveld, H. de Stigter. "Prefrontal bore mixing." *Geophys. Res. Lett.*, 44, 9408–9415, 2017, doi:10.1002/2017GL074384; 65 Adapted from Fig 3 in van Haren, H. "Challenger Deep internal wave turbulence events." *Deep Sea Research Part I: Oceanographic Research Papers*, Volume 165, 2020, 103400, ISSN 0967-0637, https://doi.org/10.1016/j.dsr.2020.103400.
Ifremer: /(2005) Rimicaris exoculata shrimp / CC BY 4.0 163BL; / (2005) Crevettes rencontrées sur le site hydrothermal Rainbow. https://image.ifremer.fr/data/00569/68076/ CC BY 4.0 193TR.
Image Quest Marine: /© Justin Marshall 17; /Kelvin Aitken / V&W 89CR; /©Peter Herring 107BL, 112, 229TR (lower); /© Peter Batson 163TL; /© Kåre Telnes 236.
Alan Jamieson: 20B, 139B, 147BL, 151Tl, 151TR, 153, 156T, 157T, 157B, 190B, 217TL.
David Liittschwager: 268.
Marine Diversity Hub: /Schmidt Ocean Institute / CC BY 3.0 28T.
Marine Institute: INFOMAR is the Department of Environment, Climate and Communications (DECC) funded national seabed mapping programme, jointly managed and delivered by Geological Survey Ireland and Marine Institute (www.infomar.ie): 60T, 132B, 134–135.
MARUM – Center for Marine Environmental Sciences, University of Bremen: 193C, 219T.
MBARI: /© 2012 MBARI 25TR; /Image courtesy of Kelly Benoit-Bird 30; /© 2004 MBARI 38B; /© 2012 MBARI 81CL; /© 2006 MBARI/NOAA 81BL; /© 2014 MBARI 85BR; /© 2018 MBARI 89B; /© 2019 MBARI 97; /© 2018 MBARI 98; /© 2021 MBARI 115TL; /© 2017 MBARI 115TR; /George Matsumoto © 1989 MBARI 119TL; /© 2017 MBARI 147TR; /© 2007 MBARI 148B; /© 2015 MBARI 152R; / © 2015 MBARI 163TR; /© 2015 MBARI 168T; /© 2017 MBARI 168B; /© 2015 MBARI 226T; © 2011 NOAA/MBARI 274B.
T. Menut/Biotope: 189.
NASA: /Timothy Marvel 31; /NASA Earth Observatory 69B.
National Institute of Water and Atmospheric Research Ltd (NIWA), New Zealand: 136, 195T.
National Oceanography Centre, UK: 226C, 226B.
Natural History Museum of Los Angeles: © Leslie Harris 87BR.
Nature in Stock: /Norbert Wu / Minden Pictures 108T, 137TL, 202B, 239; /John Holmes / FLPA 233T; /Paul Ensor / Hedgehog House / Minden Pictures 243B.
Nature Picture Library: /David Shale 2, 25TL, 83L, 91, 95, 96TL, 107BR, 111, 115BL, 115BR, 116T; 120, 121, 141T,171TL, 171BL, 171BR, 175T, 175B, 229TR (upper), 276–277; /Solvin Zankl 38T, 171TR; / Doug Perrine 96BL; /Magnus Lundgren 116B; /Sue Daly 129B; / Wild Wonders of Europe / Lundgren 145TR; /Todd Pusser 145BR; /Franco Banfi 172; /Doug Allan 197BC; /Jordi Chias 201TR; /Solvin Zankl / GEOMAR 219B.
NOAA Office of Ocean Exploration and Research: 14T; 81TR; 213T; 213B; 274C; /Discovering the Deep: Exploring Remote Pacific MPAs 6; 108B, 138; /Mountains in the Deep: Exploring the Central Pacific Basin 14B, 83CR, 90C; /2019 Southeastern U.S. Deep-sea Exploration 20T, 22T, 85CR, 89TR; /Edith Widder, Ph.D. 32B; /Image courtesy of Leah Sloan, UAF, The Hidden Ocean 2016 – Chukchi Borderlands/NOAA 33; /2017 American Samoa 58T, 83BR; /Expedition to the Deep Slope 2007, NOAA-OE 58B; /NOAA Okeanos Explorer Program, Gulf of Mexico 2014 Expedition 59T, 83TR, 270; /Image courtesy of Craig Smith and Diva Amon, ABYSSLINE Project 59B; /NESDIS 67; /NASA 71B; / NOAA Okeanos Explorer Program, Océano Profundo 2015: Exploring Puerto Rico's Seamounts, Trenches, and Troughs 74, 79BR, 113, 177T, 217TR; /North Atlantic Stepping Stones Expedition 2021 78; /Deep-Sea Symphony: Exploring the Musicians Seamounts 79TL, 139T, 141C; /NOAA Bioluminescence and Vision on the Deep Seafloor 2015 79C; /2015 Hohonu Moana: 79TR, 85CL, 100T, 139TC; /NOAA Okeanos Explorer Program, 2013 Northeast U.S. Canyons Expedition: 79BL, 130T; /Windows to the Deep 2019 79BC, 133TL, 190T; /Exploring Deep-sea Habitats off Puerto Rico and the U.S. Virgin Islands 81TL, 103T, 185TL; /NOAA OOER, Gulf of Mexico 2018 81BR, 261T; /2016 Deepwater Exploration of the Marianas /Leg 3 85T, 90T, 177B, 274T /Christopher Kelley 250; /NOAA Okeanos Explorer Program, INDEX-SATAL 2010 85BL, 89CL, 193B; /NOAA Ocean Exploration, 2021 North Atlantic Stepping Stones: New England and Corner Rise Seamounts 87TR, 184, 195B, 249L; /Image courtesy of Fisheries and Oceans Canada 87BL; /Gulf of Mexico 2017 90B; /NMFS/PIFSC 102; /Deep Connections 2019 125B, 130B; /2013 ROV Shakedown and Field Trials in the U.S. Atlantic Canyons 127; /Brooke et al. 2005, NOAA-OER Florida Deep Corals 129TL; /DEEP SEARCH 2018 – BOEM, USGS, NOAA 129TR; / Monterey Bay Aquarium Research Institute 131; /Deepwater Canyons 2013 – Pathways to the Abyss, NOAA-OER/BOEM/USGS 139BC, 185B; /NOAA Northwestern Hawaiian Islands Expedition 2003 143T; /NOAA Okeanos Explorer Program, Gulf Of Mexico 2012 Expedition 147TL, 165TL, 165B; /Exploring Atlantic Canyons and Seamounts 2014 151B; /Pacific Ring of Fire 2004 Expedition. Dr. Bob Embley, NOAA PMEL, Chief Scientist 159B; / OET/NOAA 166T; /NOAA-OER/BOEM/USGS 185TR; /Officers and Crew of NOAA Ship PISCES; Collection of Commander Jeremy Adams, NOAA Corps 186; /Deep CCZ expedition 190C; /Caitlin Bailey, GFOE, The Hidden Ocean 2016 Chukchi Borderlands 207; NOAA/Caitlin Bailey, GFOE, The Hidden Ocean 2016: Chukchi Borderlands/CC BY-SA 2.0 208; /Brooke et al, NOAA-OE, HBOI 221B; /Dr. Mridula Srinivasan, NOAA/NMFS/OST/AMD / CC BY 2.0 242; /Lophelia II 2010, NOAA OER and BOEMRE 247; /Family of Vice Admiral H. Arnold Karo, C&GS 254; /NOAA Okeanos Explorer Program, Mid-Cayman Rise Expedition 2011 258; NOAA Okeanos Explorer Program: Our Deepwater Backyard 261B; /NOAA Ocean Acidification Program 269; /NOAA / Marine Applied Research and Exploration (MARE) 271.
Nova Southeastern University: T.M.Frank, Nova Southeastern University/Global Explorer 183BL.
Monty Priede: 125CR.
Rachel Przeslawski: images collected by ROV SuBastien of the Schmidt Ocean Institute 86.
David Sandwell: 22B, 45B, 63L, 66.
Science Photo Library: /Marek Mis 28B, 51TR; /US Navy 40T, 40B; /Dennis Kunkel Microscopy 51TL; /Matt Mazloff 55; /Danté Fenolio 100B, 229TL; /Martin Jakobsson 187B, 196; /British Antarctic Survey 201B; /Gerd Guenther 217BL; /M.I. Walker 217BR; /Solvin Zankl / Nature Picture Library 229B; /Mick Harper 233B; /NOAA 273.
Alexander Semenov: 96BR, 211TL.
Shutterstock: /katatonia82 19; /Juan Gracia 49B; /Sushko Valerii 101T; /Harry Collins Photography 137TR; /Guido Montaldo 137CR; /Ian Duffield 145TL; /Ivan Hoermann 203; /AlessandroZocc 220; / Erika Kirkpatrick 223T; /V.Gordeev 249R; /FLPA 264.
Craig Smith and Mike Degruy: 89TL.
Kate Thomas: /National Science Foundation 145BL.
Wellcome Collection: Engraving by A. M. Fournier after E. Traviès 240–241
Dr. Johann Weston at Newcastle University, UK: 156B.
Edith Widder, ORCA: 109.
Wikimedia Commons: /Alexandpilat / CC BY 4.0 10–11; / Luke Thompson from Chisholm Lab and Nikki Watson from Whitehead, MIT 51BR; /NASA Earth Observatory 72; /Jay Nadeau, Chris Lindensmith, Jody W. Deming, Vicente I. Fernandez, and Roman Stocker. Image courtesy of David Liittschwager CC BY-SA 4.0 99; /Uwe Kils, CC BY-SA 3.0 100C; /Bernard Dupont from France, CC BY-SA 2.0 103B; /Espen Rekdal / Espen Rekdal Photography artsdatabanken.no/Pages/F21060, CC BY 4.0 125TL, 125CL; /Cephas, CC BY-SA 4.0 137B; /Philweb, CC BY-SA 3.0 147BR; /NOAA 159T; /Chong Chen, CC BY-SA 3.0 163C; /Magali Zbinden / University of Greifswald (CC BY 4.0) 163BR; /Charles Fisher / CC BY 2.5 165TR; /National Marine Sanctuaries, CC BY 2.0 166B; /Alf van Beem 237B; /By United States Department of Defense (either the U.S. Army or the U.S. Navy) – Library of Congress, Public Domain 255T; /WESTON, J.N.J. et al. 2020. New species of Eurythenes from hadal depths of the Mariana Trench, Pacific Ocean (Crustacea_Amphipoda). *Zootaxa* 4851(1)_151–162., CC BY 4.0 262B; /Silver Leapers, CC BY 2.0 265T.
Helena Wiklund: 87TL.